Bob Weil

Geographic Objects with Indeterminate Boundaries

D1482220

Series Editors

I. Masser and F. Salgé

Geographic Objects with Indeterminate Boundaries

EDITORS

PETER A. BURROUGH and ANDREW U. FRANK

GISDATA II

SERIES EDITORS

I. MASSER and F. SALGÉ

Taylor & Francis
Publishers since 1798

UK   Taylor & Francis Ltd, 1 Gunpowder Square, London, EC4A 3DE
USA   Taylor & Francis Inc., 1900 Frost Road, Suite 101, Bristol, PA 19007

Copyright © Taylor & Francis Ltd 1996

All rights reserved. No part of this publication may be reproduced, stored in a retrieval system, or transmitted, in any form or by any means, electronic, electrostatic, magnetic tape, mechanical, photocopying, recording or otherwise, without the prior permission of the copyright owner.

British Library Cataloguing in Publication Data

A catalogue record for this book is available from the British Library.

ISBN 0-7484-0386-8 (hard)
ISBN 0-7484-0387-6 (paperback)

Library of Congress Cataloguing Publication data are available

Cover design by Hybert Design and Type.

Typeset in Times 10/12pt by Santype International Ltd, Salisbury, Wiltshire.

Printed in Great Britain by T. J. Press (Padstow) Ltd.

Contents

The GIS Data Series

Editors' Preface

Over the last few years there have been many signs that a European GIS community is coming into existence. This is particularly evident in the launch of the first of the European GIS (EGIS) conferences in Amsterdam in April 1990, the publication of the first issue of a GIS journal devoted to European issues (*GIS Europe*) in February 1992, the creation of a multipurpose European ground-related information network (MEGRIN) in June 1993, and the establishment of a European organization for geographic information (EUROGI) in October 1993. Set in the context of increasing pressures towards greater European integration, these developments can be seen as a clear indication of the need to exploit the potential of a technology that can transcend national boundaries to deal with a wide range of social and environmental problems that are also increasingly seen as transcending the national boundaries within Europe.

The GISDATA scientific programme is very much part of such developments. Its origins go back to January 1991, when the European Science Foundation funded a small workshop at Davos in Switzerland to explore the need for a European level GIS research programme. Given the tendencies noted above it is not surprising that participants of this workshop felt very strongly that a programme of this kind was urgently needed to overcome the fragmentation of existing research efforts within Europe. They also argued that such a programme should concentrate on fundamental research and it should have a strong technology transfer component to facilitate the exchange of ideas and experience at a crucial stage in the development of an important new reseach field. Following this meeting a small coordinating group was set up to prepare more detailed proposals for a GIS scientific programme during 1992. A central element of these proposals was a research agenda of priority issues grouped together under the headings of geographic databases, geographic data integration, and social and environmental applications.

The GISDATA scientific programme was launched in January 1993. It is a four-year scientific programme of the Standing Committee of Social Sciences of the European Science Foundation. By the end of the programme more than 300 scientists from 20 European countries will have directly participated in GISDATA activities and many others will have utilized the networks built up as a result of them. Its objectives are:

- to enhance existing national research efforts and promote collaborative ventures over coming European-wide limitations in geographic data integration, database design and social and environmental applications;
- to increase awareness of the political, cultural, organizational, technical and informational barriers to the increased utilization and inter-operability of GIS in Europe;
- to promote the ethical use of integrated information systems, including GIS, which handle socio-economic data by respecting the legal restrictions on data privacy at the national and European levels;
- to facilitate the development of appropriate methodologies for GIS research at the European level;
- to produce output of high scientific value;
- to build up a European network of researchers with particular emphasis on young researchers in the GIS field.

A key feature of the GISDATA programme is the series of specialist meetings that has been organized to discuss each of the issues outlined in the research agenda. The organization of each of these meetings is in the hands of a small task force of leading European experts in the field. The aim of these meetings is to stimulate research networking at the European level on the issues involved and also to produce high quality output in the form of books, special issues of major journals and other materials.

With these considerations in mind, and in collaboration with Taylor & Francis, the GISDATA series has been established to provide a showcase for this work. It will present the products of selected specialist meetings in the form of edited volumes of specially commissioned studies. The basic objective of the GISDATA series is to make the findings of these meetings accessible to as wide an audience as possible to facilitate the development of the GIS field as a whole.

For these reasons the work described in the series is likely to be of considerable importance in the context of the growing European GIS community. However, given that GIS is essentially a global technology most of the issues discussed in these volumes have their counterparts in research in other parts of the world. In fact there is already a strong UK dimension to the GISDATA programme as a result of the collaborative links that have been established with the National Center for Geographic Information and Analysis through the United States National Science Foundation. As a result it is felt that the subject matter contained in these volumes will make a significant contribution to global debates on geographic information systems research.

Ian Masser
François Salgé

Preface

In November 1993, an Italian working in the UK (Max Craglia), a Swiss working in Austria (Andrew Frank), an Englishman working in the Netherlands (Peter Burrough) and a Frenchman working in France (Benoit David) met in Paris to plan the European Science Foundation GISDATA Workshop on 'Concepts and Languages in GIS'. This task force, which had been steamrollered into existence by Max Craglia, the GISDATA Research Coordinator, laid the foundations for a fascinating workshop that was held the following June in Baden, just south of Vienna.

The idea emerged for a meeting that would bring together a broad mix of researchers to confront the problems of dealing with real geographic phenomena that cannot easily be forced into one of the two current standard data models, namely, exact objects or continuous fields. Subsidiary aims were to explore the problems of perception and description of distinct and less distinct spatial phenomena, both natural and anthropogenic, in the context of different national and scientific cultures and means of communication. Researchers from 12 European countries and the USA (including two representatives from the United States, NCGIA) participated. The disciplines included computer science, human and physical geography, soil science, remote sensing, surveying and photogrammetry, geology and linguistics.

Current models in GIS are only useful if positions in space are known, or if it is assumed that space and spatial processes are continuous. However, spatial reasoning does not always require knowledge about position (it is topology that matters most), space and processes are not always continuous, and objects often have indeterminate boundaries. Although in the past there has been a tendency to force reality into sharp objects, there is now an increasing recognition of the need to develop appropriate new methods to represent complexity. A starting point is to distinguish between uncertainty in the description or measurement of crisp objects from uncertainty in phenomena which, by their very nature, are difficult to define (such as baldness). Human beings can cope very well with abstract concepts, while computerized systems do not. Therefore, it is important to elicit and formalize mental models. However, these models are influenced by the perception of the individual, his/her background and culture. The discussion thus needs to focus not only on uncertainty and complexity, but also on the extent to which these concepts vary for different people and cultures. In this respect, the key concern is not on what the world is, but how people see the world and how they deal with it, how they measure it. To understand the different perceptions of reality, language is critical as it is an expression of how we experience space. With this in mind, the aim of this meeting was to elicit the underlying assumptions and perceptions of reality of different groups by analyzing how they formalize or describe a common issue, then to develop a taxonomy of models, to identify current limitations of GIS in dealing with objects with no sharp boundaries, and to stimulate research to overcome these limitations.

Before the meeting we wrote a position paper which was circulated to all participants well in advance. That paper is not included in this volume, but has been published in the *International Journal of Geographical Information Systems* (1995, **9**(2): 101–16) as an extended, reviewed Guest Editorial entitled 'Concepts and paradigms in spatial information: Are current geographic information systems truly generic?'. All participants responded to the invitation to write a reply to the position paper from the point of view of their own discipline and experience. These papers were circulated to all participants before the meeting so that they could form the focus of discussion. An important part of the meeting was the 6 km walk from Baden to Gumpoldskirchen through urban, park and rural landscapes, which gave everyone the opportunity to test ideas in practice before consuming much wine at the *heurige* and a good meal. Enthusiasts walked back in the dark, continuing the discussions.

The papers were reviewed and rewritten during the summer of 1994 and form the substance of Parts 2–6 of this book. Our contributions have been two introductory chapters (Part 1) which review ideas, methods and concepts in defining, delineating and operating with indeterminate geographical objects in both natural phenomena (Chapter 1, Burrough) and anthropogenic situations (Chapter 2, Frank). These chapters set the scene for the rest of the book and provide essential supporting material and references for the reader. We have also provided a short introduction to each Part in order to draw out some of the main points that we have perceived as being noteworthy and important. Part 7 was written after the main text was complete to show that the ideas presented here do indeed relate to reality.

We believe that the material included in this book presents a fascinating cross section of up-to-date opinion from many different but related disciplines and cultures on the problems of describing 'geographic reality' and capturing that reality for applications using GIS. The different chapters raise many issues that can form the basis of future research programmes: ideas have emerged that must surely affect the way geographic data are collected and stored in future if future GIS are to deal successfully with multi-scale, polythetic phenomena in 4D space–time.

We are particularly grateful to Max Craglia and the ESF GISDATA programme for stimulating and financing this work, and to the US NCGIA for support for two participants. Most thanks are due to the participants who wrote, discussed, reviewed and rewrote chapters with enthusiasm and care within a remarkably short timespan. Thanks are also due to Peter Fisher and his team of international reviewers for their efforts to get the position paper reviewed within a few weeks just before the meeting

Peter A. Burrough
Andrew U. Frank

Contributors

Jochen Albrecht
ISPA, University of Osnabrück, Postfach 1553, D-49364 Vechta, Germany

P. Andrieux
INRA-ENSA Science du Sol, 9 Place Viala, F-34060 Montpellier Cedex, France

Thomas Barkowsky
Fachbereich Informatik, Universität Hamburg, Vogt-Kölln-Strasse 30, D-22527 Hamburg, Germany

R. Bouzigues
INRA-ENSA Science du Sol, 9 Place Viala, F-34060 Montpellier Cedex, France

Martin Brändli
Department of Geography, University of Zürich, Winterthurerstrasse 190, CH-8057 Zürich, Switzerland

Peter A. Burrough
Department of Physical Geography, University of Utrecht, PO Box 80115, 3508 TC Utrecht, The Netherlands

Irene Campari
Department of Geoinformation, Technische Universität Wien, Gusshausstrasse 27-29/127, A-1040 Vienna, Austria

Eliseo Clementini
Dipartimento di Ingegneria Elettrica, Università de L'Aquila, I-67040 Poggio di Roio, L'Aquila, Italy

Anthony Cohn
AI Division, School of Computer Studies, University of Leeds, Leeds LS2 9JT, UK

Helen Couclelis
Department of Geography and National Center for Geographic Information and Analysis, University of California, Santa Barbara, CA 93106, USA

Benoit David
Institut Geographique National, Recherche et Developpement, 2 avenue Pasteur, BP 68, F-94160 Saint-Mande, France

Paolino Di Felice
Dipartimento di Ingegneria Elettrica, Università de L'Aquila, I-67040 Poggio di Roio, L'Aquila, Italy

Giacomo Ferrari
Department of Philological Studies, University of Torino at Vercelli, Via g. Ferraris 109, I-13100 Vercelli, Italy

Peter Fisher
Department of Geography, University of Leicester, Leicester LE1 7RH, UK

Andrew U. Frank
Department of Geoinformation, Technische Universität Wien, Gusshausstrasse 27–29/127, A-1040 Vienna, Austria

Christian Freksa
Fachbereich Informatik, Universität Hamburg, Vogt-Kölln-Strasse 30, D-22527 Hamburg, Germany

Nicholas M. Gotts
AI Division, School of Computer Studies, University of Leeds, Leeds LS2 9JT, UK

Thanasis Hadzilacos
Computer Technology Institute, University of Patras, Box 1122, GR-26110 Patras, Greece

M. van den Herrewegen
Institut Geographique National, Abbaye de la Cambre 13, 1050 Bruxelles, Belgium

Marinos Kavouras
Faculty of Rural Surveying and Engineering, National Technical University of Athens, 9 H. Polytechniou Street, Zografos, GR-15780 Athens, Greece

Philippe Lagacherie
INRA-ENSA Science du Sol, 9 Place Viala, F-34060 Montpellier Cedex, France

Robert Laurini
Laboratoire d'Ingénierie des Systèmes d'Information, INSA de Lyon, Bât. 502, 20 avenue Albert Einstein, F-69621 Villeurbanne Cedex, France

Martien Molenaar
Centre for Geo-Information Processing (CGI), Wageningen Agricultural University, PO Box 339, 6700 AH Wageningen, The Netherlands

Dillon Pariente
Laboratoire d'Ingénierie des Systèmes d'Information, INSA de Lyon, Bât. 502, 20 avenue Albert Einstein, F-69621 Villeurbanne Cedex, France

Mark Poulter
Space and Communications Department, Defence Research Agency, DRA Farnborough, Hants GU14 6TD, UK

Francois Salgé
Institut Geographique National, 2 avenue Pasteur, BP 68, F-94160 Sainte-Mande, France

Tapani Sarjakoski
Finnish Geodetic Institute, Department of Cartography and Geoinformatics, Geodeetinrinne 2, FIN-02430, Masala, Finland

Christoph Schlieder
Institute for Computer Science and Social Research, University of Freiburg, Friedrichstrasse 50, D-79098 Freiburg, Germany

Markus Schneider
Praktische Informatik IV, FernUniversität Hagen, Postfach 940, D-58084 Hagen, Germany

E. Lynn Usery
Department of Geography, University of Georgia, Athens, GA 30602-2502, USA

The **European Science Foundation** is an association of its 55 member research councils, academies, and institutions devoted to basic scientific research in 20 countries. The ESF assists its Member Organizations in two main ways: by bringing scientists together in its Scientific Programmes, Networks and European Research Conferences, to work on topics of common concern; and through the joint study of issues of strategic importance in European science policy.

The scientific work sponsored by ESF includes basic research in the natural and technical sciences, the medical and biosciences, and the humanities and social sciences.

The ESF maintains close relations with other scientific institutions within and outside Europe. By its activities, ESF adds value by cooperation and coordination across national frontiers and endeavours, offers experts scientific advice on strategic issues, and provides the European forum for fundamental science.

This volume is the second of a series arising from the work of the ESF Scientific Programme on Geographic Information Systems: Data Integration and Database Design (GISDATA). This four-year programme was launched in January 1993 and through its activities has stimulated a number of successful collaborations among GIS researchers across Europe.

Further information on the ESF activities in general can be obtained from:
European Science Foundation
1 quai Lezay Marnesia
F-67080 Strasbourg Cedex, France
Tel: +33 88 76 71 00
Fax: +33 88 37 05 32

EUROPEAN SCIENCE FOUNDATION

This series arises from the work of the ESF Scientific Programme on Geographic Information Systems: Data Integration and Datahouse Design (GISDATA). The Scientific Steering Committee of GISDATA includes:

Dr Antonio Morais Arnaud
Faculdade de Ciencas e Tecnologia
Universidade Nova de Lisboa
Quinta da Torre
P-2825 Monte de Caparica
Portugal

Professor Hans Peter Bähr
Universität Karlsruhe (TH)
Institut für Photogrammetrie und
Fernerkundung
Englerstrasse 7, Postfach 69 80
D-7500 Karlsruhe 1
Germany

Professor Kurt Brassel
Department of Geography
University of Zurich
Winterthurerstrasse 190
CH-8057 Zurich
Switzerland

Dr Massimo Craglia (Research
Coordinator)
Department of Town and Regional
Planning
University of Sheffield
Western Bank, Sheffield S10 2TN
United Kingdom

Professor Jean-Paul Donnay
Université de Liège, Labo. Surfaces
7 place du XX août (B.A1–12)
B-4000 Liège
Belgium

Professor Manfred Fischer
Department of Economic and Social
Geography
Vienna University of Economic and
Business Administration
Augasse 2–6.
A-1090 Vienna,
Austria

Professor Michael F. Goodchild
National Center for Geographic
Information and Analysis (NCGIA)
University of California
Santa Barbara, CA 93106
USA

Professor Einar Holm
Geographical Institution
University of Umeå
S-901 87 Umeå
Sweden

Profesor Ian Masser (Co-Director and
Chairman)
Department of Town and Regional
Planning
University of Sheffield
Western Bank, Sheffield S10 2TN
United Kingdom

Dr Paolo Mogorovich
CNUCE/CNR
Via S. Maria 36
I-50126 Pisa
Italy

Professor Nicos Polydorides
National Documentation Centre, NHRF
48 Vassileos Constantinou Ave.
GR-116 35 Athens
Greece

M. François Salgé (Co-Director)
IGN
2 ave. Pasteur, BP 68
F-94160 Saint-Mandé
France

Professor Henk J. Scholten
Department of Regional Economics
Free University
De Boelelaan 1105
1081 HV Amsterdam
The Netherlands

Dr John Smith
European Science Foundation
1 quai Lezay Marnesia
F-67080 Strasbourg
France

Professor Esben Munk Sorensen
Department of Development and Planning
Aalborg University, Fibigerstraede 11
DK-9220 Aalborg
Denmark

Dr Geir-Harald Strand
Norwegian Institute of Land Inventory
Box 115,
N-1430 Ås
Norway

Dr Antonio Susanna
ENEA DISP-ARA
Via Vitaliano Brancati 48
I-00144 Rome
Italy

PART ONE

Introductions by the Editors

CHAPTER ONE

Natural Objects with Indeterminate Boundaries

PETER A. BURROUGH

Department of Physical Geography, University of Utrecht, The Netherlands

> To think is to forget a difference, to generalize, to abstract. In the overly replete world of Funes there was nothing but details, almost contiguous details.
> Jorge Luis Borges, *Funes the Memorious* (1962)

The aim of this introduction is to provide a backdrop for subsequent chapters, particularly those in Parts 5 and 6. It reviews the basic concepts of 'objects' and 'fields' used for modelling geographic objects and their handling using conventional discrete logic and explains how these conventional approaches are insufficient for working with complex, polythetic natural phenomena. The basic principles of geostatistics for dealing with quantitative variation in fields are explained. The main ideas of fuzzy set theory and fuzzy k-means classification are introduced and illustrated by applications that (a) demonstrate the decrisping of idealized polygon boundaries, and (b) the extraction of crisp boundaries as zones of confusion between interpolated, polythetic, overlapping classes.

1.1 Simple data models: Objects and fields

In order to model natural phenomena in terms that people can understand it has always been necessary to abstract and to generalize. Because most natural phenomena are complex, varying at many scales in space and time, geographers, soil scientists, geologists, hydrologists, climatologists and other environmental scientists have been forced to select the most important aspects of any given phenomenon and to use these as the basis for information storage and transfer.

Generalizing and abstracting complex, multiscale phenomena requires serious thought: it is far from easy. The 'best' generalization for one purpose may be unsuitable for another. Ideally, each discipline and each scientist would derive a separate generalization for every different situation as the need arises but in practice, and certainly in times before computers and remote sensing, that was simply not possible. To reduce the natural environment to terms that could be easily understood, where information could be stored and exchanged readily, required agreements and standardization. These led to a plethora of classification systems, mapping conventions and jargon that sometimes served their purpose, and sometimes failed as efficient carriers of information. The testing of classification and map quality in soil science, for example, became a major research activity in the years between 1965 and 1985 as scientists increasingly discovered that the classification and mapping systems they had established did not always provide a sound basis for information transfer (cf. Beckett and Webster, 1971; Burrough, 1993). Raper and Livingstone (1995) argue convincingly that the failure to appreciate the interactions between three-dimensional spatial and temporal extents of environmental objects and the nature of physical processes that mould them has resulted in serious philosophical and practical problems when attempting to link environmental models with GIS.

1.1.1 Geographic data models in space and time

Until recently, two diametrically opposed geographic data models were available for encapsulating the important aspects of physical phenomena as perceived by the natural environmental scientist: these are the exact object model and the continuous field model. Ideal geographic entities have crisply defined spatial boundaries and a well defined set of attributes. A typical example is the land parcel in which the boundaries have been accurately surveyed and the attributes of area, ownership, actual use, permitted use, tax value, and so on, apply uniformly to the whole object. On the other hand, there are geographical phenomena that are more often thought of as continuous fields – air pressure, elevation as represented by the hypsometric surface, hydraulic heads or pollution plumes. These are usually represented by smooth mathematical surfaces (often polynomial functions) that vary continuously and smoothly over space–time.

In the conventional exact object model of geographical phenomena, natural entities are represented by crisply delineated 'points', 'lines', 'areas' (and in three dimensions) 'volumes' in a defined and absolute reference system. Lines link a series of exactly known coordinates (points), areas are bounded by exactly defined lines (which are called 'boundaries') and volumes are bounded by smooth surfaces. Lines are linked by a defined topology to form networks which, if open, can represent rivers or blood veins, or if closed, the abstract or defined boundaries of polygons that in turn represent land parcels, soil units or administrative areas. The properties of the space at the points, along the lines or within the polygons or volumes are described by attributes, whose values are assumed to be constant over the total extent of the object. This is the *choropleth* (areas of equal value) model. Cartographic convention has reinforced these abstractions by insisting that mapped boundaries should only be represented by lines of a given style and thickness.

In the simplest form of the continuous field model there are no boundaries. Instead, each attribute is assumed to vary continuously and smoothly over space: its variation can be described efficiently by a smooth mathematical function and it can be visualized by lines of equal value (*isopleths*, or contours). In practice, these fields are often discretized to a regular grid at a given level of resolution.

Both the conventional object model and the continuous field model (Figure 1.1) are two extreme abstractions of reality that are attractive for their logical consistency and their ease of handling using conventional reasoning and mathematics.

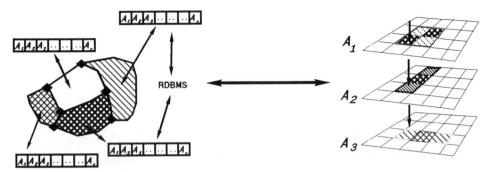

Figure 1.1 Examples of simple spatial objects and fields. Left: a simple object structure with topology and attributes for each polygon. Right: Simple continuous fields (discretized) with a separate layer for each attribute.

Both are easily implemented in computerized information systems. The object model in its simplest form has been implemented using a relational database structure for the attributes and a network data structure for the topology: the continuous field can be handled in a purely relational database structure once it has been discretized to a regular grid. Alternatively, it can be approximated by a network database structure such as a TIN (triangular irregular network) if discretized by a Delauney triangulation (Burrough, 1986). In the TIN the attribute values at data points at the apices of the triangle represent the variation of the elevation field, but each triangle carries a constant value of the local second derivatives of the field, namely, the slope and aspect. Both object and field models have become so accepted as fundamental aspects of geographic information systems that few persons question their general validity.

Both models have been implemented using the graphic models of *vector* and *raster* structures. The vector structure enhances even further the abstraction to exact objects because in the computer digitized lines are by definition infinitely thin (Figure 1.2, left), while the raster structure introduces approximations in shapes and form because of discretization on a regular grid (Figure 1.2, right). Directed pointers between objects can easily be handled in vector system but network interactions can only be handled in raster system by considering cell-to-cell neighbour interactions.

In the sciences that map natural phenomena – surveying, geology, soil science, geomorphology, vegetation, land use, climate, hydrology, etc. – general agreements have developed over the years about when either one of the two data models should be used. In general, when the phenomenon being studied is interpreted as a predominantly static, qualitative complex entity which can be mapped on external features of the landscape then the object model is used (e.g. soil series, land systems, lithologic units, ecotopes). When single quantitative attributes are measured and mapped then the continuous field model is used (e.g. isohyets for rainfall, isobars for atmosphere, contour lines for elevation, etc.).

Note that in the conventional descriptions of phenomena, entities and fields are seen as frozen in time; the possible change in form or composition as a result of a physical process is usually ignored. At best, time is also discretized in steps, by displaying entities and patterns at different points in time: snapshots of diurnal, seasonal or other temporal variation. Neglect of time-induced spatial change may be

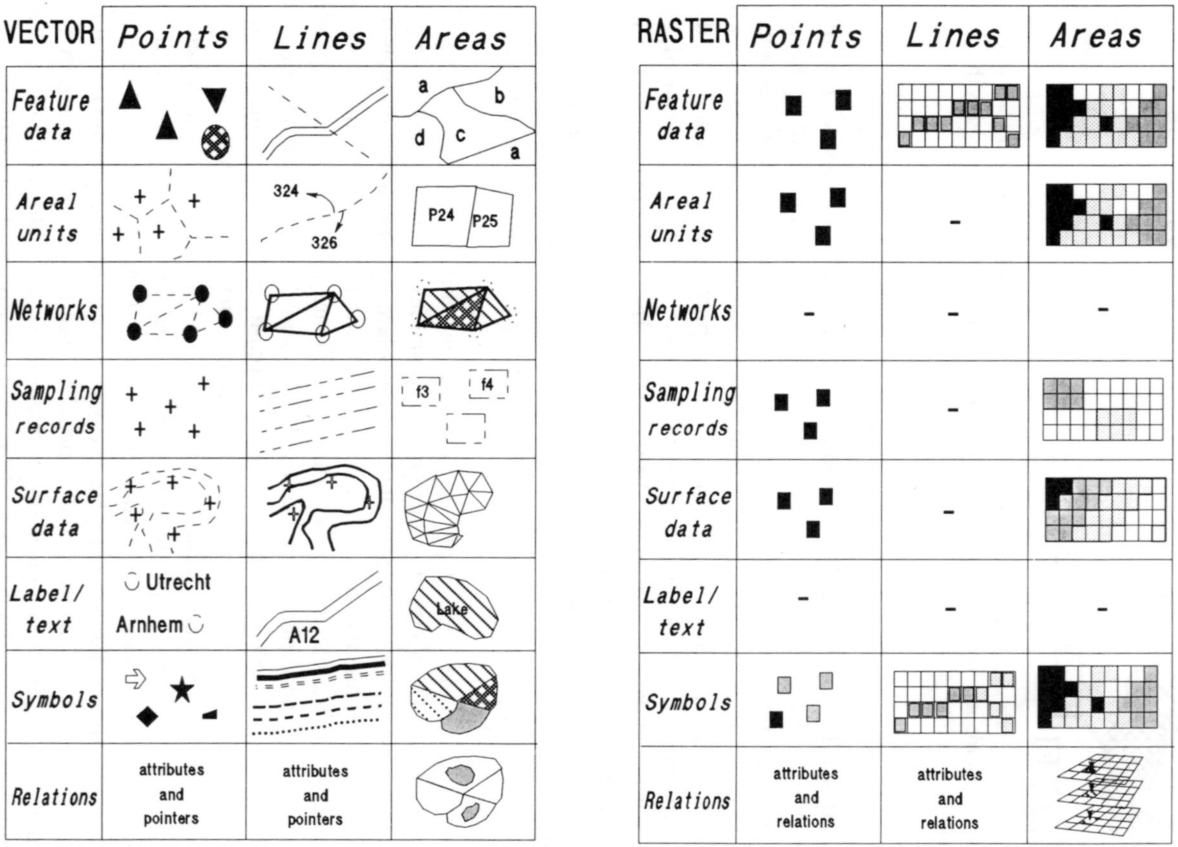

Figure 1.2   Conventional representation of objects and fields in vector and raster GIS.

a root cause of natural geographical entities, like the Aral Sea or the polar ice caps, being perceived as 'objects with indeterminate boundaries'. Note also that conventional descriptions usually relate to a single, often unspecified level of resolution or 'scale'. Linking observations of related phenomena which have been made at different scales is often seen as an exercise in uncertainty (Bierkens and Weerts, 1995; Raper and Livingstone, 1995).

1.1.2 From real world to database

Conventionally, natural resources such as soil, rock types, land use and vegetation are inventoried as static patterns. There are at least three important stages in the assessment process before data can be entered into an information system. These are (a) observation, (b) relating the observation to a conceptual data model, and (c) representation of the data in formal terms. After this follow decisions about implementation in a computer system and whether the graphic display should be raster or vector (Burrough, 1992).

Figure 1.3 shows these three stages. In the first (observation), the surveyor must decide whether the phenomenon in question belongs to the set of fully defined and definable entities (Figure 1.3a, left), or to the set of smoothly varying surfaces (Figure 1.3a, right), or to neither. If it clearly belongs to one or the other, the surveyor can place the observation in one of the two data models (Figure 1.3b) and can choose how it can best be represented (Figure 1.3c). If it belongs to neither, then there are really only two choices. The first is to assume that the phenomenon is truly exact, but obscured by noise and redundant variation, so informed judgement is used to infer and map the boundaries of the supposed objects. This is what happens in aerial photo interpretation and field survey.

If the surveyor feels that the variation is continuous and complex, but not smooth, he can resort to filtering, image enhancement and boundary extraction to extract crisp objects. These methods of pattern recognition are successful when the underlying entity has a regular form and scale, but even in skilled hands, they may produce chimeras when used on data from natural observations.

If image enhancement does not extract 'objects', the surveyor may be forced to conclude that the smooth, continuous model is more appropriate. Univariate and multivariate methods (e.g. principal components) can then be used to interpolate data by suppressing local variation and enhancing longer-range variation.

As knowledge about natural phenomena has progressed it has become clear that these simple 'object/field' models of reality are often inadequate to convey the detailed information that is required to support investigations of how different aspects of the environment interact. Many studies have shown that mapped soil units (one example where there has been much investigation) are not necessarily internally homogeneous, and neither are the boundaries between different soil, geological or vegetation units everywhere crisp and unambiguous. For example, the boundary between the Gault and Kimmeridge clays in the Vale of the White Horse in Oxfordshire, England, was originally mapped along the line of the nineteenth-century railway track! Studies have also shown that very often there can be continuous variation *within* supposedly uniform objects (cf. Burrough, 1993), and that the fields used to map 'continuous' attributes may sometimes be interrupted by abrupt

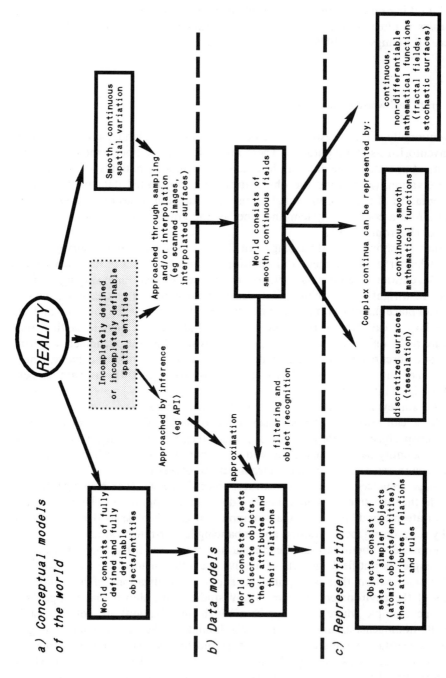

Figure 1.3 The main steps in proceeding from perceptions of reality to a representation in a GIS (adapted from Burrough, 1992).

discontinuities. This should worry managers of natural environment resource databases and their clients because they imply that the fundamental conceptual models
used to represent reality are insufficient to capture the complexities of multiscale,
polythetic phenomena, and that therefore the information they store or use is
incomplete.

1.2 Strategies for dealing with complex variation in space and time

Scientists have several different, general strategies for dealing with complex variation within the scale of the objects or fields used to model a given phenomenon.
These range from investigating hierarchies of nested objects to statistical analyses of
spatial variation.

1.2.1 Complex objects

The first approach is to suggest that an object must be made up of smaller objects:
if they can be identified and it is known how they interact to form larger objects,
then the problem will have been solved. The standard approach of reductionist
science (Cohen and Stewart, 1994) has been very successful in physics, chemistry and
biology. The approach has also been used in the mapping sciences and it has been
successful in applications dealing with complexes of man-made objects, such as in
utilities. Although the hierarchical approach has been espoused in ecology (O'Neill
et al., 1986; Allen and Starr, 1982) and in global change modelling (Kittel and
Coughenour, 1988; Mellilo, 1994), in natural resource mapping the idea of a true
hierarchy seems not to be completely convincing. For example, soil maps have been
made at all scales from the world at $1 : 5 \times 10^{-6}$ (FAO 1975), to almost $1 : 1$ in thin
sections of rocks and soil, but in all cases the data model is the same. There is a
paradox, because a map of the soils of Africa, for example, uses the choropleth
model (exact object) which ideally assumes homogeneity within boundaries, yet
every soil scientist knows that at every level of resolution at least down to the
individual clay particles in the soil (2 μm) variation within 'entities' will not only be
encountered, but can be explained in terms of a natural process of transport, erosion
or chemical change. Within-object homogeneity is therefore often a poor or even
unrealistic assumption. Also, it is rare that a landcover, soil or geological unit identified at smaller mapping scales can be uniquely divided into smaller sub-units that
interact in simple, predictable ways comparable to the sub-objects in a telephone
network. With landcover, for example, increasing the level of resolution often results
in the identification of new areas or classes, particularly in the border areas of the
higher aggregation level.

1.2.2 Exact boundaries can only be approximated

A second approach to dealing with perceived uncertainty, this time along the
boundaries of the entities mapped, is to assume that the true boundary is indeed

exact, but that because measurements are inaccurate or insufficient we make errors when locating them. The answer is to make better or more measurements, which works when the boundary is a defined cadastral or administrative boundary, but which often fails with natural phenomena. With natural phenomena, taking more observations in the boundary zone does not necessarily resolve the 'boundary' but reveals new details of variation in that zone. Mandelbrot (1982) based large parts of his fractal theory on Lewis F. Richardson's observations that the measured lengths of coasts and frontiers depend on the scale at which they are measured. For example, the definition of what is a beach, and what are the boundaries of the beach, are both scale- and context-dependent (e.g. Lam and De Cola, 1993).

1.2.3   Exact attributes can only be approximated

A third approach is to assume that true values of the attributes of an object exist, but again because of errors or difficulties in measurement we can only estimate what these values are. Attributes are sampled at locations within the boundaries (which in this case are assumed to be crisp) and average values and their standard deviations are calculated. The best map is that which yields the lowest pooled (average) within-class variance (Beckett and Webster, 1971). The standard deviations for each attribute are added to the relational data tuples as extra attributes. In a more extreme form, multivariate attribute values at point locations are subjected to numerical classification methods in an attempt to determine a classification (an alternative object model) that might be more appropriate than that derived by field observation (Webster and Oliver, 1990). The central data model of the exact 'object' remains, however.

1.2.4   Exact topology can only be approximated

A fourth approach for the object model is to examine the topology. The links between pairs of objects must be clear and unambiguous, so if they are not, detailed observations should reveal ambiguities in the form of 'pits – closed depressions' or 'saddle points', which can then be removed. Again, more detailed observations may not resolve the problem but may reveal new aspects of the variation, such as leakage from the catchment, subsurface flow, or a river that changes direction of flow according to hydrological conditions.

1.2.5   Replace 'objects' by continuous fields: geostatistics

A fifth approach is to reject the object models of areas or volumes in favour of a continuous field, but one in which the variation is assumed to be continuous, but irregular. Simple methods of interpolation use a straightforward mathematical interpolator such as inverse distance or mathematical splines to generate a smoothly varying surface. However, these methods are not necessarily optimal in the sense that they impose constraints on the interpolation weights, which can lead to bias (cf. Burrough, 1986, ch. 8).

A very successful alternative has been to attempt to examine the nature of the spatial variation in terms of values at data points before the interpolation is carried out. This has been the approach of geostatistics which considers all attributes to vary throughout 2D or 3D space as *regionalized variables* which can be modelled using stochastical theories (e.g. Journel and Huijbregts, 1978; Isaaks and Srivastava, 1989; Deutsch and Journel, 1992).

The following is a very brief introduction. Regionalized variable theory assumes that the spatial variation of any variable can be expressed as the sum of three major components. These are (a) a structural component, associated with a constant mean value or a constant trend; (b) a random, spatially correlated component, known as the variation of the regionalized variable, and (c) a random noise or residual error term. Let x be a position in 1, 2 or 3 dimensions. Then the value of a random variable Z at x is given by

$$Z(x) = m(x) + \varepsilon'(x) + \varepsilon'', \qquad (1.1)$$

where $m(x)$ is a deterministic function describing the 'structural' component of Z at x, $\varepsilon'(x)$ is the term denoting the stochastic, locally varying but spatially dependent residuals from $m(x)$ – the regionalized variable – and ε'' is a residual, spatially independent Gaussian noise term having zero mean and variance σ^2.

The first step is to decide on a suitable function for $m(x)$. In the simplest case, where no trend or 'drift' is present, $m(x)$ equals the mean value in the sampling area, and the average or expected difference between any two places x and $x + h$ separated by a distance vector h, will be zero:

$$E[z(x) - z(x + h)] = 0, \qquad (1.2)$$

where $z(x)$, $z(x + h)$ are the values of random variable Z at locations x, $x + h$. Also, it is assumed that the variance of differences depends only on the distance between sites, h, so that

$$E[\{z(x) - z(x + h)\}^2] = E[\{\varepsilon'(x) - \varepsilon(x + h)\}^2] = 2\gamma(h), \qquad (1.3)$$

where $\gamma(h)$ is known as the semivariance. The two conditions – stationarity of difference and variance of differences – define the requirements for the intrinsic hypothesis of regionalized variable theory. This means that once structural effects have been accounted for, the remaining variation is homogeneous, so that differences between sites are merely a function of the spatial covariance structure and the distance between them. If the conditions specified by the intrinsic hypothesis are fulfilled, the semivariance can be estimated from the sample data:

$$\gamma(h) = \frac{1}{2n} \sum_{i=1}^{n} \{z(x_i) - z(x_i + h)\}^2, \qquad (1.4)$$

where n is the number of pairs of sample points of observations of the values of attribute Z separated by distance h. A plot of $\gamma(h)$ against h is known as the *experimental variogram* (Figure 1.4). The experimental variogram is the key to providing a quantitative description of spatial covariance structure of the attribute, which provides useful information for interpolation, optimizing sampling and determining spatial patterns. A key feature of interpolation with kriging is that predictions of

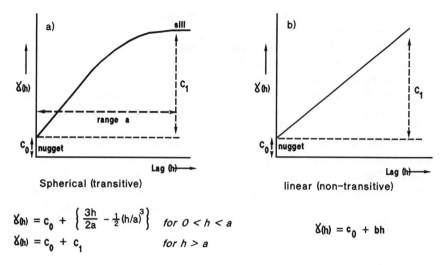

$$\gamma(h) = c_0 + \left\{ \frac{3h}{2a} - \frac{1}{2}(h/a)^3 \right\} \quad \text{for } 0 < h < a$$

$$\gamma(h) = c_0 + c_1 \qquad \qquad \text{for } h > a$$

$$\gamma(h) = c_0 + bh$$

Figure 1.4 Examples of frequently used variogram models.

attribute values at unsampled locations are always accompanied by a standard deviation, which depends on the covariance structure and the configuration of data
points – fewer points in the region of the location and a complex covariance structure leads to a greater uncertainty.

As with exact objects (above), regionalized variable theory also recognizes that
variation may occur at more than one scale, and that each scale may have its own,
different covariance structure. Figure 1.5 illustrates four of these surfaces, ranging
from pure nugget (a), through surfaces with reduced proportions of nugget to correlated variation (b, c) to a completely smooth surface (d). The *linear model of
coregionalization* allows complex variation to be modelled in terms of the sum of
variograms for each individual scale:

$$\gamma(h) = \gamma(h_1) + \gamma(h_2) + \gamma(h_3) + \cdots, \tag{1.5}$$

so that complex spatial variation can be seen as the linear sum of nested levels of
variation. There are no *completely unambiguous* ways of decomposing complex
irregular variation into unique sets of nested local covariance structures, however.

Geostatistics has developed into a major branch of spatial statistics with applications in geology, mining and mineral prospecting, soil science and hydrology.
Because of the complexities of the mathematics and the need to have sufficient data
to model the spatial covariance structures, it has been generally limited to detailed
studies of single, or small groups of related attributes in relatively small areas. Geostatistics has provided a new model for continuous data because it allows for variation at all scales to be modelled in terms of probabilistic terms that depend on the
location of data points and on spatial covariance structures. In its simplest form it
does not require that data on natural phenomena are classified as entities with crisp
boundaries, though increasingly practitioners are using the boundaries of geographic 'objects' to delineate the space in which a given spatial covariance structure
holds (Burrough, 1993). In this way extreme restrictions of the internal homogeneity
of the choropleth model can be usefully circumvented.

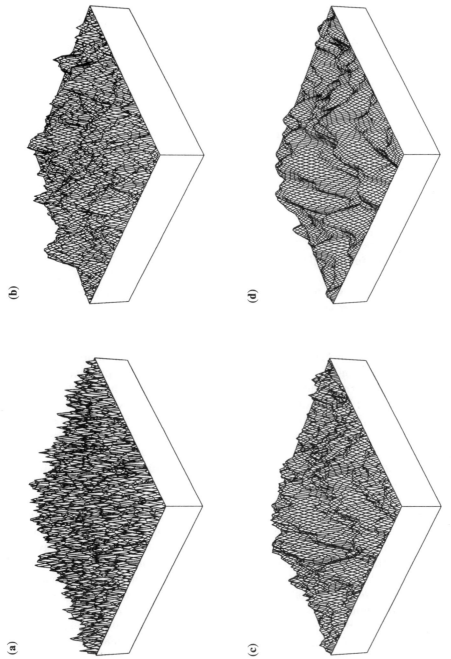

(a)

(b)

(c)

(d)

Figure 1.5   Simulated surfaces having different spatial covariance structures.

Current geostatistical methods can cope with qualitative, multivariate attribute data (Deutsch and Journel, 1992; Bierkens and Burrough, 1993), but the methods are complex and computationally intensive. Multivariate data from sets of point observations can be interpolated by any sensible method, but there is often a problem of interpreting the results in physical terms, or in terms of a perspective that is wider than the area from which the samples were taken.

1.2.6   Conclusions

Although exact data models may be used for the perception, observation and storage of data about the natural environment, it is being increasingly realized that they are insufficient to deal with the polythetic, multiscale nature of much natural variation. Geostatistical tools have gone the furthest to providing practical solutions in certain areas.

1.3   Logical tools for handling simple objects and fields

> Logic, like geometry, was to become divorced from physical reality: a complex suite of games played with symbols that created paper worlds of their own.
>
> J. D. Barrow, *Pi in the Sky* (1992)

The philosophical basis behind the division of geographical space into smaller and smaller, but exactly defined objects, is that of the reductionist approach, which is a hallmark of Western scientific thought. The reductionist approach has been extremely successful in providing a fundamental basis for thinking about the molecular and sub-atomic world, and it has also been overwhelmingly important to the development of technology (Cohen and Stewart, 1994). This approach acknowledges that the best way to solve any problem is to divide it up into a set of separate, smaller problems, which are easier to tackle. The resulting solution is then the sum of all the sub-solutions.

The axioms behind the reductionist approach, based on 'laws of thought' that were first developed by Aristotle, have so permeated Western thinking that until this century they were regarded by many, including mathematicians, as uncontentious, even though simple intuition and thought structures in other civilizations have long been less constrained (Barrow, 1992). The most important aspects of Aristotelian logic are: (a) the law of *identity* (everything is what it is – a house is a house), (b) the law of *non-contradiction* (something and its negation cannot both be true – a house cannot be both a house and a non-house), and (c) the *principle of the excluded middle* (every statement is either true or false – this house is lived in or it is not inhabited).

The principle of the excluded middle ensures that all statements in conventional logic can take only two values, true or false. This assumption lies at the heart of most of mathematics and computer science (Barrow, 1992) and although it and the other Aristotelian axioms are currently important subjects of discussion, the paradigms are very deeply embedded in our tools for computation and data retrieval. The axioms used in database languages such as SQL implicitly assume that binary logic (true/false) is necessary and sufficient for carrying out all tasks of data retrieval and transformation.

1.3.1 Inexact objects require inexact data models

Developments in both the study of fractals and geostatistics have revealed the strengths and the rigidity and poverty of conventional data models of objects and fields, but neither have provided complete answers for dealing with those situations where the natural phenomenon is both polythetic and is not self-similar. The original fractal theory requires self-similarity over a large range of scales, and though people have adopted the idea of *multi-fractals* (scale-dependent fractal behaviour) this too seems an artificial constraint, particularly when each scale-related range of variation has been caused by a different physical process. Besides, we should look not only at the spatial representation of a complex phenomenon, but also at how it should be treated logically.

Increasingly, people are beginning to realize that the fundamental axioms of simple binary logic present limits to the way we think about the world. Not only in everyday situations, but also in formalized thought, it is necessary to be able to deal with concepts that are not necessarily 'TRUE' or 'FALSE', but that operate somewhere in between.

In practice, problems arise because there are geographical phenomena that possess attributes of both 'objects' and 'fields'. These phenomena are polythetic – they are defined by more than one attribute – and while there may be general agreement about the core concept of the object, both the geographic boundaries and the boundaries in attribute space (which are not necessarily the same) around that core cannot be located exactly. A good example is the geographic 'region' – an area of a country or town that is perceived as belonging to a certain core idea, but the boundaries of which are diffuse or uncertain. Two instances will suffice: a Cockney was said to be someone who was born within the sound of Bow Bells in the City of London, but does this depend on the direction and strength of the wind, or on the background level of traffic noise? The Caribbean is usually thought of as being only the island states, but through history, culture and population does it also extend to the South American mainland in Guyana, and possibly to French Guiana and Surinam or even the north coast of Venezuela as well? Given the uncertainties about core and periphery (well understood geographical concepts) how should these be treated by exact data models and Aristotelian logic?

To go to the other extreme, some polythetic phenomena vary continuously, with a different mix of attribute values being observed at each separate sample point. We can interpolate wind fields, temperatures and pressures, but only by looking at them together do we see that a particular set of combinations represents a hurricane. Similarly, a particular combination of temperature, humidity and vegetation availability might translate into a potential danger area for locust swarms. In both cases – hurricane and locust swarm – there is an identifiable entity, which could be modelled as an exact object, or as a set of fields, yet abstraction to either extreme rejects much of the really useful information. Many other examples, such as lakes that vary in area depending on the inputs of precipitation or snowmelt, the succession of vegetation complexes, the composition of socio-economic groups, and even in the light of the recent troubles in the former Yugoslavia, the unifying concept of the nation state could have been used.

In the past, scientists and administrators have ignored or suppressed important aspects of inexact or 'fuzzy' phenomena and for their own reasons have forced them to be 'object' or 'field'. Practical scientists, faced with the problem of dividing up undividable complex continua have often imposed their own crisp structures on the raw data, as have often been the case in the aerial photo interpretation of soil or vegetation patterns. The result has been that different scientists have mapped the same area differently (Bie and Beckett, 1973) but without insight into the inter-actions of 4D patterns and processes, no one can say that one person is more correct than another.

The problems associated with binary logic for data retrieval can be seen from the following simple example taken from the discipline of land evaluation (Burrough *et al.*, 1992). With binary logic, complex land qualities and suitability classes can be defined using the operators AND, OR, NOT, XOR to specify just which com-binations of attribute values are required for any given purpose. Class membership is commonly defined by specifying the ranges of a certain number of key property values that an individual must meet. To qualify as a member of a given suitability class, an individual point, line or area must match all the specifications, which is a multiple Boolean AND, or intersection.

$$R = \text{'true' if } (A \text{ AND } B \text{ AND } C \text{ AND } D \cdots), \tag{1.6}$$

where R is the result and A, B, C, D, ..., represent the specified ranges of the key properties. The result is binary, '*true*' being represented by the character 1, '*false*' being represented by the character 0. This logic is often extended to a limited number of discrete classes (>2) that describe grades of suitability, through the prin-ciple of 'limiting conditions' (FAO, 1976). For example, if the suitability of a site for a given land use is determined from the levels of several land qualities, then the land quality with the lowest value determines the site classification.

$$R = \text{MIN } (Q_1, Q_2, Q_3, \ldots), \tag{1.7}$$

where the Q_i are the classified values of each key land quality (usually integers).

Now consider the rule:

IF SLOPE $\geq$ 10% **AND** SOIL TEXTURE

= SAND **AND** VEGETATION COVER $\leq$ 25%

THEN EROSION HAZARD **IS** SEVERE.   (1.8)

This specifies a central concept, namely, that bare sandy soils on more than gently sloping sites are prone to extreme erosion. Everyone would agree, but let us con-sider what the result would be if the slope is only 9%, or vegetation cover is 27%? Clearly, with crisp logic the erosion hazard would not be severe, although most experts would point out that it is still very great. If the rule were used in a GIS on soil polygons with simple attribute values, however, then the data retrieval oper-ation would only find those polygons that are an exact match. Clearly, polygons that for one reason or another had attributes that were just outside the class bound-aries would be rejected. The result might lead to a serious underestimate of the area of land that is prone to severe erosion (cf. Burrough, 1989).

Consider now the problem of measuring the attributes in the field. Placing a soil profile in a sand texture class may not seem too difficult, but in some places the texture might vary greatly over short distances because of differential weathering and deposition, so that 'sand' may only represent a small part of the polygon. Measuring slope to a precision of 1% is often difficult, particularly in rough country, and then there is also the difficulty of deciding the size of the area of the hillside over which slope is measured. Are we interested in an average slope over a given area, or a maximum slope measured at a point? Equally difficult is the exact assessment of vegetation cover, which can only be assessed approximately, even if the location of each plant is known. Even if we accept that vegetation cover is determined by the vertical projection of plant outlines on to the soil which protects against rain splash, we must recognize that the parts of the plant in contact with the surface of the ground will be more effective in slowing overland flow, and hence erosion, than those that are a metre above the surface. Alternatively, the presence of water dripping from leaves and branches can enhance local erosion even though the ground is fully 'covered'. In short, none of the inputs to the logical model of erosion hazard can be determined exactly. Because estimates of values that are just outside the accepted class limits lead to rejection, many sites that have the potential to be seriously eroded will not be retrieved.

The conventional approach to solve this problem is to take more measurements, map out smaller areas that are more uniform or devise more accurate ways of collecting the data. This is costly and unnecessary if we are prepared to accept a risk of being wrong part of the time and can move away from the constraining model of binary or simple discrete logic to a more flexible, continuous system that allows the ideas of class overlap, partial membership of a set and the concept of partial truths that we use in everyday speech.

In natural language (everyday speech) we have the full freedom to add modifiers to divide up the middle ground between potentially overlapping classes, and any difficult situation can be decided locally. For example, consider the discrimination between hirsute and bald men. At first, we might consider that a person who has hair growing normally on the head is hirsute and one who has no hair is bald. But is a skinhead with a shaved pate bald? Surely not. Is a middle-aged man with a receding hairline or a thinning patch on the dome, bald? He does not want to think so. Some men even grow the hair above the ears so long that they can sweep it over and pretend that the top is nicely thatched. Without generally defined and definable criteria, who is bald and who is not remains a ticklish business, even though we all have a clear notion of the concept 'bald'.

Similar problems occur with classifying and mapping natural phenomena. Many users of environmental information have a clear notion (or central concept) of what they need. Land evaluators and planners can usually define the ideal requirements for a particular kind of land use, but they are often unsure about just where the boundaries between 'suitable' and 'unsuitable' classes of land should be drawn. In many situations they also formulate their requirements generally in terms such as 'Where are areas of soil suitable for use as experimental fields', 'How much land can be used for smallholder maize?', 'Which areas are under threat of flooding?' or 'Which parts of the wetlands suffer from polluted discharges?'. Such imprecisely formulated requests must then be translated in terms of the basic units of information available. Furthermore, not all information stored in the simple data model is exact, however. Many data collected during field survey are often described in seemingly vague terms: soils can be recorded as being 'poorly drained', having 'moderate nutrient availability', being 'slightly susceptible to soil erosion'; vegetation can be described as 'vital', 'partially vital', and so on. Even though standard manuals define these terms with more precision, in practice they retain a strong flavour of qualitative ambiguity.

In natural resource inventory and modelling it is not sensible to permit users to define their classes in a totally *ad hoc* way because we need to agree on a limited number of standardized classes to facilitate training, research and general information transfer. Clearly, we need methods for specifying how we must deal with imprecise information and borderline cases.

1.4   Fuzzy sets and fuzzy objects

The problem of dealing with non-exact objects and classes is not unique to environmental phenomena but is a wider part of human experience as is shown by several other chapters in this book. In 1965, Zadeh introduced the idea 'fuzzy sets' to deal with inexact concepts in a definable way (see Zadeh, 1965). Since the 1960s the theory of fuzzy sets has been developed to the point where useful, practical tools are available for use in other disciplines (Zadeh, 1965; Kandel, 1986; Kauffman, 1975; Klir and Folger, 1988).

Fuzziness is a type of imprecision characterizing classes that for various reasons cannot have, or do not have sharply defined boundaries. These inexactly defined classes are called fuzzy sets. Fuzziness is often a concomitant of complexity. It is appropriate to use fuzzy sets whenever we have to deal with ambiguity, vagueness and ambivalence in mathematical or conceptual models of empirical phenomena.

Fuzziness is *not* a probabilistic attribute, in which the degree of membership of a set is linked to a given statistically defined probability function. Rather, it is an admission of the *possibility* that an individual is a member of a set, or that a given statement is true. The assessment of the possibility can be based on subjective, intuitive ('expert') knowledge or preferences, but it can also be related to clearly defined uncertainties that have a basis in probability theory. For example, uncertainty in class allocation could be linked to the possibility of measurement errors of a certain magnitude. Of course, geographically speaking, we still have to define what is meant by an *individual* in a spatial sense.

1.4.1   Crisp sets and fuzzy sets

Conventional or crisp sets allow only binary membership functions (i.e. TRUE or FALSE): an individual is a member or it is not a member of any given set. Fuzzy sets, however, admit the possibility of *continuous partial membership*, so they are generalizations of crisp sets to situations where the class boundaries are not, or cannot be sharply defined, as in the case of bald and hirsute men. The same is true for geographical properties such as 'vegetation cover', 'trafficability', 'urbanization', and so on. To state just what is, and what is not, 'moderate vegetation cover' requires not a strict allocation to a class, but a qualitative judgement that implies the possibility of partial membership.

It is necessary to provide some definitions of the concepts that will be used. A *crisp set* is a set in which all members match the class concept and the class boundaries are sharp. The degree to which an individual observation z is a member of the set is expressed by the *membership function*, MF^B, which for crisp (Boolean) sets can take the value 0 or 1. Formally, we write

$$MF^B(z) = 1 \quad \text{if } b_1 \leq z \leq b_2,$$

$$MF^B(z) = 0 \quad \text{if } z < b_1 \text{ or } z > b_2,$$

(1.9)

where b_1 and b_2 define the exact boundaries of set A. For example, if the boundaries between 'unpolluted', 'moderately polluted' and 'polluted' soil were to be set at $b_1 = 50$ units and $b_2 = 100$ units, then the membership function given in (1.9) defines all 'moderately polluted soils'.

A *fuzzy set* is defined mathematically as follows: If $Z = \{z\}$ denotes a space of objects, then the fuzzy set A in Z is the set of ordered pairs

$$A = \{z, MF_A^F(z)\}, \quad z \in Z,$$

(1.10)

where the membership function $MF_A^F(z)$ is known as the 'grade of membership of z in A' and $z \in Z$ means that z is contained in Z. Usually $MF_A^F(z)$ is a number in the range 0, 1 with 1 representing full membership of the set (e.g. the 'representative profile' or 'type'), and 0 representing non-membership. The grades of membership of z in A reflect a kind of ordering that is not based on probability but on admitted possibility. The value of $MF_A^F(z)$ of object z in A can be interpreted as the degree of compatibility of the predicate associated with set A and object z; in other words $MF_A^F(z)$ of z in A specifies the extent to which z can be regarded as belonging to A. So, the value of $MF_A^F(z)$ gives us a way of giving a graded answer to the question 'to what degree is observation z a member of class A?'

Put simply, in fuzzy sets, the grade of membership is expressed in terms of a scale that can vary *continuously* between 0 and 1. Figure 1.6(a) illustrates graphically the difference in membership functions between crisp and fuzzy sets. In both cases, individuals close to the core concept have values of the membership function close to or equal to 1, but in fuzzy sets membership function values decline continuously rather than abruptly away from the core zone. Note that this immediately gets around the problems of the principle of the excluded middle (truth is *not* absolute) and individuals can be members to different degrees of more than one set. The problem is to decide on how the membership function is chosen, and its shape or form.

1.4.2 Choosing membership functions

Membership functions for crisp sets are chosen either (a) on the basis of expert knowledge (e.g. boundary values of discriminating criteria in soil or vegetation classification: the key identifies the 'individuals'), or (b) by using methods of numerical taxonomy. Classes based on expert knowledge are usually *imposed* or imported classes that are set up without direct reference to the local data set. They may approximate 'natural' divisions, but they are not necessarily optimal in any statistical sense. Only two parameters, the lower and upper boundary values, are needed. These classes are used a great deal in practical science and administration.

(a)

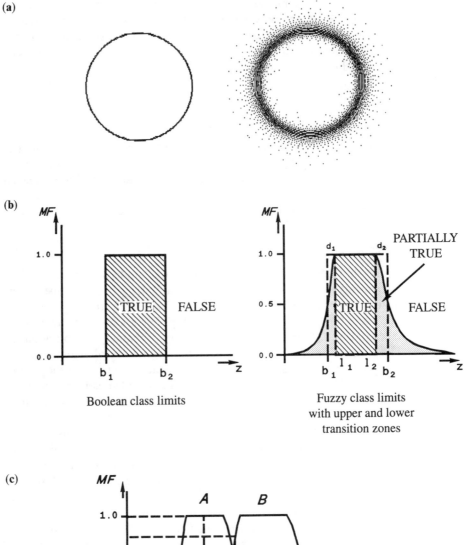

(c)

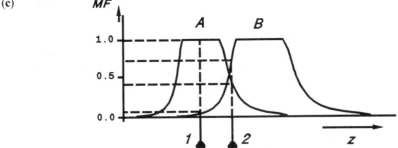

Figure 1.6 Properties of crisp and fuzzy sets. (a) Simple representation of crisp and fuzzy sets. (b) Crisp and fuzzy membership functions. (c) Partial membership in fuzzy sets.

The alternatives to imposed class boundaries are the data driven, so-called 'natural' classification methods, which are locally optimized to match the data set. In practice, the choice of classification method, parameter values, etc., can strongly affect the results of the classification.

Both options are also possible with fuzzy sets. The first and simpler approach uses an *a priori* imposed membership function with which individual objects or attribute values can be assigned a membership grade. This is known as the *semantic import approach or model* (SI).

The fuzzy class limits shown in Figure 1.6(b) are defined by the following three equations:

$$MF^F(z) = \frac{1}{1 + \left(\dfrac{z - b_1 - d_1}{d_1}\right)^2} \qquad \text{if } z < b_1 + d_1 \tag{1.11a}$$

$$MF^F(z) = 1 \qquad \text{if } b_1 + d_1 \leq z \leq b_2 - d_2 \tag{1.11b}$$

$$MF^F(z) = \frac{1}{1 + \left(\dfrac{z - b_2 + d_2}{d_2}\right)^2} \qquad \text{if } z > b_2 - d_2 \tag{1.11c}$$

where $MF^F(z)$ is the value of the continuous membership function corresponding to the attribute value z. Note that if parameters d_1 and d_2 and zero, Equation (1.11) yields the Boolean membership function (Equation (1.9)). These are by no means the only definitions of continuous membership functions, but they are among the most general because they allow for differentiation in both upper and lower transition zones.

The second method is analogous to cluster analysis and numerical taxonomy in that the value of the membership function is computed from a set of attribute data. One frequently used version of this model is known as the method of *fuzzy k-means*. There are several algorithms for computing fuzzy k-means, but that of Bezdek (Bezdek, 1981; Bezdek *et al.*, 1984) is commonly used which computes the membership μ of the *i*th object to the *c*th cluster in ordinary fuzzy k-means, with d the distance measure used for similarity, and the fuzzy exponent q determining the amount of fuzziness:

$$\mu_{ic} = [(d_{ic})^2]^{-1/(q-1)} \left/ \sum_{c'=1}^{k} [(d_{ic'})^2]^{-1/(q-1)} \right. . \tag{1.12}$$

The computation of fuzzy k-means proceeds by iteration (Figure 1.7). The membership of an individual with respect to a class mean in multivariate space is computed and it is assigned to the nearest class. Following this the class mean is recomputed. The procedure is repeated until the classes stabilize.

To sum up, fuzzy methods permit the computation of fuzzy membership values either through an imposed 'expert' model, or by a data driven multivariate procedure. In both cases the methods allow class overlap, which can be taken into account when 'objects' or 'fields' are processed with logical operations.

1.4.3   Logical operations with fuzzy sets

Logical operations with fuzzy sets are generalizations of the usual Boolean algebra (see Kandell, 1986; Klir and Folger, 1988) applied to observations that have partial membership of more than one set (Figure 1.6c). The 'AND' in binary logic is

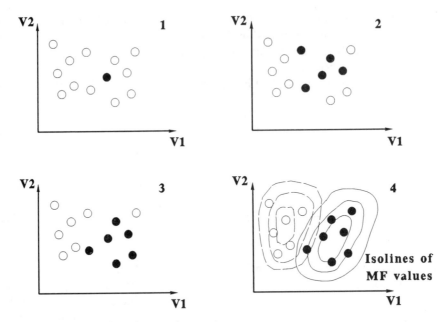

Figure 1.7   Computing fuzzy *k*-means by iteration.

replaced by a 'MIN' operation; the 'OR' by a 'MAX' operation. For example, the rule

IF SLOPE ≥ 10% **AND** SOIL TEXTURE

= SAND **AND** VEGETATION COVER ≤ 25%

THEN EROSION HAZARD **IS** SEVERE

can be rewritten as

$$SE_B = MIN\ (SL, ST, VC),\tag{1.13}$$

where SL, ST and VC are slope, soil texture and vegetation cover classes. In the Boolean case SL, ST and VC can only take values of 0 or 1, so the susceptibility to erosion SE_B can only be 0 or 1. In the fuzzy case the result is the minimum of the three values

$$SE_F = MIN\ (SL, ST, VC),\tag{1.14}$$

so if $SL = 0.9$, $ST = 0.8$ and $VC = 1.0$, the result is 0.8, or in other words, this combination of attributes fulfils 80% of the requirements to be an erosion hazard, and therefore should not be rejected. Heuvelink and Burrough (1993) have demonstrated that the fuzzy selection procedure is much less sensitive to errors in the data than the equivalent discrete procedure.

   The simple rules of fuzzy set theory should be used with care on membership values computed by fuzzy *k*-means because, unlike the SI approach where each class is scaled from 0–1, fuzzy *k*-means class memberships *sum to one*. Consequently, each class is not independent and neither do classes have equal weights.

1.5 Fuzzy geographical objects

Two examples will serve to show how fuzzy membership functions can be applied to geographical data, namely, polygon boundaries and the mapping of polythetic soil classes.

1.5.1 Decrisping polygon boundaries

If there is information about how abrupt or diffuse the boundaries of an entity really are it can be used to improve the data model of exactly bounded objects that are usually handled in geographic information systems. For example, Figure 1.8(a) shows a conventional choropleth map, of which Figure 1.8(b) shows the crisp boundaries of the largest polygon. If the boundaries were the sort that should be exact, but through measurement and poor digitizing have been located with an error, then the approach of Perkal (see Burrough, 1986, p.117) is appropriate. If the boundary indicates a degree of membership of the polygon, then it may be more appropriate to use a SI model to compute membership values of sites with respect to their geographical distance perpendicular to the digitized boundary. This means that sites well inside the digitized boundary are full members of the set of sites belonging to the polygon, whereas those near the boundary (both inside and outside) belong to the polygon in a degree that varies with distance and the chosen membership function.

Figure 1.8(c) shows the gradation of membership function values when the digitized boundary is replaced by a uniform transition zone of 500 m (10 grid cells). This map could easily be used in map algebra operations for computing weighted land suitabilities that would avoid the artificiality of the crisp polygon envelope.

In practice, there is no *a priori* reason for the whole boundary of a single polygon to have the same degree of diffuseness along its whole length. The boundaries of geological, vegetation and soil units are often sharp in some places and diffuse in others, information which is usually noted by the surveyor in the field or in aerial photo interpretation but rarely used. If this information is available and can be attached to the relevant sections of the polygon boundary then the spatial differentiation of the diffuseness of the boundary can easily be computed (Figure 1.8(d)).

1.5.2 Deriving boundaries from continuous variation

In an earlier paper (Burrough, 1989), I used the SI model to classify suitable sites for agricultural experiments on a farm in Turén, Venezuela. Soil data on 11 attributes (depth to plough pan, depth to reduction mottles and per cent organic matter, sand and clay from layers 0–20 cm, 30–40 cm and 70–80 cm) had been collected from 69 sites on a 75 m regular grid. Because of the young alluvial soil of the area it was not easy to classify the 69 sites into unambiguous classes.

For this chapter, the data were classified into four overlapping classes using the fuzzy *k*-means procedure. Figure 1.9(a–d) presents maps of the results which were obtained by ordinary point kriging of the membership values for each class at each site. Clearly, the membership values for each class vary continuously and although there are some abrupt transitions (e.g. with class 4), gradual change is more common

(a)   **(b)**

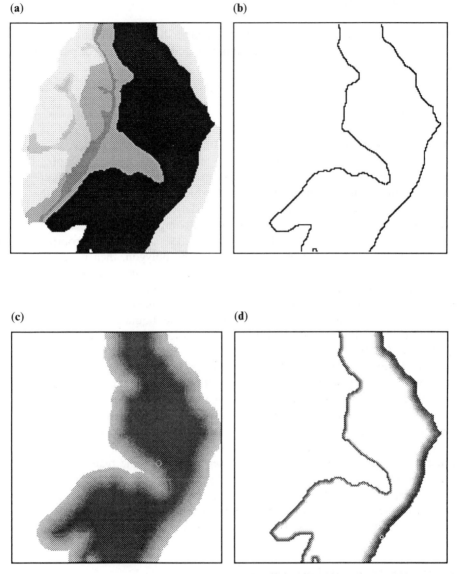

Figure 1.8 Using information on the diffuseness of boundaries to decrisp polygons. (a) A choropleth map. (b) Crisp polygon boundaries. (c) Two-sided diffuse boundary resulting from applying membership functions to a polygon. (d) Sharp, medium and diffuse fuzzy boundaries on one polygon.

than sharp boundaries. This is in line with the nature of the soil-forming process at the site.

The grey scales on these four maps show also that the maximum membership values for classes 2 and 4 are larger than those for classes 1 and 3; in other words,

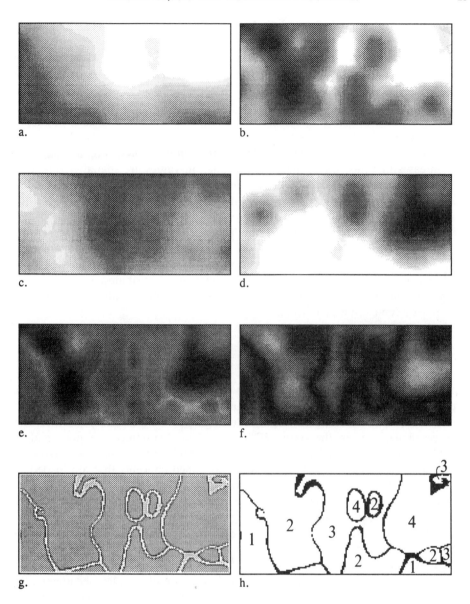

Figure 1.9 Fuzzy *k*-means classification followed by interpolation and intersection of surfaces can give rise to extractable boundaries. (a) Membership values class 1. (b) Membership values class 2. (c) Membership values class 3. (d) Membership values class 4. (e) Maximum membership values for each cell. (f) Confusion index per cell. (g) Edge filtering to extract boundaries. (h) Extracted boundaries and classes.

some classes are less ambiguous and therefore more important than others. This can be seen more easily in Figure 1.9(e), which shows the maximum interpolated membership value for each cell. This figure also suggests that the maximum membership values for sites in between two classes are lower than average. It is as if at these sites

there is the greatest confusion about the class to which a given cell should be allo-
cated.

We can define a 'confusion index' (*CI*) for each cell in a number of ways, either as
the ratio of the second largest membership value to the largest, or as one minus the
difference between these two:

$$CI = 1 - (\text{Maxfuz} - \text{maxfuz2}). \tag{1.15}$$

In either case, the *CI* value is largest when the difference in membership function
values of the two most important classes for the cell is smallest. Figure 1.9(f) shows
that in cases like this, where membership function values are spatially contiguous,
the zones of maximum confusion divide areas with relatively homogeneous membership values. In fact, these confusion zones indicate the presence of boundaries, which
can be enhanced by conventional edge filtering and boundary extraction procedures
(Figures 1.9(g,h). Note that these boundaries are not absolute, but are entirely
dependent on the attributes used in the classification, the data recorded at the
sample sites, the nature of the soil variation in the area and, of course, the method
used and the values of its parameters. A different set of attributes, a different set of
samples or, indeed, another method could produce other boundaries, which might
be better by some criterion of quality.

1.6   Discussion

Unlike Borges' character Funes, who was unable to reduce the complexities of his
experience to manageable proportions, most people find it all too easy to generalize
and abstract. Human beings are highly skilled at classification and pattern recognition, particularly when the static, 2D entities being examined are man-made or
capable of being examined in isolation (cf. Burrough and Frank, 1995). However,
exact data models and discrete logic are over-refined concepts that are inadequate
for dealing with complex, polythetic, multiscale phenomena. It is time to relax our
formal requirements, to think naturally and accept the need for a formalized compromise. Fuzzy logic is a way of thinking that seems particularly appropriate for
this purpose, the more so because it is easy to implement on current computer
systems, as the two examples showed.

The examples showed that the concepts of a continuous membership function
and overlapping sets permit a more realistic description of complex phenomena that
fall between the conventional extremes of 'objects' and 'fields' than do either of these
alone. In both examples, the characteristics of 'boundaries' and of gradual variation
were both handled easily, and were capable of being handled as the context
demands.

The concept of a boundary as a zone of confusion has been received by many
persons as being counter-intuitive. Surely, they say, boundaries are geographical
objects that we really must be certain about: they are features in their own right and
we must specify them and map them accurately. Personally I do not entirely agree
with this view. Certainly there are situations where clear features indicate abrupt
spatial change (and these are extremely valuable for locating an agreed boundary),
but common experience shows that the zone around a boundary, unless it is clearly
indicated, can give rise to disputes and argument. For example, when the ice
mummy was found in the Ötztaler Alpen a few years ago, a dispute between Austria

and Italy ensued over the ownership of the body. Had he been found in Austria or in Italy – there were insufficient markers on the ground to be sure? The ownership of the body and the right to undertake prestigious scientific research depended on the supposed position of a hypothetical line in a rocky wilderness about which previously no one had cared. Another example is Poland. As a country, its borders have varied from once encompassing the largest national entity in Europe (during the sixteenth and early seventeenth centuries) to being totally non-existent as a nation state 200 years later (Zamoyski, 1987) with a modest expansion to the present day. What then is 'Poland'? I am assured that Poles know, but there is no doubt that over the centuries cartographers, historians and, above all, politicians have had strongly differing views. So much for certainty.

I believe that it is much more likely that our passion for boundaries stems from a territorial need to concentrate confusion in the smallest zone possible so that we have a firm basis for settling disputes. If not, then everywhere is a transition zone and everyone is confused about to which geographical unit they belong. The problem, then, is to agree on how and where to concentrate the confusion. The two examples demonstrate that we do not have to force geographical objects with indeterminate boundaries into exact moulds, nor do we need to remain confused about how to deal with indeterminacy. Indeterminacy is a way of life in many areas and it is sensible to have formal ways for handling it, which is what the chapters in this book are about.

Acknowledgements

Thanks are due to Jonathan Raper, Andrew Frank and Michael Goodchild for constructive comments to a first draft of this chapter.

REFERENCES

ALLEN, T. F. H. and STARR, T. B. (1982). *Hierarchy: Perspectives for Ecological Complexity*, Chicago: University of Chicago Press.

BORGES, J. L. (1962). *Funes the Memorious*, in: *Ficciones*, transl. Anthony Kerrigan, London: Weidenfeld and Nicolson.

BARROW, J. D. (1992). *Pi in the Sky*, Harmondsworth: Penguin.

BECKETT, P. H. T. and WEBSTER, R. (1971). Soil variability: a review, *Soils and Fertilizers*, **34**, 1–15.

BEZDEK, C. J. (1981). *Pattern Recognition with Fuzzy Objective Function Algorithms*. New York: Plenum Press.

BEZDEK C. J., EHRLICH, R. and FULL, W. (1984). FCM: the fuzzy *c*-means clustering algorithm, *Computers and Geosciences*, **10**, 191–203.

BIE, S. W. and BECKETT, P. H. T. (1973). Comparison of four independent soil surveys by air-photo interpretation, Paphos area (Cyprus), *Photogrammetria*, **29**, 189–202.

BIERKENS, M. F. P and BURROUGH, P. A. (1993). Stochastic Indicator Simulation: Parts I and II, *J. Soil Science*, **44**, 361–368, 369–381.

BIERKENS, M. F. P. and WEERTS, H. J. T. (1995). Block hydraulic conductivity of cross-bedded fluvial sediments, *Water Resources Research*, **30**, 2665–2678.

BURROUGH, P. A. (1986). *Principles of Geographical Information Systems for Land Resources Assessment*, Oxford: Oxford University Press.

BURROUGH, P. A. (1989). Fuzzy mathematical methods for soil survey and land evaluation, *J. Soil Science*, **40**, 477–492.

BURROUGH, P. A. (1992). Are GIS Data Structures too simple minded?, *Computers and Geosciences*, **18**, 395–400.

BURROUGH, P. A. (1993). Soil variability: A late 20th century view, *Soils and Fertilizers*, **56**(5), 531–562.

BURROUGH, P. A. and FRANK, A. U. (1995). Concepts and paradigms in spatial information: Are current geographic information systems truly generic?, *Int. J. Geographical Information Systems*, **9**(2), 101–116.

BURROUGH, P. A., MACMILLAN, R. A. and VAN DEURSEN, W. (1992). Fuzzy classification method for determining site suitability from soil profile observations and topography, *J. Soil Science*, **43**, 193–210.

COHEN, J. and STEWART, I. (1994). *The Collapse of Chaos: Discovering Simplicity in a Complex World*, New York: Viking Press.

DEUTSCH, C. V. and JOURNEL, A. J. (1992). *GSLIB: Geostatistical Software Library and User's Guide*, Oxford: Oxford University Press.

FAO (1975). *FAO–UNESCO Soil Map of the World*, Paris: UNESCO.

FAO (1976). A framework for land evaluation, *Soils Bulletin No. 32*, FAO, Rome.

HEUVELINK, G. B. M. and BURROUGH, P. A. (1993). Error propagation in cartographic modelling using Boolean logic and continuous classification, *Int. J. Geographical Information Systems*, **7**, 231–246.

ISAAKS, E. H. and SRIVASTAVA, R. M. (1989). *An Introduction to Applied Geostatistics*, New York: Oxford University Press.

JOURNEL, A. J. and HUIJBREGTS, Ch. J. (1978). *Mining Geostatistics*, New York: Academic Press.

KANDEL, A. (1986). *Fuzzy Mathematical Techniques with Applications*, Reading, MA: Addison-Wesley.

KAUFFMAN, A. (1975). *Introduction to the Theory of Fuzzy Subsets*, New York: Academic Press.

KITTEL, T. G. F. and COUGHENOUR, M. B. (1988). Prediction of regional and local ecological change from global climate model results: A hierarchical modeling approach, in: Pielke, R. A. and Kittel, T. G. F. (Eds), *Monitoring Climate for the Effects of Increasing Greenhouse Gas Concentrations*, Cooperative Institute for Research in the Atmosphere, Fort Collins, CO, pp. 173–193.

KLIR, G. J. and FOLGER, T. A. (1988). *Fuzzy Sets, Uncertainty and Information*, Engelwood Cliffs, NJ: Prentice Hall.

LAM N. and DeCOLA, L. (1993). *Fractals in Geography*, Englewood Cliffs, NJ: Prentice Hall.

MANDELBROT, B. (1982). *The Fractal Geometry of Nature*, San Francisco: Freeman.

MELLILO, J. M. (1994). Modeling land–atmosphere interactions: A short review, in: Meyer, W. B. and Turner, B. L. II (Eds), *Changes in Land Use and Land Cover: A Global Perspective*, 1991 OIES Global Change Institute, Cambridge: Cambridge University Press.

O'NEILL, R. V. DeAngelis D. L., WAIDE, J. B. and ALLEN, T. F. H. (1986). *A Hierarchical Concept of Ecosystems*, Monographs in Population Biology No. 23. Princeton, NJ: Princeton University Press.

RAPER, J. and LIVINGSTONE, D. (1995). Development of a geomorphological spatial model using object-oriented design, *Int. J. Geographical Information Systems*, **9**(4), 359–383.

WEBSTER, R. and OLIVER, M. A. (1990). *Statistical Methods in Soil and Land Resources Survey*, Oxford: Oxford University Press.

ZADEH, L. A. (1965). Fuzzy sets, *Information and Control*, **8**, 338–353.

ZAMOYSKI, A. (1987). *The Polish Way: A Thousand-year History of the Poles and their Culture*, London: John Murray.

CHAPTER TWO

The Prevalence of Objects with Sharp Boundaries in GIS

ANDREW U. FRANK

Department of Geoinformation, Technische Universität Wien, Vienna, Austria

2.1 Introduction

The debate on vector- versus raster-based models is nearly as old as the concept of GIS itself (Dutton, 1979). It was restated as a debate between GIS with an object concept (not to be confused with object-oriented as used in software engineering; Worboys (1994) uses the term object-based GIS), where objects have sharp boundaries delimited by vectors, and GIS which model the continuous variations of attributes over space using a regular tessellation (e.g. a raster; Frank, 1990). Efforts to merge the two representations have been attempted (Peuquet, 1983). This has been a very fruitful debate as it has forced us to consider and reconsider the epistemological bases of our work and has led to an extensive discussion of fundamental questions (Chrisman, 1987; Frank and Mark, 1991; Mark and Frank, 1991, 1995). The debate has prompted the development of ever-more powerful software, achieving a nearly complete integration of vector and raster data (Herring, 1990).

It has been repeatedly pointed out that only few objects in geographic space have natural boundaries which are sharp and well determined (Couclelis, 1992). Most geographic objects seem to be an abstraction of things which have unclear, fuzzy boundaries, if they have boundaries at all. They include most natural phenomena, from biotopes to mountain ranges, and extensive research efforts have centred around soil type data (Burrough, 1986, 1993), and often use the techniques of fuzzy logic (Zadeh, 1974). Nevertheless, many practically used GIS model reality in terms of crisply delimited objects. Cadastral systems, GIS used for facility management and automated mapping (AM/FM), and communal information systems are all appropriately oriented towards distinct objects with well defined boundaries. The same systems, with the same models, are also used to manage soil maps and land use data, where the fiction of sharp boundaries contrasts with our view of reality.

Depending on the area of application or profession, one or the other spatial concept is more appropriate and allows the aspects of interest to be captured succinctly. Users are accustomed to seeing a process described in a particular spatial framework; for example, a legal discussion is most often cast in the framework of spatial objects with crisp and determined boundaries, whereas a discussion of the application of fertilizer might consider soil types as having indeterminate boundaries. Both Burrough and Frank (1995) and Couclelis (Chapter 3, this volume) classify applications or phenomena according to an evolving taxonomy of object types and of boundary types. This leaves two questions: where do these models of well-defined objects come from; and why are object models with sharp boundaries very often used in GIS applications despite all the obvious problems with delimiting geographic objects?

This chapter attempts to address these two questions, starting from the point of view of experiential realism (Lakoff, 1987). Experiential realism argues that the natural categories of human cognition are based on our experience of the world, and that the way in which we perceive the world is influenced by our cognitive apparatus (Couclelis, 1992; for a practical application to geography, see Mark, 1993). There are two different environments for our experience, namely small-scale and large-scale (geographic) space. Our daily experience with handling small objects in small-scale space, e.g. apples or stones, leads to a concept of objects with well-defined boundaries. This contrasts with the direct experience of large-scale space, which leads to a different conceptual structure of space, mostly without dividing large-scale space into delimited objects. These are both fundamental human experiences, which are deeply embedded in our thinking and give rise to the concepts of 'object' and 'field'.

Experiential realism also assumes that metaphorical mapping (Lakoff and Johnson, 1980) can be used to conceive of situations in terms of previous experiences in different circumstances. For example, the experience with small prototypical objects with well-determined boundaries is metaphorically translated to conceive of the large-scale space situation, typically by creating conceptual objects with more or less clear boundaries in the landscape. The mechanism of mapping geographic objects from large-scale space onto small-scale space or, more specifically, onto 'pictorial space' (Montello, 1993) is, in a wider sense, an application of this method of metaphorical mapping.

The tendency of structure the world in terms of objects can be observed in many areas of application. Scientific methods often need to construct distinct objects, which can be the object of discourse, of measurement and formal description. Technical tools, e.g. computer-aided design (CAD), lead to the same pressure to 'objectivize' the world. Most important is the use of small-scale objects in all situations where things are manipulated and managed, as indicated by verbs like 'fetch' and 'buy' (Couclelis, 1992). This is highly visible in the legal and administrative processes dealing with large-scale space, again forcing the creation of (legal) objects in geographic space. These metaphorical transformations of experiences with small objects applied to large-scale space are in contrast to the direct experiences with large-scale space and the concepts that follow from them.

This chapter is structured as follows. Section 2.2 briefly reviews the recent literature and indicates where in-depth reviews can be found. Section 2.3 discusses the prototypical object concept as experienced in small-scale space and links it to Euclidean geometry. Section 2.4 contrasts the small-scale object view with the experience of large-scale space, which is discussed in Section 2.5. The metaphorical use of objects in science to organize large-scale space is pointed out in Section 2.6. The legal and administrative 'object' concept is developed in Section 2.7. Finally, Section 8 summarizes why the object view is so prevalent despite its problems, and argues for systematic research on the experience of large-scale space.

2.2   The recent object versus field debate

Initially, the choice between raster and vector representation of geometry was regarded as a technical issue of implementation. The early discussion often mixed conceptual and implementation considerations (in an early and extreme example, fundamental mathematical considerations from topology were expressed in the assembler code of a particular computer; see Corbett, 1979). An attempt to summarize and bridge the gap was made by Peuquet (1983).

With the attempts to clarify the underlying abstract models and to provide for their formalization (Frank, 1987, 1990; Goodchild, 1990), there was a move away from implementation issues and to concentrate on the conceptual level. One implicit goal was to find a universal representation into which all other representations could be translated (Frank and Mark, 1991; Mark and Frank, 1991, 1995; Mark et al., 1989). In this process the terminology moved from 'raster versus vector', stressing an implementation point of view, to 'field versus objects', pointing to the conceptual issue. Worboys (1994) proposes to call them 'object-based' GIS and to separate them from GIS using object-oriented technology for their implementation.

Most authors contrast objects with distinct boundaries with continuous fields. Objects have prototypical sharp boundaries, but often the limits are not so well defined or are not precisely determined or measured. Objects have properties and are homogeneous within their boundaries (at least with respect to some properties). The field concept assumes a function in such a way that for each point a property value results (Goodchild, 1990):

$$f(x, y) = a.$$

A rich discussion about the errors that can occur when abstracting from a field concept to a concept with sharp boundaries started from this point (Goodchild and Gopal, 1989).

The distinction between objects and fields is not a complete taxonomy. Further distinctions are necessary; for example, with respect to object conceptualization, Burrough and Frank (1995) considered

- whether objects are complex (i.e. further subdivided into objects which are composed to form complex objects; Haerder et al., 1987) or simple, i.e. existing only on a single level of resolution. This issue has previously been treated under the more general heading of 'multiple representation' (Beard and Buttenfield, 1991), or

- if the model includes processes and allows changes over time. Many of the difficulties of determining boundaries relate to the changing nature of reality and the way in which a GIS models temporal aspects (Goodchild and Gopal, 1989).

Couclelis, points out that there are many different types of objects with indeterminate or unclear boundaries and any analysis must note precisely which particular case is being discussed (Couclelis, this volume). She has also extensively surveyed the psychological tradition in this area, extending the position detailed by Couclelis and Gale (1986).

The 'object versus field' debate is intimately linked with the two prototypical cases of spatial perception. Summarizing the discussion of spatial cognition related to GIS (see Couclelis, 1992; Montello, 1993; Mark *et al.*, 1989), one can see agreement to differentiate between at least two major different kinds of space from a human perception point of view: small-scale and large-scale (or geographic) space (Kuipers, 1978). *Small-scale space* describes a situation in which objects are smaller than the human body and can be moved; the configuration can be perceived as a whole at once. *Large-scale space* contains objects much larger than the human body and we can move among them; the configuration can not readily be perceived at once, but knowledge must be integrated over time. The intent of this chapter is to show that these two types of experience are relevant for the distinction between objects and fields.

2.3 Experiential realism and spatial image schemata

Experiential realism (Lakoff, 1988) posits that human cognition is based on practical experience. The physiological similarities between all human bodies lead to the similarity of most basic aspects of human life and thus to similar experiences, independent of culture or language. All children have essentially the same experiences in their early years (eating, grasping things, letting things drop, etc.), and cultural and individual differentiation follows later. If the conceptualizations are appropriate for a purpose they will be retained for future usage.

Many of these early experiences common to all humans are spatial, leading to the fundamental concepts of 'in'/'out', 'up'/'down', etc. Following the framework of experiential realism, meaning is associated with abstract concepts through repeated experience. The concepts of space (and other similar concepts) are thus dependent on the physiological construction of the human body in a similar way that the values of the primary colours are dependent on the sensibility of the retina (Rosch, 1973). The kind of experience differs according to the senses involved; for example, Couclelis and Gale (1986) differentiate between sensori-motor and visual experience, etc.

The spatial experiences are aggregated into schemata, which abstract the essence of a prototypical situation. Of particular interest here are image schemata, which describe a number of fundamental situations, many of them spatial. Johnson (1987, p. 126) lists them as follows:

> ... Much of the structure, value, and purposeness we take for granted as built into our world consists chiefly of interwoven and superimposed schemata ... *My chief point has been to show that these image schemata are pervasive, well-defined, and full of sufficient internal structure to constrain our understanding and reasoning.* [italics in original] To give some idea of the extent of the image-schematic structuring of our understanding (as our mode of being-in-the-world or our way of having-a-world), consider the following partial list of schemata, which includes those previously discussed:

Container	Balance	Compulsion
Blockage	Counter force	Restraint removal
Enablement	Attraction	Mass–count
Path	Link	Center–periphery

Cycle	Near–far	Scale
Part–whole	Merging	Splitting
Full–empty	Matching	Superimposition
Iteration	Contact	Process
Surface	Object	Collection

These image schemata are then metaphorically transformed to structure other situations where they partially match the situation (Kuhn and Frank, 1991; Martin, 1990). This mechanism is so powerful that it can be used to explain completely abstract situations, such as in the sentence 'the candidate is midway to her PhD'. Here no spatial position is intended, but the spatial metaphor is used to convey the abstract non-spatial meaning. Often the spatial experience also includes a value assessment, as in 'up' is 'more' or 'better', leading to metaphors such as 'upper class'.

2.4   Objects in small-scale space

Human experience can be roughly divided into experiences in small-scale space and experiences in large-scale situations. Small-scale space is characterized by a layout which can be perceived at a glance; objects are smaller than human beings and can easily be moved. The list of spatial image schemata can be divided into those where the (likely) prototypical situation is one of small-scale space, and those where the prototypical situation is taking place in large-scale space. Large-scale space experience often involves the movement of the observer in space. The spatial image schemata at the small scale include: container (in/out), object, link, up/down, left/right, before/behind, surface, support, part/whole, contact.

Experience in small-scale space primarily takes place with movable objects which have sharp boundaries, but there are other types of objects that do not have well defined boundaries, such as fluids, grains, or balls of cotton. The generic properties of prototypical objects can be analyzed, e.g. a piece of fruit lying in a bowl can be picked up and handled. What operations can be performed? What are the generally available properties, independent of the particulars of the fruit in this case? These properties can be described in terms of a universal algebra (Birkhoff and Lipson, 1970), as is now customary for the specification of semantics in software engineering (Liskov and Guttag, 1986). The properties of an object are described as operations which can be performed with it:

- An object has an *identity*; each object is an individual and is differentiated from any other. The identity of an object is immutable during its lifetime (Al-Taha and Barrera, 1994). One may form classes of objects which are interchangeable, but this is already a second level of abstraction.

- An object has properties that are not dependent on its current location (*invariance of properties* under movement).

- An object has sharp boundaries and a *geometric form and size*, which are not variable (again invariant under movement).

- An object *can be moved* around and remains where it is until it is moved (this is known as the 'frame problem' in artificial intelligence (Hayes, 1985).

- Objects can be joined: two similar objects are twice as long (or high) as a single one.

- An object cannot be placed at the same location as another one.
- An object falls until it rests on something; it obeys the effects of gravity (although there are some non-prototypical objects, e.g. birds and smoke, which are not subject to gravity).
- One object can be placed on top of another.
- An object can be put into or taken out of a container.
- An object can be split into parts and put together again.

Experience in small-scale space leads directly to abstraction in Euclidean geometry. Euclidean geometry is structured after the experience with rigid, movable objects, i.e. objects which allow solid body motion. The classical geometric instruments, ruler and compass, are rigid objects that can be moved in space to produce marks where they are placed. Theoretical investigations demonstrate that the requirement of 'rigid body motion', meaning that material objects can be moved around without change of shape, leads to a class of geometry of which the Euclidean is the simplest case (curvature 0) and respects our daily experience with relatively small objects at rest (Adler *et al.*, 1965).

Small-scale space is in principle static – nothing moves until a force is applied – and our perception is primarily one of a static situation. The image schemata describe situations, not motion. Changes are brought about swiftly, nearly instantaneously; things are moved from one place to another without consideration of the path nor the time the move requires. The image schemata describing force and movement, e.g. compulsion, blockage, restraint, belong to experience in small-scale space.

Of particular interest in our context is the aspect of a sharp boundary: if an object can be moved, its boundary is determined by lifting it up. What is moved is the object; what remains is not part of it. The boundary, which might not have been clearly visible at first, is sharp and becomes determined when the object is moved and can be measured with any precision desired.

There are also experiences with other things that do not have all these properties. Some classes of unbounded objects are so fundamental that they have found expressions in language; for example, liquids and other 'mass nouns', which do not refer to identified objects but to indefinite or measured amounts of a particular material. Words like 'water', 'sand', etc., usually do not have plural forms, and many languages have particular constructs to indicate amounts (the Latin partitive case).

From an experiential point of view, objects are a basic experience of small-scale space. Object identity and sharp boundary, invariance of shape and properties under movement, etc., are the most salient characteristics of objects. They are significantly different from the properties found in large-scale space.

2.5 Experience in large-scale space

> *Q*: What goes uphill and downhill and yet always stays in the same place?
> *A*: A road.
>
> (Children's riddle)

The typical experience in large-scale space is one of moving around in a landscape only part of which is visible. The most important operation is navigation, to find the way back and to enter the 'cave' after a day of hunting or collecting nuts and berries. Extensive psychological literature exists about how people navigate, mainly relating to an artificial environment. Kuipers and Levit (1990) detail the most fundamental elements.

Places can be recognized, but they are neither objects with definite boundaries, nor abstract points from Euclidean geometry. Places are connected by *paths*, which can be followed to get from one place to another. Paths and links are examples of two very similar image schemata, one from the small-scale, the other from the large-scale space. The abstract formulation for the relations between places and paths is found in graph theory (Deo, 1974), which deals with a bipartite set of objects (nodes and edges) and the relations between them (adjacency).

Graph theory does not capture all experiences of large-scale space, since it does not cover the area between nodes and edges: a path goes through a wood or across a field. This is most often expressed in language as 'fictive motion', described as if a person were actually walking along the path: the path winds along the valley (Talmy, 1983). Fields, woods and other areas which extend to the right and left of the path may be visible, but they do not have boundaries and can gradually change from one to the other (from a wood to a grazing area). Woods and fields do not typically have set boundaries and are not perceived in their totality as objects. The experience of the local neighbourhood, as it is experienced in large-scale space, leads to the theory of topology (Spanier, 1966).

Along the path there are intermediate locations which are not sharply determined and are not even named. The path also goes up and down gradually, with high and low points that might qualify as places, but all intermediate points are just relatively higher or lower than others. The flow of water – extremely important for human activities – follows the same pattern. One might assume that this is the prototype for the mathematical field $f(x) = h$, which leads to calculus, discussing gradual changes of a surface embedded in space.

This fuzzy organization is sufficient because things do not move or remain stable over time. Natural change in landscapes is usually very slow. Changes in vegetation or even snow cover or light changes, are slow in comparison with movement in small-scale space. Thus society can create places through conventions, typically selecting landmarks of importance which do not change rapidly (Lynch, 1960). Topographic maps then depict stable objects in the landscape which can be useful for navigation.

The movement of animals in a landscape is still slow enough that gradual movement along a path is perceived. Of great importance, however, is the deduction of a temporal sequence from spatial clues; a good example is provided by North American Indians who, by reading from marks on the ground, could deduce which animals or people had passed, and when (for the application of the same concept to geology, see Flewelling *et al.* (1992).

Experiences of small-scale and large-scale space are substantially different, but there is a strong tendency to carry over the experiences gained from one, to structure experiences in the other. Most obvious is the application of the 'container' metaphor in large-scale space applied to landscape elements like mountains, lakes, swamps, good hunting areas, fields for planting and finally territory. Even if they lack the 'clear boundary' property (recall that for metaphorical transformation not all the properties of the source domain must be present in the target domain; Martin, 1990), they can be treated as objects. For example, one can say 'the deer is in the wood', without implying that the wood has well-defined boundaries.

2.6 Scientific use of objects to structure large-scale space

The most successful model for modern science is physics. In particular, mechanics –
like Euclidean geometry – is an abstraction of the movement of ideal objects in
small-scale space. Classical mechanics is an example of a 'rational' science, and is
firmly based on mathematics, algebra and Euclidean geometry. It deals ideally with
mass points (the abstraction of regular balls, i.e. prototypical objects) moving on a
frictionless, ideally plane and horizontal surface. The image schemata used are
'force', 'counter force', 'blockage', 'link', 'restraint', etc. It uses infinitesimal calculus
to model fields of forces which are not directly visible, covering the image schema of
'attraction'.

Concepts from mechanics are used extensively in science. Mechanics embodies
the image schemata of causal thinking, which are fundamental to science. As a consequence, the things discussed in a scientific discourse must be converted to objects
('reified'); in speaking of a 'deer population' or of 'rainfall' one applies the same
operations one would apply to simple objects like an apple. Geography (and other
sciences) also uses fields to model influences, often in the form of a 'gravity law'
where attraction spreads out from a centre towards a periphery.

Much human reasoning is not quantitative but qualitative, as are most arguments in science (Kuipers, 1994). The quantitative revolution, which attempted to
move geography and other sciences towards full quantitative modelling as used in
physics, has not been successful in completely converting the discipline. Much scientific reasoning remains qualitative, but even in applications of physics, qualitative
reasoning is introduced to solve problems which are too complex to be modelled
quantitatively.

2.7 Legal objectification

Management in the widest sense requires the manipulation of objects and action
applied to objects. Human social interaction calls for rules pointing out what is
permitted and what is not. Most cultures have developed a concept of ownership by
individuals (or by small groups) of small-scale space objects: fruits collected, tools
prepared are 'privately' 'owned'. The owner is free to determine what to do with the
objects, even whether to destroy them or to give them away. Every culture has
developed its own rules on what can be owned, how ownership is established and
what rights it grants. Ownership is linked to the 'container' image schema: I own
what I have in my hand and I can exclude all others from using it.

From this simple ownership relation between a person and an object many other
legal relations can be deduced. Ownership of a (mobile) thing was a fundamental
concept of Roman law, which was uniformly used by analogy for ownership of
animals, slaves and real estate. Similarly, civil law in continental Europe is based on
the concept of the ownership of things (and similar rights). From this prototype
came the extensions to the ownership of real estate (called *Immobilar-Sachenrecht* in
German, *vastgoed* in Dutch, betraying its origin in the ownership of things), and to
the ownership of rights of immaterial 'things' such as intellectual property (patents,
copyright), etc. In all cases a 'metaphorical' transformation of the basic concept of
ownership of small-scale objects to a new situation is used to structure the legal
situation of, say, the ownership of a piece of text.

In order for this metaphorical transformation to work, the fields and woods must become objects, so they can be dealt with like tools or cattle – sold as individualized items: I sell you this parcel of land, these cattle. Alternatively, land can be seen as a mass and is sold by quantity; for example, I can sell you 10 hectares of land.

It should be noted that the original German law, as found in the *Sachsenspiegel*, made particular provisions for the ownership of land and how it could be sold, which were different from the rules for buying and selling movable goods. But even then the thing had to be individualized and bounded. There have been extensive and detailed descriptions of how boundaries are to be created – an obviously difficult endeavour and the details (religious ceremonies, wars, etc.) indicate how 'artificial' (i.e. man-made) these boundaries are. Similarly, Roman mythology relates in detail how the original boundaries of the town were created with a plough, and counsels respect of the boundaries, even if they could be physically easily crossed.

As ownership rights are the 'fundamental' schema of law, all administrative law relates to objects and the relations between them. For any legal decision, a person or an individualized object must be determined; if the administrative action applies to land, the land must be an 'object' with an identity and a clear boundary. Thus any GIS relating to administrative action affects bounded land-objects; areas of application like planning, where phenomena without clear boundaries must be dealt with, encounter problems in the administrative process.

2.8 Conclusions

This chapter has considered the basic experiences of mankind and of individuals, leading to an understanding of the fundamental schemata of spatial cognition. Two major situations for spatial experience have been discussed. First, from experiences in small-scale space and with movable objects smaller than human beings, emerges the fundamental concept of an object, which has the following properties:

- an object has an identity, which is independent of minor changes in attributes and properties,
- an object can be moved,
- an object can have a geometric form and other properties, which are not affected by movement of the object,
- an object cannot be placed at the same location as another one,
- one object can be placed on top of another, etc.

The second environment for spatial experience is the human moving in large-scale (or geographic) space: visiting recognized places and wandering along a path through woods and fields over an undulating landscape. From these two experiences emerge the fundamental spatial image schemata which are used by all humans to structure their experiences with space. They can be separated into small-scale and large-scale image schemata:

- *small scale*: container (in/out), object, link, surface, support, part/whole, contact; and
- *large scale*: place, path, near/far, centre/periphery.

The two basic conceptual models for GIS are directly linked to these two experiences: the object (vector) model, which deals with individual parcels of land with sharp boundaries, uses the 'small-scale' object experience; and the field (raster) model, which models variable properties without individualizing objects or determining boundaries, builds on the 'landscape' experience.

Humans have the ability to use metaphorical transformation through the image schemata from one experience, to structure experience in another situation. Two trends to transport the object metaphor to the landscape and to bound ('objectivize') the unbounded phenomena are found:

- Science often requires individualized objects which can be measured, counted and described (but science also uses the 'field' metaphor extensively).
- Law and administration use the object concept, in particular the concept of the ownership of an object, to structure other legal situations. It is thus required that all things become 'objects' with identities and boundaries.

This explains why GIS for administrative use, as well as for many other scientific uses of GIS, enforce an 'object' view. It also resolves part of the puzzle of why, from the immense variability of objects with definite boundaries to all sorts of fuzzy limits, the sharp boundary variant covers so many applications.

However, this should not be construed as a conclusion that all the models we need have been fully developed. On the contrary, it indicates that there is a need for more complex models that will allow the true integration of large-scale image schemata and related processes in a GIS. This becomes even more pressing when process models are required, as is currently the case in environmental applications. The experience of small-scale space does not provide the necessary tools.

Acknowledgements

Comments from Helen Couclelis, Catherine Dibble, Werner Kuhn and Stephen Hirtle have helped me to sharpen the arguments in this paper. I appreciate their contributions. Many thanks also go to Peter Burrough for his thoughtful review. Grants from Intergraph Corp. and from the Austrian Science Foundation have supported this and related work.

REFERENCES

ADLER, D., BAZIN, M. and SCHIFFER, M. (1965). *Introduction of General Relativity*, New York: McGraw-Hill.

AL-TAHA, K. and BARRERA, R. (1994). Indentities through time, in: ISPRS Working Group II/2, *Workshop on the Requirements for Integrated Geographic Information Systems*, New Orleans.

BEARD, K. and BUTTENFIELD, B. (1991). *Visualization of the Quality of Spatial Data*, Initiative 7 Position Paper, NCGIA.

BIRKHOFF, G. and LIPSON, J. D. (1970). Heterogeneous algebras, *J. Combinatorial Theory*, **8**, 115–133.

BURROUGH, P. A. (1986). *Principles of Geographical Information Systems for Land Resource Assessment*, Monographs on Soil and Resources Survey, Oxford: Oxford University Press.

BURROUGH, P. A. (1993). Soil variability: A late 20th-century view, *Soils and Fertilizers*, **56**, 529–562.

BURROUGH, P. A. and FRANK, A. U. (1995). Concepts and paradigms in spatial informa-tion: Are current geographic information systems truly generic?, *Int. J. Geographical Information Systems*, **9**(2), 101–116.

CHRISMAN, N. (1987). Fundamental principles of geographic information systems, in: Chris-man, N. R. (Ed.), *Proc. Auto-Carto 8*, Baltimore, MD, ASPRS-ACSM, pp. 32–41.

CORBETT, J. P. (1979). *Topological Principles of Cartography*, Technical Paper 48, Bureau of the Census, US Department of Commerce.

COUCLELIS, H. (1992). People manipulate objects (but cultivate fields): Beyond the raster–vector debate in GIS, in: Frank, A. U., Campari, I. and Formentini, U. (Eds), *Theories and Methods of Spatio-Temporal Reasoning in Geographic Space*, pp. 65–77, Berlin: Springer.

COUCLELIS, H. and GALE, N. (1986). Space and spaces, *Geografiske Annaler*, **68B**, 1–12.

DEO, N. (1974). *Graph Theory with Applications to Engineering and Computer Science*, Engle-wood Cliffs, NJ: Prentice Hall.

DUTTON, G. (Ed.) (1979). *Proc. 1st Int, Study Symposium on Topological Data Structures for Geographic Information Systems*, Reading, MA: Addison-Wesley; Harvard Papers on GIS, Cambridge, MA: Harvard University Press (1979).

FLEWELLING, D., EGENHOFER, M. J. and FRANK, A. (1992). Constructing cross geo-logical sections with a chronology of geologic events, in: *Proc. 5th Int. Symposium on Spatial Data Handling*, Charleston, SC, IGU Commission on GIS, pp. 544–553.

FRANK, A. U. (1987). Towards a spatial theory, in: Aangeenbrug, R. T. and Schiffmann (Eds), *Int. Geographic Information Systems (GIS) Symposium: The Research Agenda*, Crystal City, VA.

FRANK, A. U. (1990). Spatial concepts, geometric data models and data structures, in: Maguire, D. (Ed.), *GIS Design Models and Functionality*, Leicester, Midlands Regional Research Laboratory.

FRANK, A. U. and MARK, D. M. (1991). Language issues for geographical information systems, in: Maguire, D., Rhind, D. and Goodchild, M. (Eds), *Geographic Information Systems: Principles and Applications*, London: Longman.

GOODCHILD, M. F. (1990). A geographical perspective on spatial data models, in: Maguire, D. (Ed.), *GIS Design Models and Functionality*, Leicester, Midlands Regional Research Laboratory.

GOODCHILD, M. F. and GOPAL, S. (Eds) (1989). *Accuracy of Spatial Databases*, London: Taylor & Francis.

HAERDER, T. *et al.*, 1987, PRIMA: A DBMS prototype supporting engineering applica-tions, *Proc. 13th Int. Conference on Very Large Databases*, Brighton, UK, pp. 433–442.

HAYES, P. J. (1985). The second naive physics manifesto, in: Hobbs, J. R. and Moore, R. C. (Eds), *Formal Theories of the Commonsense World*, pp. 1–36, Norwood, NJ: Ablex.

HERRING, J. R. (1990). TIGRIS: A data model for an object-oriented geographic informa-tion system, in: Maguire, D. (Ed.), *GIS Design Models and Functionality*, Leicester, Mid-lands Regional Research Laboratory.

JOHNSON, M. (1987). *The Body in the Mind: The Bodily Basis of Meaning, Imagination and Reason*, Chicago: University of Chicago Press.

KUHN, W. and FRANK, A. U. (1991). A formalization of metaphors and image schemata in user interfaces, in: Mark, D. M. and Frank, A. U. (Eds), *Cognitive and Linguistic Aspects of Geographic Space*, pp. 419–434, Dordrecht: Kluwer.

KUIPERS, B. (1978). Modeling spatial knowledge, *Cognitive Science*, **2**(2), 129–154.

KUIPERS, B. (1994). *Qualitative Reasoning: Modeling and Simulation with Incomplete Know-ledge*, Cambridge, MA: MIT Press.

KUIPERS, B. and LEVIT, T. S. (1990). Navigation and mapping in large-scale space, in: Chen, S.-S. (Ed.), *Advances in Spatial Reasoning*, pp. 207–251, Norwood, NJ: Ablex.

LAKOFF, G. (1987). *Women, Fire, and Dangerous Things: What Categories Reveal about the Mind*, Chicago: University of Chicago Press.

LAKOFF, G. (1988). Cognitive semantics, in: Eco, U., Santambrogio, M. and Violo, P. (Eds), *Meaning and Representations*, pp. 119–154, Bloomington, IN: Indiana University Press.

LAKOFF, G. and JOHNSON, M. (1980). *Metaphors We Live By*, Chicago: University of Chicago Press.

LISKOV, B. and GUTTAG, J. (1986). *Abstraction and Specification in Program Development*, Cambridge, MA: MIT Press.

LYNCH, K. (1960). *Image of the City*, Cambridge, MA: MIT Press.

MARK, D. M. (1993). Toward a theoretical framework for geographic entity types, in: Frank, A. U. and Campari, I. (Eds), *Spatial Information Theory: A Theoretical Basis for GIS*, pp. 270–283, Berlin: Springer.

MARK, D. and FRANK, A. U. (Eds) (1991). *Cognitive and Linguistic Aspects of Geographic Space*, Vol. 63, NATO ASI Series D, Dordrecht: Kluwer.

MARK, D. and FRANK, A. U. (1995). Experiential and formal models of geographic space, *Environment and Planning* B (in press).

MARK, D. M. *et al.* (1989). *Languages of Spatial Relations*, Initiative 2 Specialist Meeting Report, Technical Report 89-2, National Center for Geographic Information and Analysis.

MARTIN, J. H. (1990). *A Computational Model of Metaphor Interpretation*, Boston, MA: Academic Press.

MONTELLO, D. R. (1993). Scale and multiple psychologies of space, in: Frank, A. U. and Campari, I. (Eds), *Spatial Information Theory: A Theoretical Basis for GIS*, pp. 312–321, Berlin: Springer.

PEUQUET, D. J. (1983). A hybrid structure for the storage and manipulation of very large spatial data sets, *Computer Vision, Graphics, and Image Processing*, **24**(1), 14–27.

ROSCH, E. (1973). Natural categories, *Cognitive Psychology*, **4**, 328–350.

SPANIER, E. (1966). *Algebraic Topology*, New York: McGraw-Hill.

TALMY, L. (1983). How language structures space, in: Pick, H. and Acredolo, L. (Eds), *Spatial Orientation: Theory, Research, and Application*, New York: Plenum.

WORBOYS, M. (1994). Unifying the spatial and temporal components of geographical information, in: Waugh, T. C. and Healey, R. G. (Eds), *Proc. 6th Int. Symposium of Spatial Data Handling*, Edinburgh, pp. 505–517.

ZADEH, L. A. (1974). Fuzzy logic and approximate reasoning, *Synthese*, **30**, 407–428.

PART TWO

Objects versus Fields: Contrast in Concepts

ANDREW U. FRANK

Introduction

The four chapters in Part 2 of this book discuss the difficulty in achieving comprehensive descriptions of objects with uncertain boundaries in general terms. The authors concentrate on the concepts we use to understand the world and how they lead to objects and boundaries. A reasoned discourse is hindered by too many cases, which are similar in that they do not result in crisp boundaries, but in other respects they are very different. The observations and conclusions drawn in one case cannot be generalized and applied to another; the findings of different researchers contrast with each other. Because reality is immensely complex, many different abstractions may be meaningful, resulting in contrasting observations of essentially the same reality. Terminological confusion is often added into the picture.

In Chapter 3 Helen Couclelis attempts a taxonomy of different kinds of objects with uncertain boundaries. She observes first that there is only a single concept of a spatial object with a crisp boundary, but there are many different types of objects with uncertain boundaries. How should one proceed to organize these many different types of object with uncertain boundaries in a taxonomy? She proposes three different dimensions: the empirical nature of the object, the mode of observation, and the user's purpose.

The nature of the object (e.g. whether it is continuous or discontinuous, homogeneous or inhomogeneous, fixed or moving) influences how we perceive the boundaries and their level of crispness. The mode of observation, and the resolution and the error, affect how well the position of the boundary is known. Finally, the purpose for which a model is constructed and data are collected, leads to a preference for one or the other. Management and administration demand defined objects; at the other extreme, scientists want to include the uncertainty of the boundary in their models. Couclelis concurs here with Burrough and Frank (1995, p. 151): different user categories tackle the uncertainty in the boundaries in different ways, which could be a first result of these investigations. Her taxonomy gives more detail and should be very helpful for checking whether descriptions of objects with uncertain boundaries are complete.

In Chapter 4 Irene Campari applies some of the characteristics identified by Couclelis to the urban environment. It is surprising to see how many uncertain boundaries exist in this small and seemingly well defined environment. Administrations subdivide space and create boundaries; these boundaries are not always easy to translate into an urban reality, and different services, e.g. the post, census, or telephone, translate essentially the same conceptual units into different boundary networks. Cultural tradition subdivides space in yet another way, for example by identifying sacred places. Finally, the artifacts in the city have sharp boundaries, but these boundaries interact with their usage. The road system is used for transportation, for water runoff, etc., and each use results in the creation of a different boundary.

The modelling of geographic objects in a feature-based geographic information system must consider the problem of crispness of object boundaries. In Chapter 5 Lynn Usery analyzes the conceptual model underlying this popular method. Features are characterized by space, theme and time – essentially a subset of the observation modes in Couclelis' chapter. The concept of feature exploits the extensive cartographic tradition; features thus depend on the intended scale or the resolution of observation. Usery then shows that the 'basic level' categories from cognitive science research apply as well: road, river and lake are basic level categories, where transportation network or hydrography are the superordinate concepts and local road, highway, etc., are the subordinate ones. Such categories are best described with prototypes, which the logic of today's databases does not provide. Usery then gives an example of the representation of specific features with undetermined boundaries using fuzzy logic. Later, Usery's theoretical treatment is nicely balanced by Sarjakoski's contribution (Chapter 20), which reports efforts in Finland to count well-understood geographic features (lakes, islands, rivers, etc.) and discusses the specific and practical problems influencing the results, many of which echo the theoretical points raised here.

In Chapter 6 Peter Fisher continues with a discussion of a particularly important aspect of objects with uncertain boundaries, namely, the difference between Boolean and fuzzy regions, which has often led to confusion. Certain areas are delimited by Boolean criteria, which may be influenced by probability, such as the area within five minutes' (i.e. 300 sec) drive. Particular circumstances affect the decision as to whether a point lies within a given area or not; a probability statement can adequately describe the situation. Fisher contrasts this probability-based Boolean region with areas delimited by an uncertainty in the criteria for membership, for example, land use determination based on remote sensing data. His extensive example of a Boolean region with probability is complemented by the detailed description of fuzzy boundaries for soil types given by Lagacherie et al. in Chapter 18.

Chapters 3–6 differentiate between kinds of uncertainty in the boundaries of objects, and the sources of this uncertainty. They all stress that the ways in which we confront the world, the purpose and method of observation, and our cultural backgrounds influence how we divide the world into objects, and the kind of boundaries these objects have. Couclelis, Campari and Usery explicitly, and Fisher implicitly, use methods and results from cognitive science. Of particular importance are the methods of forming categories, and the relations between the categories. Usery and Fisher both apply fuzzy logic to construct a formal model of prototype-based categories.

Part 2 provides a comprehensive schema into which the different specific cases described in Parts 5 and 6 can be classified. Each of them presents a particular view from an application perspective, and therefore a particular selection of the aspects listed here. To emphasize one or two points as examples: Poulter concentrates on the influence of scale on land use classification methods. In highly developed scientific and technological efforts similar problems are encountered to those that human beings experience in everyday life in a town. This suggests that generic solutions, as discussed in Parts 3, 4 and 5, based on mathematical formalization, can be found which will have wide applicability. It also predicts that a complete formalization will require a differentiation of cases and limitation to specific scopes – a point borne out in Chapters 11–17.

REFERENCES

BURROUGH, P. A. and FRANK, A. U. (1995). Concepts and paradigms in spatial information: Are current geographic information systems truly generic?, *Int. J. Geographical Information Systems*, **9**(2): 101–116.

CHAPTER THREE

Towards an Operational Typology of Geographic Entities with Ill-defined Boundaries

HELEN COUCLELIS

Department of Geography and National Center for Geographic Information and Analysis, University of California, Santa Barbara, USA

3.1 Introduction

Disciplines create their own problems, just as they help solve those of others. Few outside the community of GIS researchers (and perhaps also cartographers) would think that the representation of geographic objects with undetermined boundaries is a problem at all, or that the distinction of geographic entities into well-bounded and ill-bounded recognizes an important division of things in nature. Yet the issue *is* a very important one in the context of a problem-solving technology that happens to be very good at dealing with geographic objects of the one kind but not the other. Significantly, the concept of geographic object, itself a product of that technology (for who was thinking of mountains, cities and shorelines as 'objects' before GIS?), is largely responsible for the creation of the problem discussed in this volume.

To an extent much greater than most scientists would care to admit, the models and theories of science have their roots in the intuitive cognitive schemes and models humans form of the empirical world around them (Lakoff, 1987; Casti, 1989). Of these schemes, because of its significance for both the species and the individual, the concept of object is particularly fundamental. In the commonsense, intuitive meaning of the term, the prototypical object is small, solid, permanent, manipulable, and definitely bounded. An object with undefined boundaries in particular is a 'bad' object, a conceptual chimera, almost a contradiction in terms: something that is bound(!) to cause trouble. In experiment after experiment, subjects asked to name examples of objects will mention things such as book, cup, chair and pen, but not fire, graffiti, ship or shadow. On such lists, geographic entities like bays, meadows, rivers, mountains and streets are also highly unlikely occurrences. Yet GIS technology has 'objectified' such entities by rendering them small enough to fit within a computer screen by the dozens; it has given them a reassuring concreteness, expressed in sharp lines and bright, solid colours; it has captured them as permanently as the magnets of computers will allow; it has tamed them to the point of making them manipulable by anyone who can type in the right keystroke sequences; but it hasn't yet solved the problem of the bad, unbounded objects – the objects that shouldn't be.

The specialized use of the term 'object' in object-oriented programming (OOP), widely popular by now within the GIS community, further corrupts the meaning of the original natural-language concept. Just recently a colleague received a comment from a technical journal editor asking him not to use the term object in the ordinary-language sense as it may confuse the OOP-oriented readership. As Alice learned in Wonderland, words can be made to mean whatever you want, but it is difficult to make them un-mean it. Also, more ominously, the word can become the thing.

There is a problem here of confusing the thing represented with the representation, the thing modelled with the model, or in more geographic terms, the territory with the map. Let us be clear from the start: within the context of geographic-scale things and events, *objects* belong in the GIS representation (and any OOP implementation thereof), whereas the world contains *entities* that may or may not have the properties of objects (including their typical boundedness). I have argued elsewhere (Couclelis, 1992) that, with the exception of the two man-made categories of engineered artifacts (such as roads, bridges, dams, and runways) and some mostly immaterial spatial objects of social control (such as national, political, administrative, or property boundaries), entities with crisp boundaries are very rare in the real, one-to-one scale geographic world. The problem discussed in this volume arises because, in the mapping from the territory to the GIS, there is a discrepancy between the boundary-poor empirical world on the one hand, and the boundary-happy technology that is strongly influencing our ways of thinking on the other.

Thus the answer to the problem of undefined boundaries should be sought neither in the geographic world alone ('let's make more precise measurements to find out where the boundaries of things really lie'), nor internally within the GIS representation ('let's figure out in how many different ways we can fuzz up our object boundaries'), but in the cognitive act that transforms geographic *entities* into database *objects*. That mapping, I will argue, is a function of three distinct though not altogether unrelated things: 'objective' reality, mode of observation, and human intentionality or purpose. Certainly, empirically speaking, some geographic entities are bounded, and others are not; some modes of observation yield boundaries, and others do not; but most importantly perhaps, we must appreciate that the existence or not of well-defined boundaries is a problem for some categories of GIS users and not for others, and that it is not the same kind of problem for everyone. To improve the representation of geographical entities with undetermined boundaries it is essential that we understand *for whom* this is an issue, and why this is so.

In the following I will first examine this question from the three perspectives of empirical reality, mode of observation (or representation), and user purpose. This discussion is based on the intuitive meaning of boundary (as that which delimits, as that which divides) rather than on any mathematical definition, although there should be no conflict between the two. The three perspectives are then woven into a rough typology, and the paper ends with the tentative formulation of an approach that may allow a systematic and comprehensive practical treatment of the boundary problem in GIS.

3.2 Dimension one: Empirical nature of the entity

Boundaries and the entities they bound are dual concepts: each takes at least part of its meaning from the other, which is why discussions of boundaries couched in abstract terms of lines and polygons, interiors and exteriors, fuzzy zones and probability surfaces, can never go far enough. What kinds of boundaries things may have depends on a host of material, topological, functional, temporal, and other empirical considerations pertaining to the bounded entities themselves. It is thus necessary to consider the relevant characteristics of the different kinds of entities that are part of the geographic world.

Talking about what the world is 'really' like is always tricky. Thankfully, the geographic world has, for the most part, a reassuring empirical concreteness to it, by which I mean that it is reflected in fairly robust and practically testable cognitive models. That concreteness can also be misleading, because not every phenomenon of interest is tangible, visible, permanent, actual, or measurable. The following considerations regarding the nature of geographic entities illustrates the variety of factors that may affect the empirical properties of their boundaries.

(a) *Atomic or plenum*

Is the world ultimately made up of discrete, indivisible elementary particles, or is it a continuum with different properties at different locations? This question, already debated by the ancient Greeks, remains one of the major unanswered problems in the philosophy of physics (Hooker, 1973; see also Couclelis, 1992). At the level of the empirical, one-to-one scale geographic world however, the answer is, trivially, 'it depends'. Houses, runways, trees, cars and humans are 'atoms' which, as the Greek etymology of the word indicates, are non-subdivisible smallest units for most purposes of geographic-scale modelling and study; extensive entities such as oceans, prairies, forests, and geological formations, on the other hand, may be subdivided within very wide limits and still maintain their identity. Somehow boundaries are intrinsic to the notion of atom, whereas in the case of extensive entities they are contingent. In other words, the notion of boundary *a priori* sits better with the atomic view of things (and vector GIS) than with the plenum view (and raster GIS), whereas the real geographic world forces us to consider both discrete and extensive entities.

(b) *Homogeneous or inhomogeneous*

This criterion is at the root of the classic soils map problem. Boundaries draw binary distinctions (same – different), whereas inhomogeneous entities vary only by degree from place to place. Inhomogeneity, and the resulting boundedness problem, is an emergent property of many common extensive geographic entities defined on collectivities of well-bounded individuals (a house is well-bounded, but it is not a small town; a pine tree is relatively well-bounded, but it is not a small pine forest). Thus the soils map problem is also the vegetation map problem, the urban land use map problem, the population structure map problem, the species distribution map problem, the Jewish/Palestinian territory map problem...

(c)   *Discontinuous or continuous*

Often the transition from one geographic entity to another is smooth and continuous, so that any boundary between them is conventional rather than empirically
real. Where do the foothills end and the valley begins? Where at last do we leave
urban sprawl behind and reach the open country? Many geographic entities are
well defined at their most typical, most intense, most prominent, most central, etc.,
part, but de-grade away from there, gradually merging with the next one. In other
cases the discontinuity between two geographic entities is as clear as the difference
between a wet and a dry shoe on the seashore, and the binary treatment afforded by
the notion of boundary is in principle appropriate (but see below for the case of
moving fluid entities such as water).

(d)   *Connected or distributed*

Some geographic entities consist of several non-contiguous pieces that 'work'
together in some way (functional, aesthetic, etc.). Because of these connections
across space (which may link together similar or very dissimilar things), this case
differs from the inhomogeneous distributions discussed earlier. A dispersed settlement, an incompletely preserved historic downtown, and a regional system of health
care provision are of that nature. It is unclear whether and how such entities should
be (considered) bounded. Most people would perceive a boundary around the historic downtown, to a lesser degree around the dispersed settlement, but most probably not around the health-care provision system.

(e)   *Solid or fluid*

Material entities differ with respect to boundedness in obvious ways. Generally,
solids are well-bounded, but fluids not in rigid containers are not. Fluids moving
inside other fluids (as in oceanic and atmospheric phenomena) are the worst, calling
into question the very notion or boundary.

(f)   *Two- or three-dimensional*

We can get away with representing much of the geographic world as a flat surface,
and most volumes in it as polygons, for two reasons: because the vertical dimension
is usually negligible in magnitude compared to the horizontal, and because the projection of significant geographic volumes (mountains, tall buildings) is usually contained within their footprint. Problems with ill-defined boundaries can arise
however whenever the three-dimensionality of the world must be taken into
account. Thus a tree projected at the level of the trunk will yield a neatly bounded,
fairly stable polygon, but at the level of the canopy it will not. A bridge, on the other
hand, being a man-made object, will yield well-defined boundaries both at the level
of the span (above ground) and of the foundations (below ground), as well as at
many levels in between, but it may be unclear which boundary is the 'proper' one.
Geographic-scale examples of entities that may have ill-defined 2D boundaries
because of their higher dimensionality (among other things) are geologic and
meteorological formations.

(g)   *Actual or non-actual*

Temporality is a long-neglected aspect of the representation of geographic entities that has been receiving quite a bit of attention lately (Langran, 1993). Tensed exis-tence (is, was, will be) is a basic aspect of temporality that has implications for the boundary problem. If it is often difficult to determine the boundaries of *actual* entities, representing those of entities that have been but no longer are (a vanished settlement, an ancient lake), or of those that may be, but are not yet (a future city expansion, post-global-warming sea levels), can be a formidable challenge.

(h)   *Permanent or variable*

Temporality also means change, and nothing is permanent at the grand scale of things. Within the temporal scales of interest to those dealing with geographic phe-nomena some entities may be considered stable, and many others not. Geographic entities may change their boundaries (along with any of their other characteristics) in a regular, periodic, or irregular manner, and such change may be along any of the dimensions discussed above (solid/fluid, homogeneous/inhomogeneous, continuous/ discontinuous, etc.), including change from non-being to being or vice versa (actual/ non-actual).

(i)   *Fixed or moving*

While not drastically different from some of the variable entities in the category above (e.g. a toxic plume expanding from a source), many dynamic entities changing their location as well as their shape through space (a tornado; a herd of migrating animals; a traffic jam) pose such difficult representational problems that they deserve separate mention. It is questionable whether the very notion of boundary makes sense in such cases.

(j)   *Conventional or self-defining*

Although many geographic-scale natural and man-made entities have boundaries, the most precise boundaries existing in the geographic world are immaterial. These are the national, political, administrative, and property boundaries that express social and political control over space, to the extent that they are defined through geographic coordinates (or some other convention) rather than through natural fea-tures. At the other extreme are the self-defining boundaries of social, cultural, and biological territories and socio-economic regions, which are usually very difficult to pin down. In between is the class of conventional constructs dependent on empirical measurements, such as isolines and the boundaries of watersheds, which are concep-tually extremely well defined, though usually not practically.

The above categories do not purport to exhaust the distinctions that may be made in the empirical world with respect to the well-boundedness of geographic entities. But they do help illustrate the futility of dividing things into those with determined and those with undetermined boundaries – rather like dividing animals into eleph-ants and non-elephants, and hoping to say something meaningful about the latter. The problem is compounded by the fact that, next to the empirical one, the boundedness of geographic entities depends on two further dimensions: the mode of observation and representation, and the purpose the representation is to serve.

3.3 Dimension two: Mode of observation

Regardless of the empirical characteristics of geographic entities, these may *appear* well- or ill-bounded depending on both the chosen conventions of representation and measurement, and the (objective or subjective) quality of the observation. Indeed, the mode of observation actively determines what kind of boundaries an entity will have.

(a) *Scale*

Scale is the great boundary maker of geography. The increasing generalization and abstraction entailed by a decreasing scale of representation turns scattered entities into points, ill-defined bands into lines, and regions with undetermined boundaries into neatly bounded polygons. The reason why bounded geographic objects appear to many to be the rule, and ill-bounded ones the anomaly, is because so much geographic reasoning and manipulation is geared towards small-scale *map* representations of geographic entities (as opposed to, say, mathematical representations). In many ways, the boundary problem in GIS is the legacy of the continuing influence of the cartographic perspective. Like other abstractions the cartographic abstraction is legitimate and often necessary, up to the point where the properties of the model are mistaken for those of the reality represented. Also, the emphasis on cartographic scale and its properties should not detract from the more fundamental problem of scale in the context of scale-dependent geographic phenomena and entities. As an example, the question of when a collection of (well-bounded) atomic entities such as buildings becomes an (ill-bounded) extensive entity such as a settlement acquires a new twist in the context of a technology that allows easy back-and-forth changes in cartographic scale.

(b) *Resolution*

In general, the higher the resolution, the more difficult it will be to obtain objects with well-defined boundaries, since the irregularities, discontinuities and inhomogeneities of entities will be more difficult to ignore. Because of the strong linear correlation between resolution and cartographic scale, these two issues have generally been treated together (e.g. the work on map generalization: see Buttenfield and McMaster, 1991). GIS, by allowing descriptive detail on entities to be stored independently of map scale, restores resolution as a separate consideration within the problematic boundary (as when two entities registered at different levels of resolution appear on the same map layer).

(c) *Perspective*

A classic example of how perspective creates boundaries is in the different outlines of the Earth globe generated by different map projections – all of them extremely well-defined, though totally spurious boundaries imposed on a continuous unbounded surface. 'Viewshed' functions in GIS have a similar effect of creating non-existent boundaries through perspective. Along with other kinds of perspective boundaries produced by the increasingly popular 3D visualization techniques, the latter can change widely and discontinuously with small changes in input data.

(d)   Time

If scale is the great boundary-maker of geography, time is the great boundary-breaker. Most well-defined boundaries that may be visible in a static snapshot of the world will be invalidated a moment later – a moment being a second, a day, a year, a decade, a season, an epoch – whatever the applicable time-step of observation might be.

(e)   Error

Measurement and representation error is not an issue with respect to object boundaries unless the observer (or representer) is aware of it, or unless the user suspects or fears it. Closely related is the notion of *uncertainty*, which, in the technical sense, is an admission of possible error expressed quantitatively. Statements about accuracy margins, confidence limits, epsilon-bands, probability surfaces and the like, all have the effect of casting doubt on the positional and/or categorical validity of different kinds of boundaries represented.

(f)   Theory

Many geographical entities are theoretical constructs depending on definitions, some of which imply boundaries while others do not. Market areas in a Christaller hierarchy are virtually pure boundary; at the other end of the spectrum, it is difficult to conceive of boundaries in connection with interaction flows. In between are constructs such as zone productions and attractions in spatial interaction models, which, by emphasizing properties of the interior of zones, shift the focus away from boundaries, but are in fact very sensitive to them.

3.4   Dimension three: User purpose

The boundaries of geographic objects are, at the very least, the complex products of representational convention and empirical fact. The combination of the several empirical cases and modes of observation outlined above yields several dozens of different classes of geographic objects that may have well-defined or undefined boundaries in several dozens of different ways. The problem is compounded by the fact that, in general, one is dealing with *intersections* of entities rather than isolated entities, and therefore with intersections of boundaries that may be of several different kinds. To impose some order on this proliferation we must go back to the original question posed in this paper: *who cares* if geographical objects have well-defined or ill-defined boundaries, and in what context?

Different users of geographic information work with different conceptual models of their domains of interest, populated with entities (or objects, for some) that may have similar names but very different properties, and are subject to very different kinds of operations. Burrough and Frank (1995) offer a typology of users of geo-graphic information that captures some distinctions of procedure and purpose that are critical to the issue discussed: The *managers of defined objects* use GIS in areas such as cadastral mapping and utility management; *the planners and resource mana-gers* deal with mostly extensive geographic entities, the evolution of which they try to control; the *modelers of space–time* (scientists for short) also deal with extensive entities, the behaviour of which they try to describe, explain and predict, rather than control. The *politicians and entrepreneurs* and the *public at large* constitute addi-tional categories of GIS users.

This classification could be further refined on the basis of either user purpose or of the GIS operations and functions most commonly used in its pursuit. Such an analysis would reveal what kinds of boundaries matter to whom and what manipu-lations render the lack of well-defined boundaries problematic for users. Since this would be well beyond the scope of the present paper, I propose the following simpli-fied scheme, which also summarizes the discussion in the previous two sections:

- some geographic entities are well-bounded (1), but others are not (0);
- some modes of representation and observation yield well-bounded objects (1), but others do not (0);
- some categories of users require well-bounded objects (1), but others do not (0).

This gives eight distinct cases, coded as follows (Table 3.1). The first digit represents empirical conditions, the second, the mode of observation and representation, and the third, whether or not the user purpose calls for well-defined boundaries. Clearly, only the top half of these cases, indicating a user need for well-bounded objects, deserve further investigation. I claim that case 000, where clear boundaries neither exist in reality nor appear in the representation nor are needed by the users, is that of the space–time modelers (scientists). Indeed, physical science has been largely oblivious to geometric boundaries, often concentrating on centres of mass, equi-librium points, and linear trajectories, and more generally emphasizing the interiors of entities, with the boundaries implied. Managers and controllers of entities, on the other hand, tend to emphasize the boundary and imply the interior. Interestingly, this same distinction has been made in the literature (Langran, 1993) with respect to raster versus vector GIS.

Case 111 (well-bounded empirical entities, well-bounded representations, manip-ulations requiring well-defined boundaries) is not problematic. CAD applications in industrial design are the best examples of that category, which generally comprises Burrough and Frank's (1994) user group of managers of defined objects.

Table 3.1 A rough typology of perspectives on boundaries

111	*managers of defined objects*
011	agricultural land value assessors
101	pilots of fighter planes
001	*planners and resource managers*
110	users of (man-made) facilities
010	graphic artists
100	bus and subway route map users
000	*space–time modelers*

Case 011 (empirical entities with undefined boundaries, well-bounded representations, requirement for well-bounded objects) is exemplified by the (large-scale) soils map used for agricultural land value assessment. Here the problem arises from the clash between the need for accurate boundaries for revenue control purposes and the elusiveness of boundaries in an inhomogeneous entity such as soil. The same soils map used in a descriptive vegetation study for scientific rather than management purposes may not be problematic (case 010).

Case 101 (well-bounded entities that cannot be adequately observed or represented, and users who require sharp boundaries) is the classic predicament of military operations needing to hit enemy targets from afar. No more needs to be said about this case.

Case 001 (entities that neither have, nor can easily be assigned well-defined boundaries, users who require such boundaries) is probably the one deserving the most attention, as it corresponds to one of the most important categories of GIS users – the planners and resource managers. These professionals deal with entities that often defy boundaries in both reality and representation (ecosystems, estuaries, wetlands, present and future land uses, air quality, fisheries), but need to control and manipulate them as some kind of objects.

3.5 The next step: Combining ill-bounded entities

The preceding rough analysis has helped focus the problem of ill-defined boundaries to particular kinds of users who need to apply operations conceptually appropriate for objects (control, manipulate, label, retrieve, move, exchange, subdivide, buy, sell, price, replace, restore, renovate, destroy, etc.) to geographic entities that more often than not lack most of the properties of (prototypical) objects. Such operations 'objectify' geographic entities conceptually in ways that make object-like representations appropriate and even necessary (Couclelis, 1992). Still, the question of how to deal with this very real problem remains open. The one thing that is clear is that given the wide array of possible ways for boundaries to be ill-defined either in fact or in representation, no single approach will do. Fuzzy boundaries, fractal boundaries, multiple boundaries, movable boundaries, *ad hoc* boundaries, flashing boundaries, probability surfaces, buffer zones, bands, colour gradations, aural signals, textual warnings, multiple representations, and so on, may all be appropriate for different cases. Finding what is appropriate for each case should be the next challenge.

Thus far in this chapter we discussed boundaries relative only to isolated geographic entities. There is a natural way to extend the discussion to intersecting entities and intersecting boundaries of different kinds, which is by far the more relevant case in practice. The boundedness conditions outlined in Sections 3.2 and 3.3 lend themselves to a simple representation of geographic entities as binary codes indicating which empirical or observational factors may be causing a boundary problem. Thus, regarding the empirical nature of the entity, we have defined the following binary variables:

$x1$ atomic (plenum)
$x2$ homogeneous (inhomogeneous)
$x3$ discontinuous (continuous)

x4   connected (distributed)
x5   solid (fluid)
x6   two-dimensional (higher-dimensional)
x7   actual (non-actual)
x8   permanent (variable)
x9   fixed (moving)
x10 conventional (self-defining)

The first characteristic listed in each case is conducive to well-defined boundaries, and is coded as 0, while the second one (in parentheses) is not, and is coded as 1, as being the more critical with regard to the boundary issue. Thus, the more 1's the code of an entity contains, the more it is likely to have undefined boundaries. For example, a lake would be represented as:

x1	x2	x3	x4	x5	x6	x7	x8	x9	x10
1	0	0	0	1	1	0	1	0	1

indicating the existence of five distinct factors potentially causing boundary problems (plenum, fluid, three-dimensional, moving, self-defining).

The extension to the case of intersecting entities is straightforward. Here is the case where a river flows into the lake through an area covered with sparse seasonal vegetation, crossing a property boundary defined through its coordinates:

	x1	x2	x3	x4	x5	x6	x7	x8	x9	x10
Lake	1	0	0	0	1	1	0	1	0	1
River	1	0	0	0	1	1	0	1	1	1
Landcover	1	1	0	0	0	0	0	1	0	0
Property	0	0	0	0	0	0	0	0	0	0

The resulting table is a representation of a mathematical structure known as a *simplicial complex*, with the individual entities composing it being the associated *simplices*. The incidence relation is: 'Geographic entity Y_n has property X_m', where Y is the set of intersecting entities and X is the set of empirical properties conducive to ill-defined boundaries. Here, the simplicial complex contains a lot of information about the intersection of the four entities as regards the properties of the resulting boundaries. There is a heavy analytical apparatus in algebraic topology for finding the properties of simplicial complexes, and a methodology known as Q-analysis or polyhedral dynamics (see Atkin and Casti, 1977) that is familiar to many geographers and others. Questions that may be asked include the dimensionality of the boundary problem, the possibility of dealing with different boundaries with different techniques, the effect of adding entities, and so on. A similar view can be taken of the mode of observation/representation dimension, which comprises at least six critical variables. A systematic treatment of the boundary issue should combine both the empirical and observational dimensions, as well as being sensitive to the different kinds of GIS operations that are most important for different groups of users.

3.6 Conclusions

Boundaries constitute the outer limits of individual entities, but also the locus where two or more different entities meet; they enclose and separate, divide and join, distinguish and juxtapose, contain and include, create interiors and exteriors, help tell same from other. Janus-like, they are partly inward-oriented envelope, partly outward-oriented interface. Even where boundaries are not actual or observable in the world, human intentionality often needs to create them. People kill each other over boundaries of both the well-defined and the undefined kind. The purpose of this paper has been to highlight the irreducible complexity of the issue, just as we strive to find relatively simple technical solutions that will help us get on with the work.

REFERENCES

ATKIN, R. and CASTI, J. (1977). *Polyhedral Dynamics and the Geometry of Systems*, RR-77-6, IIASA, Laxenburg, Austria.

BURROUGH, P. A. and FRANK, A. U. (1995). Concepts and paradigms in spatial information: Are current geographical information systems truly generic?, *Int. J. Geographical Information Systems*, 9(2), 101–116.

BUTTENFIELD, B. P. and MCMASTER, R. B. (Eds.) (1991). *Map Generalization: Making Rules for Knowledge Representation*, London: Longman.

CASTI, J. (1989). *Alternate Realities: Mathematical Models of Nature and Man*, New York: Wiley.

COUCLELIS, H. (1992). People manipulate objects (but cultivate fields): beyond the raster–vector debate in GIS, in: Frank, A. U., Campari, I. and Formentini, U. (Eds), *Theories and Methods of Spatio-Temporal Reasoning in Geographic Space*. Lecture Notes in Computer Science 639, pp. 65–77, Berlin: Springer.

HOOKER, C. A. (1973). Metaphysics and modern physics: A prolegomenon to the understanding of quantum theory, in: Hooker, C. A. (Ed.), *Contemporary Research in the Foundations and Philosophy of Quantum Theory*, Dordrecht: Routledge and Kegan Paul.

LAKOFF, G. (1987). *Women, Fire, and Dangerous Things: What Categories Reveal about the Mind*, Chicago: University of Chicago Press.

LANGRAN, G. (1993). *Time in Geographic Information Systems*, London: Taylor & Francis.

CHAPTER FOUR

Uncertain Boundaries in Urban Space

IRENE CAMPARI

Department of Geoinformation, Technische Universität Wien, Vienna, Austria

4.1 Introduction

> The understanding is associated with the segmentation of non-discrete space.
>
> J. M. Lotman, *Kul'tura I Vzryz*

Urban space is usually thought of as an *ensemble* of objects with determined boundaries. It is composed of human artifacts. Human artifacts are assumed *per se* to have determined boundaries. However, urban space and its structure are determined by many factors that provoke uncertainty in the definition of boundaries of artifacts. These factors are associated with the presence of many boundaries belonging to artifacts of different types. In urban space there are administrative artifacts, sacred and religious artifacts and places, social and territorial entities, physical artifacts whose nature is determined by some associated functions. The representation of urban spatial information forces to the definition of exact boundaries for each urban entity. This is due to the use of the isotropic model of space and Euclidean geometry in representing spatial information with GIS. The modelling of urban information must include the concept of urban space as structured, anisotropic space. The experiential point of view provides a more concrete account of the uncertainty in the definition of the objects that populate the structured space. In this case the uncertainty of boundaries of urban entities becomes a crucial issue for the modelling. This chapter discusses some of the main factors that make the boundaries of artifacts of urban space uncertain, using examples from actual urban contexts.

GIS manages urban spatial information using the isotropic model of space and Euclidean geometry, both of which are based on experience with solid bodies with determined boundaries. To model urban spatial information through Euclidean geometry means to interpret urban space as being populated with objects with determined boundaries. For each spatial entity of urban space a corresponding solid body in Euclidean geometry and a corresponding primitive form in isotropic space is formed. GIS modelling has inherited this way of thinking about urban space from architecture and surveying, which fields are used to seeing urban space as isotropic. This way of reasoning is rather a professional deformation than a unique way of looking at urban space.

There is a general tendency to think of artifacts as objects with determined boundaries. Artifacts are discrete objects *par excellence*. Urban space is populated with artifacts. Even some natural elements of urban space might be thought of as artifacts; even trees and rivers are manipulated and adapted to urban contexts. Therefore it is quite obvious to think of urban space as populated with objects with determined boundaries. Each element is thought as distinguishable from all others in shape, form, dimension, position, function, colour, etc., and appears definable, identifiable, describable, and determinable. Although uncertainty might arise from observing elements with fuzzy boundaries, their shape, dimension, location, function and configurations appear to be determined. Notwithstanding, the determinacy of urban artifacts is mostly appearance rather than evidence. Artifacts are not definable, identifiable, and describable *per se*, but must be defined, identified and described within a specific context of observation. The context affects the determination of artifacts as objects by attributing to them a shape, a form, a function and a location. In spite of some generally recognizable properties, these may not have the same corresponding meanings for everyone (Wittgenstein, 1961). The context for artifacts builds a whole configuration that transforms undetermined entities into determined objects, and determined objects into other determined objects. The notion of context leads us to see any boundary of urban space as an artifact boundary rather than as a boundary of artifacts.

We are trying 'to see' urban space as a realistic space accounting for some of its experiential aspects. The observation based on the experiential point of view follows the approach introduced by *experiential realism*, in particular by Lakoff (1987, 1988) and Johnson (1987). In mathematics the definition of 'reality' closest to the experiential concept was expressed by Rucker (1984), who aimed to build a model of reality based on the notion that 'everything that exists is the perception of the various observers'. He called the space of our perception a 'space of facts'. The experiential approach has been applied to the understanding of geographic space by Tuan (1977), Couclelis (1988), Campari (1991) and Mark and Frank (1995). Experiential realism offers a different view from that of professionals of land measurement. From a realistic approach urban space appears to be populated with elements and entities whose boundaries may be identified only in a defined context of observation (Campari and Frank, 1994). Realistic space like urban space is anisotropic and structured, it is a 'space of facts'. What determines boundaries in anisotropic space is the connection between elements, their functions, and the processes that give rise to particular shapes and particular configurations of different objects. The definition of boundaries of objects depends on the variables considered to describe urban space and, in particular, the type of approach used to describe it. Realistic space has a strong component of experientialism. In describing realistic space one cannot ignore how urban contexts are perceived by inhabitants and by planners. Planners share with residents the experiential perception of urban space, although planners are influenced by the professional view based on the isotropic model.

This chapter presents a view of urban space as an *ensemble* of elements and entities with uncertain boundaries. It is assumed that most of the boundaries in urban space are uncertain and undetermined *per se*, and that their determination depends on the context in which the elements and entities are observed. Some different contexts from reality are discussed which observe artifacts of urban space. In particular, in Section 4.2 some of the main artifact administrative boundaries are analyzed e.g. districts, and postal, telephone service boundaries. In Section 4.3 religious space and holy places are described as belonging to administrative and territorial contexts. Section 4.4 provides an analysis of uncertain boundaries of physical urban artifacts and how they are perceived with respect to their functions and the processes they serve.

4.2 Administrative artifact boundaries

In reality, most urban administrative boundaries are invisible, and can only be traced on maps. Their existence depends on bureaucratic decisions. Administrative boundaries discretize the continuous physical and social components of urban space; they do not exist in the reality we 'see' (in this context, the English verb 'to see' is synonymous with 'to understand'; Tuan, 1977). People make administrative boundaries visible by other means, like attributing to them a tangible and understandable value. In practice, people understand the existence of administrative boundaries when they interact with the local administration either privately or publicly.

The GIS field has always dealt with administrative fragmentation of geographic space, and most applications have been based on the representation of administrative subdivisions at all levels (i.e. local, regional, national, etc.) In the recent theoretical debate in the field of GIS, administrative subdivisions have been ignored as objects of geographic space, and the focus has been on human experience with structured space. There is some dualism in the treatment of administrative boundaries. On the one hand, one considers administrative boundaries as the exact subdivision of bureaucratic space onto which to map statistical data, in which case the certainty of boundaries is always taken for granted. The data are usually taken from other sources, and, at most, one has to define the measurement accuracy. On the other hand, from the geographical point of view, one tends to liquefy those boundaries as the expression of some form of territoriality, in which case the uncertainty of boundaries of the territorial subdivision is well recognized. In many cases the territoriality is not expressed by tracing exact boundaries on the land or as an equivalent representation on maps, but often by a sense of belonging to a land or a social group. The sense of territoriality is not always quantifiable and measurable on the land.

The difficulty in accounting for administrative and territorial boundaries from an experiential point of view is evident in the literature. Zubin (in Mark *et al.*, 1989; Couclelis, 1992) and Montello (1993) proposed a classification of four different types of space based on human perception and sensori-motor experience. This classification does not account for administrative spaces. The boundaries of administrative spaces are not visible, they do not constitute a barrier or a relevant element in human sensori-motor experience. There is an evident difficulty in defining administrative boundaries as belonging to human experience of space, yet these boundaries exist, and involve contexts of everyday life. It could be argued that administrative boundaries belong to the space of type A and D as described by Zubin. Space of type A is that of manipulable objects. Administrative boundaries are usually only represented, identified and described on maps as manipulable objects. However, in common experience, administrative boundaries may be associated with the space of type D, whose whole structure and entities are beyond the range of direct experience and are not directly manipulable.

4.2.1   District boundaries

The subdivision of urban space into districts is the most common way of bounding administrative areas. Boundaries of districts are the boundaries of public and private administration *par excellence* to (i.e. the municipality, telephone service companies, etc.). The subdivision into districts is typically associated with administration of large towns.

Making invisible boundaries visible

The boundaries of districts are the bureaucratic means of discretizing urban space into functional areal units. Entities and artifacts within them are identifiable through other visible and perceivable elements, such as the particular social group that lives in a district, the configuration of buildings and streets, the style of buildings, the main function attributed to the district within the larger urban context, etc. Some other secondary aspects coincide to identify one district with respect to another, such as the cost of houses, their structure and style, quality of shops, status of the residents, the buildings in which the bureaucratic functions of a district are concentrated, etc. The invisibility of bureaucratic boundaries is transformed by people into tangible values that visibly differentiate urban space. The character and value attributed to a district by the residents may differ from those selected by planners who make the official subdivision of urban space. Their needs are different and the functions residents or planners attribute to the various areas of a city are consequently different too. Residents reason more about immediate needs than planners; for a planner the cost of houses is not a criterion for bounding a district, but for a resident this would be a way of identifying an urban area. For example, Vienna is subdivided into 23 districts (*Bezirke*). They are identified through some of the aspects described above, i.e. the first district is the richest, the oldest and the core of the city with the cathedral; the second district Leopoldstadt, was formerly identified by its Jewish community, but now by the *Prater*; and the thirteenth district is identified with Schönbrunn and Hietzing, high-class residential areas.

The uncertain boundaries of districts

The uncertainty in the identification of district boundaries has various causes. From the experiential point of view, one cause of this uncertainty is the fact that they are invisible. From a realistic approach the two other relevant causes of this uncertainty are the morphology of the city, and the function of the district.

   Districts are areal units, whereas the urban artifacts they interact with are nets, streams, points, crosses, irregular tessellations and areal units other than the blocks. In towns and cities with a regular morphology (i.e. rectangular grids) it is easy to identify the boundaries of a district by the limits of a set of blocks defined by straight streets and 90° corners. In towns and cities with an irregular morphology (like most European towns) blocks and streets have irregular shapes. The easiest way to define the boundaries of areal administrative units is to ignore the real structure of urban space, imposing boundaries based on quantitative criteria (such as an equal number of residents and houses in each district).

The district is supposed to administer the residents and to provide services. However, residents move and are not fixed spatial objects, so the public administration chooses fixed spatial elements, such as buildings and streets. One might argue that the district area is identifiable with the physical boundaries of buildings at the edge of the blocks. However, this number is not reported in the numbering scheme of buildings. Therefore, one may argue that it is not the buildings that delimit the districts, but more probably the streets. On the map of Vienna, the boundaries of districts cross streets or divide them in two, not so well defined parts. Only the first district is thought of as bounded by a street (the *Ring*), but on its eastern side the boundary is the Danube Canal, located beyond the *Ring*, and on the south it is bounded by the River Wien.

The uncertainty of boundaries can hardly be avoided. The local adminstration is forced to choose fixed elements to control the mobile entities in urban space – the people themselves. Public adminstration fragments physical entities of different types because it cannot identify exactly where people actually live.

4.2.2   Postal service boundaries

The space of the postal service is usually subdivided into areal units or districts, based on streets and buildings where the street is only a segment of building that 'indicates' where the post is to be delivered. The concept of street as a limit is indeterminate. It is not the street that identifies a mail user, but the building in which the addressee is supposed to live. Postal services and administrative districts serve different needs. The district identifies social units and people, whereas the postal district identifies dwellings and buildings. A letter may not be delivered to a person because he or she has changed address.

4.2.3.   Telephone service boundaries

The telephone service also subdivides urban space into areal units, but it is obviously thought of as a net. The service does not delimit areas but connects points in the net, which are then aggregated within areas. While the view of the telephone service as a net implies that each point in the net has equal weight, the subdivision into areas gives each area a different weight. Users perceive the boundary of the telephone service area when they dial a prefix to reach a number outside the urban area. This influences the perception of distances between places on the basis of the cost of the call.

4.2.4   Embassy boundaries

Embassies are territories in urban space that do not come under the administration of the countries in which they are situated physically. Their physical structure belongs to the urban space, while their legal space belongs to other countries with other laws, rules and traditions. The behaviour of people who interact with embassies is usually that of the country represented by the embassy. When an Iranian woman goes to an Iranian embassy wherever it is, she wears a *chador* and speaks Persian.

Usually embassies are considered enclaves in urban space but, unlike other types of administrative enclaves, urban enclaves are not represented on maps. An administrative city map does not indicate whether a building is under the sovereignty of another country, but shows the building as belonging to a district as the host country has organized them. Actually, embassy and consulate boundaries are quite difficult to represent on maps, since they are often housed in flats in buildings with other flats with different functions.

4.3   Uncertain boundaries of religious space and holy places

'Then you go down there! You will be far!'
– 'Far from where?'

Antoine de Saint-Exupéry

Many religions discretize space, although the ideas they express are universal and continuous. This discretization follows the hierarchical organization of their secular structure. In cities and towns in regions with a prevalent religion (i.e. Catholicism or Islam) the whole urban space and its structure are organized on a religious basis. Most Islamic towns developed from the *Hejira*, dictated by Mohammed, to settle the nomad population in urban areas (Fusaro, 1984; Hourani and Stern, 1970; Tjahjono, 1989). The European towns founded in the Middle Ages had a church or cathedral as the core. In Islamic, as well as in Christian towns, the discretization of space serves to distinguish the areas for adherents of the religious groups present in these towns (e.g. Jewish communities in North African and European towns).

In urban space, within the boundaries of the Catholic parishes there are groups belonging to other religions, their places of worship and their cemeteries. In Western cities Catholic, Jewish, Orthodox, Evangelical and Muslim communities share the same space but discretize it in different ways. One has to identify boundaries for sacred or administrative religious entities by choosing a context in which to select them. The highest level contexts are the secular/religious administrative subdivision of space, and the territoriality of sacred places. These two contexts are now specified further in the context of each religion.

4.3.1   The secular/religious administrative discretization of space

Many religions have traditionally subdivided space into units following the clerical hierarchy, particularly in Catholic countries. Urban space is subdivided into parishes, the smallest secular administration of the Catholic church. The administrative subdivision of urban space is used as the basis for statistical information about adherents. In some cases these data are used by the civil administration to map official population data. In Portugal, for example, one of the main administrative subdivisions of the Portuguese territory is based on the *Freguesias* (Catholic parishes). In many countries the parishes were the only places in which data about population was collected. They guaranteed the accuracy of the information collected. This is often reflected in the representation of administrative space with GIS, although the original basis for such a discretization is lost in the technological management of data (Campari, 1990, 1991; Campari and Frank, 1995).

Vienna provides an example of the strong administrative character of the parishes. All residents of the city, must declare their religion to the local authorities, and Catholics are obliged to pay a tax to the parishes.

4.3.2 The territoriality of sacred places

The sacred subdivision of urban space expresses a strong sense of territoriality. In large towns this subdivision is mostly based on the location of temples, churches and cemeteries:

> Church buildings ... are not simply things located in space. They are places set apart by boundaries and within which authority is exerted and access is controlled. In other words they are territories. (Sack, 1986, p.93)

The urban spatial religious organization is a crucial element in defining space through the identification of places. However, there is a distinction between 'monuments', considered as 'historical' monuments, and temples and churches, which are used for worship. Sacred buildings have boundaries, but they determine only themselves within the context of the religion they serve. A synagogue may be located within a social context that is Catholic, or *vice versa*. The Jewish temple of Vienna is 50 metres away from the oldest Catholic church and is located in the same sector of the district as the main Greek Orthodox church. The temple and the oldest Catholic church are located in the same parish. The Evangelical church of the 22nd district is only a flat in a residential building in an area of the Viennese suburbs that the Catholics consider under the administration of their own parish.

Cemeteries are sacred places that are usually managed by the public administration in Western societies. The may appear to be objects with determined boundaries. In most British towns they are easily recognizable as open places and are used as public gardens (except perhaps for the Jewish cemetery in Hampstead, London, which is surrounded by a wall). In Europe under Napoleon many cemeteries were moved beyond the border of the towns, but as the towns expanded over their ancient border the cemeteries were again included in the urban fringe. In most European towns these cemeteries are now within the urban borders, usually separated by a wall from residential areas beyond. Inside, other forms of physical or symbolic separation (walls, fences, streets, etc.) were erected between areas of different religions. In Vienna, for example, the *Zentralfriedhof* is partly dedicated to the Jewish cemetery (from 1800 to 1938; doors I and IV). On the map this seems a sharp separation, but in reality, only a narrow street separates the old Israelite cemetery from the Catholic one.

To determine the boundaries of historical cemeteries is even more difficult. For example, the oldest Jewish cemetery in Vienna, is located in Seegasse (Alsergrund, ninth district). On the map of Vienna it is shown as the courtyard of the building, whereas on the map of Jewish places in Vienna the courtyard is indicated as a cemetery (Heinmann-Jelinek and Kohlbauer-Fritz, 1993). While the residents of the building may regard it as their garden, the Jewish community of Vienna regard it as belonging to their sacred territory.

4.4   Boundaries of urban physical artifacts

The better part of my affections, would
Be with my hopes abroad. I should be still
Plucking the grass to know where sits the wind,
Peering in maps for ports, and piers, and roads;
And every object that might make me fear
Misfortune to my ventures, out of doubt
Would make me sad.

Shakespeare, *The Merchant of Venice*, I.1

Physical artifacts are always understood as discrete objects: they are visible, and their supposed determined shape makes us aware of the nature objects. However, the ideas they express, or what we perceive when we look at them, are continuous. Trying to see in artifacts a bounded idea leads to account only for their usefulness. But how can we see and understand them if the ideas they express are not confined within their supposed determined form? Why are they considered to be objects with determined boundaries? And why should we question the certainty of these boundaries? We often seem to confuse sharp boundaries with a shape that 'matches a type' that is accepted within a specific experiential framework (Gould, 1991). Turning to the realistic urban space, the determination of boundaries of artifacts becomes more complex than the simple objects of isotropic space. It is enough to consider the *function* as an element of definition of objects that artifacts become difficult to define as objects with determined boundaries, predefined shape and form. When the context in which the function is observed changes, the identification of the objects changes as well, and they turn out to be other objects.

For instance, the spatial definition of urban infrastructures like street networks and runoff systems require the definition of physical boundaries that delimit them (see Figure 4.1). The street network and the runoff system may have the same shape, the same form, share the same space, and may be bounded by the same physical components, but they do not have the same infrastructure. They serve different functions and different processes. Their identification as different 'things' occurs through different identifications of the physical components that delimit them. For instance, an outdoor staircase can be a part of the boundary of the street if the context of observation is that of the street network, or can be a part of the boundary of the runoff system if the context of observation is water drainage. Each definition is based on the identification of the function of the object at any given time and in a specific context. Therefore, objects may be identified differently according to their function or the process they serve. The context is always crucial in determining and identifying objects. The discussion so far of the street network and the runoff system is a professional appraisal. The residents experience directly everyday the function of the street, by walking in it. The street is an element of social and individual life, while the runoff system is only an element in the infrastructure of urban space.

Thinking of urban space being populated with artifact objects fits very well with the concept of objects in GIS. The professional view of urban artifacts is that of architects and planners, which is filtered by Euclidean geometry and the isotropic model of space. Buildings and other urban components are treated as manipulable objects, and are usually designed and traced on map before they come to populate actual urban space. Architects and planners use the isotropic model of space to represent urban artifacts on maps as well as in a GIS. Urban artifacts in a GIS must be exactly measurable, they must be at a precise distance, they must have a form and a shape that fit with a supposed unique function. This fact limits the freedom of architects and planners in seeing and interpreting urban space with a complex structure. Some painters who played with objects with determined boundaries and isotropic space, but who were not bounded by the relation form–shape–position–function, transformed objects into things with an identity, and isotropic space into anisotropic and structured space.

4.4.1 Streets and blocks

Streets and blocks are components of the configuration of towns. They define and identify the spatial extension of infrastructures and the spatial dimension of processes that occur in urban space. The geometry of the street network is based on nodes and edges. It is topologically a connected graph. The components of the inside street network are usually defined according to their functions; streets may have a local function only or a regional function.

The configuration of street connections derives from the origin of the town. In towns built in a valley close to a navigable river, for example, the internal main streets are directly connected with the outside streets leading to the river to facilitate the transportation of goods from the river harbour to the stores in the town. An example of street network with only a local function is provided by small towns with circular shape developed on the top of hills around a castle or church. The streets connected with the outside network are circular, while the secondary streets are radial.

Town gates

Town gates limited access to the town, served as fiscal check points and were part of the town walls. They often formed part of the continuous structure of buildings (i.e. as archways between them) at the edges of towns. In large towns they connect the non-local inside street network and the regional one (e.g. Lucca, Pisa). Gates are symbols of both the continuous relationship of the town with the space outside, and of separateness from the outside.

Open spaces and town squares

The local street network includes open spaces and in particular, the town square. Open spaces are often bounded by buildings of particular importance for social life, like a municipality building or a church, or installations such as public fountains. In small towns the square is often only a component of the local inside street network, and is not directly connected with the regional street network. It serves only local needs, such as to provide access to water or to the church.

In small southern European towns the square sometimes represents a continuum with the structure of the buildings that surround it. The differentiation between outside space (the square) and the inside space (the building interior) is not visible. People living in small towns tend to look at the outside space as belonging to the interior of their home.

Water runoff system

The structure and organization of water runoff systems depend on the topography. The runoff system often shares the same urban space of the street. The streets may have been designed to carry water out of the town, into a river or lake, but sometimes the runoff system is a separate structure, organized beneath the buildings, and it is then not represented on maps. The geometry of the runoff system is linear. It may be connected with and contained within the structure of buildings.

4.4.2   Blocks

A block is a set of buildings limited by the street network. The boundaries of segments of streets are the boundaries of blocks and *vice versa*. For instance, in semicircular towns with radial secondary streets, a block is a set of buildings between two main streets and two radial secondary connections. In towns with compact morphology (a building occupies an entire parcel, with no open space around it) a block is like a continuous construction. In towns with a multi-level street network on a slope a block is located at one level and is isolated from neighbouring blocks by streets located at the upper and lower levels.

4.4.3   Buildings

> The construction of a building involves the definition and delimitation of space. Therefore it is important to analyze boundaries that indicate how spaces – here and there, inside and outside, upstairs and downstairs – are separated and linked. The concepts of boundary, threshold, and transition are useful in examining how different spaces or domains are associated with or demarcated from each other. (Lawrence, 1989)

Lawrence introduces the concept of a boundary as a separation between spaces at the building level. He considers a building as an object that may be decomposed into other objects on the basis of their function and use, but the building is the smallest unit of the town configuration considered in this chapter.

The front of a building is a segment of the street and an element of the block. It may have open space around it, but not in compact towns. The geometry of the building is areal, usually a polygon with four or more corners (at least in the class of towns under consideration), typically with right angles. The relations between neighbouring buildings are affected by the internal structure of the dwellings. A dwelling may belong physically to different buildings, having rooms in different buildings, or it may have rooms in different blocks. This complex structure is not observable from the exterior; two buildings may have a boundary line that divides their exterior fronts, but they may share dwellings and rooms in the interior.

4.4.4   The street as access to buildings and the water runoff system

In Figure 4.1 (a view of Rio nell'Elba) the outdoor staircases of the two buildings may be functionally identified as: part of the boundaries of the street, part of the boundaries of the water runoff, access to the building, access to a dwelling, or a constituent part of the building. The exterior of the building may be a functional

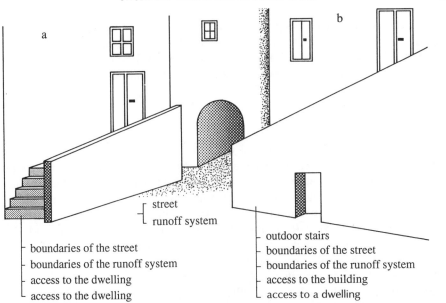

Figure 4.1   Sketch of a street scene in a small town in southern Europe.

component of the infrastructure or a boundary of its spatial extension. The archway between the two buildings may be seen as: a boundary of the runoff system structure, a boundary of the street structure, the floor of the building, the floor of a dwelling.

4.4.5   Definition and identification of components

The spatial identification of infrastructures requires the definition of physical boundaries as well as the identification of the function of the components. The definition of the physical boundaries of processes requires the identification of the components that delimit them. Therefore, components may be identified differently according to their functions or the process they serve. For instance, the spatial boundaries of the water runoff system and street network may be observed as being the same with regard to the function of the components chosen to delimit them (Figure 4.1). If the identification of a component changes with respect to an attributed function, the properties of the component may also change, together with the relations that describe the whole structure.

4.4.6   Differences in definition and identification of components

There are substantial differences between the bureaucratic and administrative subdivision of towns and the experiential and social subdivision. For instance, the administrative subdivision into blocks does not usually correspond to the social subdivision. From the experiential point of view, urban space is subdivided into a set of neighbouring buildings that may not correspond to a physical block (Hillier and Hanson, 1988). This situation has consequences for the identification by names of sites and places within the towns. Buildings belonging to different blocks may be in the same set of neighbours and may be identified by the same reference name. In Figure 4.1 the two buildings belong to different blocks. The street that divides them is covered by the archway, although the two buildings belong to the same set of neighbouring buildings in the social identification of urban places.

4.5 Conclusions

This chapter has highlighted some of the issues associated with the uncertainty of boundaries in urban space, in particular, administrative boundaries and boundaries of urban artifacts. The range of issues discussed here is far from exhaustive, but an element of uncertainty has been introduced in a matter that is not usually regarded as being affected by uncertainty. Urban space and its components are usually regarded as artifacts with determined boundaries. To determine the boundaries of artifacts and to identify objects in urban space it is necessary to specify the context of observation. The contexts of observation are many, and all of them may lead to a different definition and identification of urban artifacts.

From an experiential approach further investigations are needed to define a conceptual spatial domain for objects bounded by invisible boundaries, such as administrative ones. The modelling of objects with undetermined boundaries may benefit from a realistic approach because substantial variables can be introduced to build a structured space for those objects.

The approaches presented in this chapter open a line of research for incorporating views in the conceptual and formal modelling of geographic objects that are different from those usually considered by professional planners. These alternative views may be valuable in helping our understanding of the complex structures of urban spaces, and how these structures are understood by people from different cultural backgrounds.

REFERENCES

CAMPARI, I. (1990). Accuracy vs spatial statistical data: The Mediterranean region *Proc. EGIS'90*, I: pp.110–121, Amsterdam: EGIS Foundation.

CAMPARI, I. (1991). Some notes on geographic information systems: The relationship between their practical application and their theoretical evolution, in: Mark, D. and Frank, A. U. (Eds), *Cognitive and Linguistic Aspects of Geographic Space*, pp.35–44, Dordrecht: Kluwer.

CAMPARI, I. and FRANK, A. U. (1994). *Modeling Realistic Space: Urban Development as a Case Study*, ESPRIT Workshop on Advances in Geographical Information Systems, Ascona, March.

CAMPARI, I. and FRANK A. U. (1995). Cultural aspects and cultural differences in GIS, in: Nyerges, T. (Ed.), *Human–Computer Interaction for GIS*, NATO-ASI Series, Dordrecht: Kluwer.

COUCLELIS, H. (1988). The truth seekers: Geographers in search of the human world, in: *A Ground for Common Search*, pp.148–155, Santa Barbara, CA: Santa Barbara Geographical Press.

COUCLELIS, H. (1992). People manipulate objects (but cultivate fields): Beyond the raster–vector debate in GIS, in: Frank, A. U., Campari, I. and Formentini, U. (Eds) *Theories and Methods of Spatio-Temporal Reasoning in Geographic Space*, Lecture Notes in Computer Science 639, pp.65–77, Berlin: Springer.

FUSARO, F. (1984). *La citta' islamica*, Bari: Laterza.

GOULD, S. J. (1991). *Bully for Brontosaurus*, New York: Penguin.

HEINMANN-JELINEK, F. and KOHLBAUER-FRITZ, G. (1993). *Jüdischer Stadtplan Wien: Einst und Jetzt*, Vienna: Freytag-Berndt und Artaria.

HILLIER, B. and HANSON, J. (1988). *The Social Logic of Space*, Cambridge: Cambridge University Press.

HOURANI, A. H. and STERN, S. M. (1970) *The Islamic City*, Oxford: Oxford University Press.

JOHNSON, M. (1987). *The Body in the Mind*, Chicago: University of Chicago Press.

LAKOFF, G. (1987). *Women, Fire and Dangerous Things: What Categories Reveal about the Mind*, Chicago: University of Chicago Press.

LAKOFF, G. (1988). Cognitive semantics, in: Eco, U., Violi, P. and Santambrogio, S. (Eds), *Meaning and Mental Representation*, pp.119–154, Indianapolis: Indiana University Press.

LAWRENCE, R. L. (1989). Translating anthropological concepts into architectural practice, in: Low, S. M. and Chambers, E. (Eds), *Housing, Culture, and Design*, pp.89–114, Philadelphia: University of Pennsylvania Press.

MARK, D. and FRANK, A. (1995). Experiential and formal models of geographic space, *Environment and Planning B*.

Mark, D. M. *et al.* (Eds) (1989). *Languages of Spatial Relations*, NCGIA Initiative 2 Specialist Meeting Report.

MONTELLO, D. R. (1993). Scale and multiple psychologies of space, in: Frank, A. U. and Campari, I. (Eds), *Spatial Information Theory: Theoretical Basis for GIS*, Lecture Notes in Computer Science 716, pp.312–321, Berlin: Springer.

RUCKER, R. (1984). *The Fourth Dimension*, New York: Houghton Mifflin.

SACK, R. D. (1986). *Human Territoriality*, Cambridge: Cambridge University Press.

TJAHJONO, G. (1989). Center and Duality in the Javanese Dwelling, in: Bourdier, J-P. and Alsayyad, N. (Eds), *Dwellings, Settlements and Tradition*, pp.213–236, Lanham: University Press of America.

TUAN, Y.-F. (1977). *Space and Place: The Perspective of Experience*, Minneapolis: University of Minnesota Press.

WITTGENSTEIN, L. (1961). *Tractatus Logico-Philosophicus*, London: Routledge and Kegan Paul.

CHAPTER FIVE

A Conceptual Framework and Fuzzy Set Implementation for Geographic Features

E. LYNN USERY

Department of Geography, University of Georgia, Athens, USA

5.1 Introduction

Geographic entities can be conceptualized along spatial, thematic, and temporal dimensions. This conceptualization provided advances toward quantification of geographic analysis (Berry, 1964) and later a basis for computer representation of geography (Dangermond, 1983; Nyerges, 1991). Unfortunately, the computer-based techniques often represent all aspects of geography using the spatial dimension with little regard for the thematic and temporal dimensions. This tendency toward the spatial dimension is based primarily on a map view of geographic reality. Historically, the map has served as both the repository of spatial data and the display vehicle for conveying spatial relationships. Logically, designers of computer-based systems for representation of geographic phenomena turned to the map model as a primary source.

It has been argued that of the three dimensions, one is fixed, one is controlled, and the third is variable (Sinton, 1978). This argument holds for map representations in which time is usually fixed, theme is controlled, and spatial location is variable. However, the conventional map framework based on layers of geography appears inadequate to handle sophisticated geographic process models, spatial analysis approaches, and features with undetermined boundaries (Goodchild, 1987; 1991; Nyerges, 1991). For example, a moving oil spill cannot be adequately represented with the map model because it varies along its spatial (size and shape) and thematic dimensions (concentration of oil) as a function of time. Similarly, a modelling of urban transportation requires a dynamic model to account for the changing patterns of movement and congestion as a function of time along a fixed set of spatial routes.

One approach to representing geographic entities is to develop a set of representational objects which maintains a one-to-one correspondence with the geographic entities. For example, a hill is a geographic entity which may be represented with a unique numeric object identifier to which are attached spatial location coordinates, elevation values, soil types, and mathematical formulation from various datasets, and the range of characteristics which define the hill. Such an approach requires the spatial, thematic, and temporal dimensions of each object to be represented to provide for capabilities of examining interaction. The development of computer-based geographic information systems (GIS) has advanced to the point that such an object-based model can now be implemented. Few would argue the need to develop better interfaces to GIS in which users are able to name geographic entities and perform analysis on those named entities. The difficulty occurs in whether the underlying data model need reflect these entities directly or whether, through sophisticated data processing, these entities can be presented to the user through conventional layer-based map models. This chapter is developed from the assumption that the user's perception of geographic reality is one of geographic entities, such as roads, buildings, hills, and ethnic immigration areas, rather than layers of data, and that it is desirable for the data model to directly reflect this perception.

Given this assumption, the design for an object-based model in which objects are a direct representation of the geographic entities rather than geometric elements such as point, line, and area, is developed. This design is referred to as a feature-based geographic information system (FBGIS) since the term feature encompasses both the geographical entity and its object representation (NCDCDS, 1988). Formally, a feature is defined as an entity with common attributes and relationships (Guptill *et al.*, 1990). Unfortunately, only one type of feature, those defined by mathematical equations or legislative edicts, for example, a county boundary, is free from human perception and cognition. Thus, implementation of a FBGIS is a non-trivial task. To address the problem of representation of features including those with undetermined boundaries, and to better support spatial analysis and geographic process models, this chapter presents a conceptual framework developed from three basic abstraction levels, concepts, data model, and data structure using the spatial, thematic, and temporal dimensions of geographic phenomena. This chapter further provides details of a fuzzy set implementation construct for the spatial dimension of the conceptual framework.

To model geographic phenomena from reality some subset of that reality, i.e. an idea of geographic reality, must be developed which reflects a particular application context and resolution of representation. For example, a trafficability application in terrain analysis requires terrain data modelled as hills, valleys, and other features. Since an infinite complex of geographic entities with infinite attributes must exist, any given spatial analysis must work with a subset of these infinities (Berry, 1964). Given the basic dimensions of geographic phenomena as space (location), theme (classification or attribute), and time, to specify an idea of geographic reality one must include spatial, thematic, and temporal attributes and relationships. The attributes define the characteristics of the feature along the particular dimension and the relationships describe the linkages among features along the dimensions (Table 5.1). Thus if one describes spatial, thematic, and temporal dimensions for a geographical entity such as a road, one has described an object and thus a feature. A feature is similar to the region concept in geographic research (Grigg, 1965; NCDCDS 1988; Usery 1993).

The FBGIS model focuses on spatial entity modelling and explicitly contains

Table 5.1 Dimensions, attributes and relationships of the FBGIS conceptual model

	Space	Theme	Time
Attributes	ϕ, λ, Z point, line, area, surface, volume, pixel, voxel, ...	colour, size, shape, ph, ...	date, duration period, ...
Relationships	topology, direction, distance, ...	topology, is_a, kind_of, part_of, ...	topology, is_a, was_a, will_be ...

thematic and temporal attributes and relationships for three- and higher-dimensional data. A feature specified only by spatial location is incomplete; both theme and time are required as a part of the representation (object) and while these may be fixed attributes of the feature, they are always present for actual geographical entities. Thus the FBGIS model includes explicit thematic and temporal dimensions for each feature as well as the spatial location. Furthermore, since these dimensions can be described by their inherent characteristics and/or by their association with other characteristics, both attributes and relationships of each of the dimensions are required to specify a feature.

While the FBGIS approach is similar to the geographical matrix (Berry, 1964) with possibility of supporting infinite numbers of attributes and relations, the problem of selecting a subset of these infinities for a given application and resolution remains to be solved. As in current GIS, one guide to that solution can be taken from cartography, in which maps have always been produced for specific applications and at specific resolutions or scales. From the work in cartography, the most relevant research concerns conceptual models and theories of generalization of geographic phenomena (McMaster, 1991; Nyerges, 1991).

Since maps were designed to serve specific purposes, a base of knowledge was developed to guide which features are appropriate for those purposes. FBGIS can use this knowledge as a starting point to guide the types of features to be structured in databases and knowledge bases and provide users with the tools to generate new features which do not exist. The digital line graph-enhanced (DLG-E) of the U.S. Geological Survey (USGS) is a feature-based dataset built from this approach (Guptill *et al.*, 1990).

In Section 5.2 the concepts of a FBGIS are explored in the context of data model theories. Section 5.3 briefly examines potential tools for feature determination. Section 5.4 explores fuzzy sets as a possible implementation mechanism for features, and Section 5.5 discusses the potential and limitations of this model.

5.2 FBGIS and data model theories

To develop the concept of FBGIS, an analogy to cartography proves useful. The layered model of GIS is analogous to the colour separations of a lithographic map such as the black, blue, brown, green, and red separates of a standard USGS topographic quadrangle. On each separation, one can view the attributes and relationships of the features in that layer but attributes and relationships of features which require two or more separations cannot be viewed. The FBGIS model is analogous to the final lithographic copy of the topographic map in which all features are shown in relation to each other. This analogy is accurate to a point but neither the separations nor the lithographic copy can represent dynamic temporal phenomena. The FBGIS model includes explicit representation of these temporal attributes and relationships to permit dynamic features such as oil spills or storms to be analyzed through mathematical equations with time as a variable in the GIS.

The basic concept of a FBGIS is presented in Table 5.1, but the model must be examined in the context of representing geographic phenomena in a way which provides capabilities for producing geographic information useful for spatial decision-making. A conceptual framework for features can be developed through abstraction levels which have been defined in GIS research and through this framework the logical parts of the model can be explored and implemented.

Geographic data modelling abstraction levels from the real world to the computer implementation including ideas about reality, data model, data structure, and file structure have been developed by Peuquet (1984), Guptill et al. (1990) and Nyerges (1991). For the FBGIS model, an implementation framework which addresses each part of the spatial, thematic, and temporal aspects of features is proposed (Figure 5.1). The spatial dimension of features can be viewed as consisting of three abstraction levels, spatial concepts, spatial data models, and spatial data structures (Egenhofer and Herring, 1991). Spatial concepts are used to organize and structure human perception and cognition of space. For example, one might use a traditional Euclidean geometry model to move about a room but a spatial network concept to navigate through a city. These spatial concepts are intuitively used and based on image schemata which relate to linguistic terms (Talmy, 1983; Herskovits, 1986; Johnson, 1987; Mark, 1989; Frank and Mark, 1991).

Formalization of the spatial concepts leads to spatial data models. The most frequent formalization of space for geographic phenomena is based on the Euclidean metric although other space models, such as the taxicab metric, may be more appropriate for specific analytical purposes (Atkin, 1981; Gatrell, 1983; 1991). Often the spatial data model is used to refer to the combination of spatial and nonspatial or thematic parts of geography (for example, see Morehouse, 1989; Egenhofer and Herring, 1991), but in the proposed FBGIS model a clear separation of space, theme, and time is used since all three are required dimensions of a feature. This separation allows each dimension of a geographic phenomenon to be modelled with appropriate concepts, structures, and implementation mechanisms and still be viewed as component parts of a holistic description of the feature.

The formalization includes the basic elements of the spatial concepts such as the point, line, area, volume, pixel, and volume element or voxel objects (see Table 5.1). Attributes of these objects such as location in some spatial reference system (Euclidean or otherwise) and their interrelations such as topology are formalized through axioms and rules. For the FBGIS framework, specific formalizations building from existing vector-based topological systems and raster-based object systems are used (Tang, 1991; Usery, 1994a). The implementation of the spatial model in a computer·system requires development of spatial data structures which can accept the dimensions of the spatial data model.

Applying the three abstraction levels to the thematic dimension, the thematic concepts are placed at an equal level with spatial concepts in contrast to layer-based

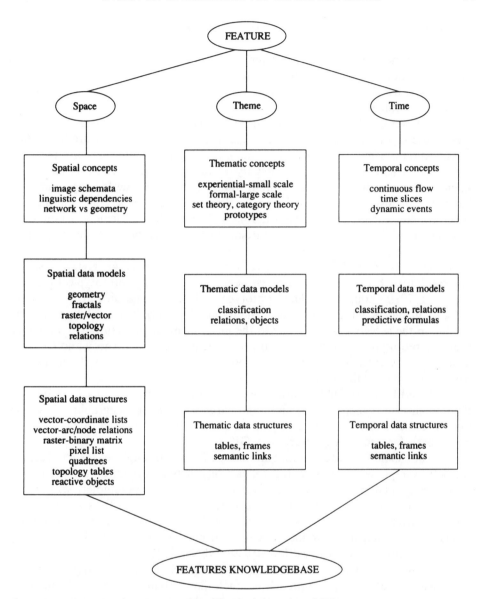

Figure 5.1 Dimensional concepts and models of a feature-based GIS.

systems which assume a space dominant model (Figure 5.1). This distinction in which a feature requires spatial, thematic, and temporal dimensions to exist at equal levels, is perhaps the most fundamental difference between the FBGIS model and conventional layer-based models. Much of the research in GIS has focused on development of databases for representing the spatial conceptual model and only recently have thematic conceptual models been explored (Nyerges, 1991). Thematic concepts include basic theories concerning categorization of phenomena which have impacts on which attributes define the spatial objects and lead to the formation of features. Basic research in cognition has led to category theory and the concept of prototypes which offer a basis in experiential models (Rosch, 1978; Lakoff and Johnson, 1980; Lakoff, 1987).

Based on the principle that the material structures in the perceived world possess high correlational structure, cognitive theory indicates that for categorical systems, such as feature classifications, a basic level of abstraction exists which possesses common attributes, common motor movements, similarities in shape, and identifiability of average shapes (Rosch, 1978). This level of abstraction logically forms the most basic category and provides a single level for devising a feature-based GIS classification scheme. Using a Rosch example, chair, table, and lamp are the basic level objects in a superordinate category of furniture. Subordinate elements are kitchen chair, living room chair, dining table, coffee table, desk lamp, and floor lamp. Analogously, road, railroad, and canal are basic level geographic features in a superordinate category of transportation with subordinate elements state road, interstate highway, standard gauge railroad, narrow gauge railroad, shipping canal, and irrigation canal (Usery, 1993).

Recognition of the basic level objects for categorization is not always a simple process, but each category may be best defined in terms of a prototypical member (Rosch, 1978; Lakoff, 1987). This prototype will not possess all attributes of the category members, but is a single element about which there is no question of membership in the category. It is similar to the mean or mode in quantitative data and, thus, the prototype can serve as the defining member of the category and elements which share greater similarity to the prototype of one category than another become members of that category.

Mark and Frank (1990) have used category theory research as a basis to differentiate small-scale (small areal space) and large-scale (large geographic space) as experiential and formal models, respectively. Usery (1993) has used category theory as one basis for structuring features in a GIS.

Thematic concepts have traditionally been explored through set theory. Perhaps the best example which uses set theoretical constructs to structure thematic groupings of geographic phenomena is multispectral classification of remotely-sensed data into land cover categories. Almost all current GIS are built on the concepts of set theory (Gatrell, 1983; 1991) with the table of the relational data model being the formal model often used for the thematic attributes (Codd, 1970). This relational model generally has not been used to structure thematic relationships.

The FBGIS thematic data model includes relational and object-oriented approaches. The relational approach is much like current attempts with the addition of true thematic relationships. It relies on the ability to construct features as collections (sets) of thematic attributes through which features may be related. The object-oriented approach to thematic concepts is detailed by Mark (1991) in an examination of phenomenon-based (feature-based) approaches to generalization. While terminology is not standardized, object-oriented usually connotes a computer science perspective and in this chapter is used to mean a computer-based implementation concept in which objects include inheritance, polymorphism, and encapsulation while feature-based is used to refer to a geographic conceptual model and is independent of the computer-based implementation strategy. For example, Usery (1994b) details a feature-based prototype system implemented using relational database techniques. Obviously, the logical implementation strategy for a feature-based conceptual model is an object-oriented approach. In the object-oriented approach, objects are constructed based on either sets or category theory concepts of prototypes. Fuzzy sets offer possible techniques to describe prototype features such as hills and valleys (Katinsky, 1994).

Temporal concepts are the least researched aspect of geographic phenomena. Only recently through the work of a few researchers such as Langran and Chrisman (1988), Langran (1989, 1992), Kelmelis (1991), and Greve et al. (1993) has progress occurred in temporal database development and implementation. Similar to the other feature dimensions, temporal concepts involve human perception and schemata of the mind. In general, the human view of time as continuous flow from birth to death dictates temporal organization. For example, the traditional view of each spatial representation as a snapshot in time accounts for our ideas of change detection in geographic phenomena. Change over time becomes the differences between two of these snapshots each at a specific instant of time. These time slices are inadequate to handle dynamic features such as oil spills. The change over time must be modelled in a continuous fashion to capture the geographic phenomena. To accomplish this, a temporal model is incorporated in the FBGIS. The temporal model is not a static attribute in a relational database but an object-oriented predictive function which models change in spatial configuration and thematic attributes and relationships over time. This modelling construct allows dynamic geographic entities to be included in the FBGIS, with static entities temporally modelled with traditional time slice techniques such as time stamp attributes.

5.3   Feature determination

Since an infinite number of features is possible based on various combinations of space, theme, and time, methods to determine features which are appropriate for inclusion in the knowledge and databases are needed. That determination is usually dependent on the particular application and has traditionally been performed in spatial analysis studies through the experience of the analyst. While FBGIS approaches cannot replace the analyst and judicious selection of geographic entities for a particular study, these systems can aid the user in two specific ways: (1) determination of a rudimentary basic feature set; and (2) construction of features in the knowledge base from geographic data and user specification. As a starting point, a set of base features such as those commonly found on topographic maps may be automatically developed from the system databases. This allows use of the accumulated knowledge in cartography on which features are important as base category information. For example, DLG-E is one such feature database currently being developed.

The preparation of features from existing databases and user input specifications require a set of tools for data extraction and feature construction. Many of the tools required for this purpose can be drawn from existing GIS. Techniques such as aggregation, clumping, buffering, and relational algebra provide the basis to use a set of user parameters against geographic data to extract specific feature information. The actual methods of feature determination from databases require a combination of database retrieval and data processing routines, such as those contained in a relational database management system, tightly interfaced to a set of analytical processing tools such as multispectral classification, factor analysis, multidimensional scaling, and q-analysis. Data reduction and presentation of the results to the user for final confirmation of the feature's correctness are prerequisite tasks. The extracted feature information can then be used to build the features in the model of the FBGIS.

The variety of applications requires a dynamic feature determination system capable of learning (Smith *et al.*, 1987). For example, in a terrain analysis application, a user may want to build features which determine the trafficability of a surface by particular types of vehicles. The FBGIS can provide the base set of terrain features to the user for review. If this is insufficient, then the user can specify requirements including spatial, thematic, and temporal attributes and relationships, for features to be added to the system. Assuming appropriate data exist in the databases, these new features can be generated and added to the set of system features. The next user requesting features for a similar application will be shown the base set plus the new features defined by this application.

5.4 Features as fuzzy sets

Fuzzy set theory is an extension of classical set theory and as such fuzzy set operations provide identical results to classical set operations when the sets involved are *well-defined*. The mathematics of fuzzy sets are a solved problem and well-documented in the literature, however their use with spatial data is still a non-trivial application (Zadeh, 1965; DuBois and Prade, 1980; Leung, 1988; Katinsky, 1994). To date most applications to geographic data have modelled the thematic dimension rather than the spatial dimension (for example, see Burrough, 1989; Wang, 1990). The approach in this brief discussion will be to model the spatial dimension of geographic features using fuzzy sets to define the spatial extent of the thematic dimension. In all cases the temporal dimension will be assumed to be constant.

This approach requires the differentiation of features with undetermined boundaries from those features with determined boundaries and measurement uncertainty. The latter type of uncertainty, or fuzziness, is error in the measurement and can be modelled as probabilities based on epsilon distances (Chrisman, 1982) or as fuzzy sets. This chapter is concerned with the former, i.e. features with undetermined boundaries which are themselves fuzzy and, regardless of the accuracy of measurement, remain fuzzy based on the concept of the feature. Examples include hills, valleys, wetlands, and many other geographic features which cannot be rigorously bounded by a mathematical line. The treatment here provides only sufficient definitions to illustrate the concept of fuzzy features. Required definitions and theorems of fuzzy set operations such as intersection, union, and complement are available in the literature (for example, see Dubois and Prade, 1980).

> Definition 1: Given a universe, V, of objects, a *fuzzy set* $A^* \subset V$ is a mapping, denoted f_{A^*} from V to the unit interval $[0, 1]$, where $f_{A^*}(x)$ is the *membership value* of x in A^* for any $x \in V$.
> Definition 2: A *map space* V is a bounded subset of R^2.
> Definition 3: A *fuzzy feature* is a fuzzy set whose universe is a map space.

Example 1: Let $V = [0, 1000]^2$ be a map space and let $A^* \subset V$ be a fuzzy feature representing an air pollution danger zone around a city at location (450, 500).

Define the feature with the following membership function:

$$f_{A^*}(v) = \begin{cases} 1 & : & \|v - (450,\ 500)\| < 30 \\ \dfrac{60 - \|v - (450,\ 500)\|}{30} & : & 30 < \|v - (450,\ 500)\| < 60 \\ 0 & : & 60 \leq \|v - (450,\ 500)\| \end{cases}$$

where the dual vertical bars indicate the Euclidean distance between the points.

The example defines any location within 30 units of the city centre as definitely within the air pollution zone (Figure 5.2). Any location farther than 60 units from the city centre is definitely outside the air pollution zone. Locations greater than 30 units and less than 60 units have membership values which linearly relate to distance from the city centre. Although this is a simplistic model, it illustrates the concept of a fuzzy feature and the related ideas of core and boundary, as defined below (Leung, 1988). A second simple example of a hill is shown below using elevation as the membership determinant. For more complex examples, see Katinsky (1994).

Example 2: Let $V = [0,\ 1000]^2$ be a map space and let $A^* \subset V$ be a fuzzy feature representing a hill with a peak elevation of 1300 m. Define the feature with the following membership function:

$$f_{A^*}(v_e) = \begin{cases} 1 & : & f(v_e) \geq 1300 \\ \dfrac{f(v_e) - 1100}{200} & : & 1100 < f(v_e) < 1300 \\ 0 & : & f(v_e) \leq 1100 \end{cases}$$

where $f(v_e) =$ elevation in metres of the pixel in a digital elevation model. This example is shown in Figure 5.3.

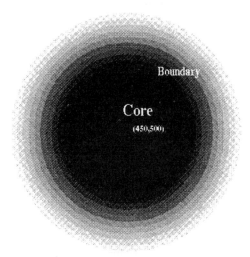

Figure 5.2   Air pollution feature defined as a fuzzy set with membership values based on distance from city centre (from Katinsky, 1994).

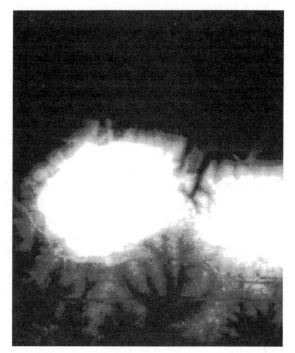

Figure 5.3   A fuzzy set representation of a hill with elevation as the controlling variable for the membership function.

Definition 4: Let $A^* \subset V$ be a fuzzy feature. Then
(1) $core(A^*) = \{v \in V : f_{A^*}(v) = \sup_{\omega \in v} f_{A^*}(\omega)\}$,
(2) $boundary(A^*) = v \in V : 0 < f_{A^*}(v) < \sup_{\omega \in V} f_{A^*}(\omega)\}$.
In general, $\sup_{\omega \in V} f_{A^*}(\omega) = 1$, so $core(A^*) = \{v \in V : f_{A^*}(v) = 1\}$.

Conceptually, the ideas of core and boundary relate to category theory. For the hill example, the core defines the prototype elevation values, that is, all values which are unquestionably a part of the hill. The periphery is defined by the boundary and is less representative of the hill. As the complexity of the feature increases, the complexity of the fuzzy set membership function also increases. The difficulty in using fuzzy sets is definition of the membership functions.

The boundary between two fuzzy features can also be defined after Leung (1987). Given two fuzzy features, X^* and Y^*, the boundary between these is:

Definition 5: Let X^*, $Y^* \subset V$ be fuzzy features. Let

$$M = \sup_{v \in V} \min\{f_x^*(v), f_y^*(v)\}.$$

Then define $boundary(X^*, Y^*) = v \in V : 0 < \min\{f_x^*(v), f_y^*(v)\} < M$.

The boundary of two fuzzy features is a well-defined set and the boundary definition only applies to fuzzy features. If either X^* or Y^* is a well-defined set, then boundary $(X^*, Y^*) = \phi$.

Once fuzzy set membership is defined, one can define fuzzy operations in a GIS environment. Using established fuzzy set operations such as intersection, union, and complement, a fuzzy overlay operator can be developed. Fuzzy buffering, fuzzy overlay including intersection, union, and complement, and fuzzy boundary operators have been developed by Katinsky (1994). For example, forest and field features extracted with a fuzzy set classifier from a Landsat thematic mapper (TM) image (Figures 5.4a and b) share a fuzzy boundary (Figure 5.4c) determined by a GIS fuzzy boundary operator constructed from definition 5 (Katinsky, 1994). Since fuzzy fea-

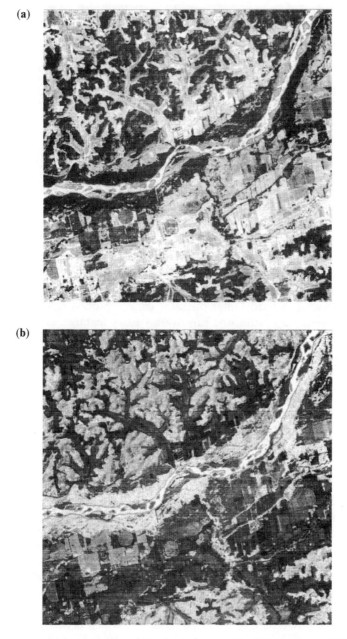

(a)

(b)

Figure 5.4 Fuzzy set representation of (a) a forest feature, (b) a field feature, and (c) (*overleaf*) the boundary between them as defined by fuzzy boundary concepts (from Katinsky, 1994).

(c)

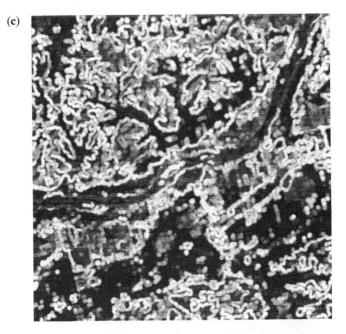

Figure 5.4 *(cont.)*

tures cannot be processed with hard operators, only specific combinations of features and operators are possible. For example, a fuzzy buffer operator can be used
with hard features but a hard buffer operator cannot process fuzzy features. Also,
with the fuzzy overlay operator, one can overlay fuzzy features with hard features.
The combinations allow processing of many geographic phenomena with indefinite
boundaries and provide extensions to complement the capabilities of current GIS.
The utility of these concepts in practical applications appears promising but remains
to be explored.

5.5   Conclusions

A conceptual modelling framework for geographic features has been presented. The
model explicitly includes spatial, temporal, and thematic dimensions and is firmly
grounded in region theory from geography, category theory from cognitive psychology, and data modelling theories developed in cartography and GIS. This model
holds potential for effectively representing geographic entities. The model is not
constrained to map and layered representations of geography and can represent
three- and higher-dimensional entities and temporal events. Because the model
structures features from geographical entities, multiple spatial representations, such
as raster and vector geometries, are possible. Difficulties of the model include the
selection of the appropriate entities to be modelled, which depend on applications
and resolutions of representation, and the lack of a defined set of critical relations
among features. Implementation difficulties occur because features other than those

defined mathematically or legally, are based on human perception and cognition. Fuzzy sets provide one possible mechanism to capture some of the ambiguity of specific features and fuzzy operators allow analysis with fuzzy features.

REFERENCES

ATKIN, R. H. (1981). *Multidimensional Man*, Harmondsworth: Penguin.

BERRY, B. J. L. (1964). Approaches to spatial analysis: A regional synthesis, *Annals of the Association of American Geographers*, **54**, 2–11.

BURROUGH, P. A. (1989). Fuzzy mathematical methods for soil survey and land evaluation, *Journal of Soil Science*, **40**, 477–492.

CHRISMAN, N. R. (1982). A theory of cartographic error and its measurement in digital databases, *Proc. Auto-Carto 5*, Crystal City, VA, pp. 159–168.

CODD, E. F. (1970). A relational model of data for large shared data banks, *Communications of the ACM*, **13**(6), 377–387.

DANGERMOND, J. (1983). A classification of software components commonly used in geographic information systems, in: Peuquet, D. J. and O'Callaghan, J. (Eds), *Design and Implementation of Computer-Based Geographic Information Systems*, IGU Commission of Geographical Data Sensing and Data Processing, Amherst.

DUBOIS, D. and PRADE, H. (1980). *Fuzzy Sets and Systems: Theory and Applications*, Academic Press: New York.

EGENHOFER, M. J. and HERRING, J. R. (1991). High-level spatial data structures for GIS, In: Maguire, D. J., Goodchild, M. F. and Rhind, D. W. (Eds), *Geographical Information Systems: Principles and Applications*, Vol. 1, pp. 227–237, London: Longman.

FRANK, A. U. and MARK, D. M. (1991). Language issues for GIS, in: Maguire, D. J., Goodchild, M. F. and Rhind, D. W. (Eds), *Geographical Information Systems: Principles and Applications*, Vol. 1, pp. 147–163, London: Longman.

GATRELL, A. C. (1983). *Distance and Space: A Geographical Perspective*, Oxford: Clarendon Press.

GATRELL, A. C. (1991). Concepts of space and geographical data, in: Maguire, D. J., Goodchild, M. F. and Rhind, D. W. (Eds), *Geographical Information Systems: Principles and Applications*, Vol. 1, pp. 119–134, London: Longman.

GOODCHILD M. F. (1987). A spatial analytical perspective on geographic information systems. *Int. J. Geographical Information Systems*, **1**(4), 327–334.

GOODCHILD M. F. (1991). Spatial analysis with GIS: Problems and prospects, *Proc. GIS/LIS '91*, Atlanta, pp. 40–48.

GREVE, C. W., KELMELIS, J. A., FEGEAS, R., GUPTILL, S. C. and MOUAT, N. (1993). Investigating U.S. geological needs for the management of temporal GIS data, *Photogrammetric Engineering and Remote Sensing*, **59**(10), 1503–1508.

GRIGG, D. (1965). The logic of regional systems, *Annals of the Association of American Geographers*, **55**, 465–491.

GUPTILL, S. C., BOYKO, K. J., DOMARATZ, M. A., FEGEAS, R. G., ROSSMEISS, H. J. and USERY, E. L. (1990). *An Enhanced Digital Line Graph Design*, U.S. Geological Survey Circular 1048, Reston, VA.

HERSKOVITS, A. (1986). *Language and Spatial Cognition: An Interdisciplinary Study of the Prepositions in English*, Cambridge: Cambridge University Press.

JOHNSON , M. (1987). *The Body in the Mind: The Bodily Basis of Meaning, Imagination, and Reason*, Chicago: University of Chicago Press.

KATINSKY, M. H. (1994). *Fuzzy Set Modeling in Geographic Information Systems*, Unpublished Master's Thesis, Department of Geography, University of Wisconsin–Madison, Madison, WI.

KELMELIS, J. A. (1991). *Time and Space in Geographic Information: Toward a Four-Dimensional Spatio-Temporal Data Model*, Unpublished PhD Thesis, Pennsylvania State University.

LAKOFF, G. (1987). *Women, Fire, and Dangerous Things: What Categories Reveal about the Mind*, Chicago: University of Chicago Press.

LAKOFF, G. and JOHNSON, M. (1980). *Metaphors We Live By*, Chicago: University of Chicago Press.

LANGRAN, G. (1989). A review of temporal database research and its use in GIS applications, *Int. J. Geographical Information Systems*, **3**(3), 215–232.

LANGRAN, G. (1992). *Time in Geographic Information Systems*, London: Taylor and Francis.

LANGRAN, G. and CHRISMAN, N. (1988). A framework for temporal geographic information, *Cartographica*, **25**(1), 1–14.

LEUNG, Y. (1987). On the imprecision of boundaries, *Geographical Analysis*, **19**(2), 125–151.

LEUNG, Y. (1988). *Spatial Analysis and Planning Under Imprecision*, Amsterdam: Elsevier.

MCMASTER, R. B. (1991). Conceptual frameworks for geographical knowledge, in: Buttenfield, B. P. and McMaster, R. B. (Eds), *Map Generalization: Making Decisions for Knowledge Representation*, pp. 21–39, London: Longman.

MARK, D. M. (1989). Cognitive image-schemata for geographic information: Relations to user views and GIS interfaces, *Proc. GIS/LIS '89*, Vol. 2, Orlando, FL, pp. 551–560.

MARK, D. M. (1991), Object modelling and phenomenon-based generalization, in: Buttenfield, B. P. and McMaster, R. B. (Eds), *Map Generalization: Making Decisions for Knowledge Representation*, pp. 103–118, London: Longman.

MARK, D. M. and FRANK, A. U. (1990). *Experiential and Formal Models of Geographic Space*, NCGIA Technical Report 90-10, University of California, Santa Barbara, CA.

MOREHOUSE, S. (1989). The architecture of Arc/Info, *Proc. Auto-Carto 9*, Baltimore, MD, pp. 388–397.

NCDCDS (1988). A proposed standard for digital cartographic data: National Committee on Digital Cartographic Data Standards, *American Cartographer*, **15**(1).

NYERGES, T. L. (1991). Representing geographical meaning, in: Buttenfield, B. P. and McMaster, R. B. (Eds), *Map Generalization: Making Decisions for Knowledge Representation*, pp. 59–85, London: Longman.

PEUQUET, D. J. (1984). A conceptual framework and comparison of spatial data models, *Cartographica*, **21**(4), 66–113.

PEUQUET, D. J. (1988). Representations of geographic space: Toward a conceptual synthesis, *Annals of the Association of American Geographers*, **78**(3), 375–394.

ROSCH, E. (1978). Principles of categorization, in: Rosch, E. and Lloyd, B. B. (Eds), *Cognition and Categorization*, pp. 27–48, New York: Halstead Press.

SINTON, E. (1978). The inherent structure of information as a constraint to analysis: Mapped thematic data as a case study, in: Dutton, D. (Ed.), *Harvard Papers on Geographic Information Systems*, Vol. 6, Reading, MA: Addison-Wesley,

SMITH, T., PEUQUET, D., MENON, S. and AGARWAL, P. (1987). KBGIS II: A knowledge-based geographical information systems, *Int. J. Geographical Information Systems*, **1**(2) 149–172.

TALMY, L. (1983). How language structures space, in: Pick, H. and Acredolo, L. (Eds), *Spatial Orientation: Theory, Research, and Application*, pp. 225–282, New York: Plenum.

TANG, A. Y. (1991). *Data Model Design for a Feature-Based GIS*, Unpublished Master's Thesis, Department of Geography, University of Wisconsin–Madison , Madison, WI.

USERY, E. L. (1993). Category theory and the structure of features in GIS, *Cartography and GIS*, **20**(1), 5–12.

USERY, E. L. (1994a). Implementation constructs for raster features, *Proc. ASPRS/ACSM Annual Convention*, Reno, Nevada, pp. 661–670.

USERY, E. L. (1994b). Display of Geographic Features from Multiple Image and Map Databases, *Proc. International Society for Photogrammetry and Remote Sensing, Commission IV Symposium on Mapping and Geographic Information Systems*, Athens, GA.

WANG, F. (1990). Improving remote sensing image analysis through fuzzy information representation, *Photogrammetric Engineering and Remote Sensing*, **56**(8), 1163–1168.

ZADEH, L. A. (1965). Fuzzy sets, *Information and Control*, **8**, 338–353.

CHAPTER SIX

Boolean and Fuzzy Regions

PETER FISHER

Department of Geography, University of Leicester, Leicester, UK

6.1 Introduction

There are two principal types of logic which are used and applied in spatial data processing: Boolean and fuzzy logic. There seems, however, to be a relatively poor understanding of the distinctions between the two, and there have been several cases of misrepresentation of the distinctions between fuzzy sets and probability in the GIS literature. Howard and Barr (1991) in their discussion of the British countryside change database, state that:

> ITE is investigating the application of fuzzy logic (Chang and Burrough, 1987) to present such boundaries. Although ARC/INFO is designed to present crisp boundaries, it is possible to include probablistic information to present parcels as part of a continuum (Wang *et al.*, 1990). (p. 218).

Similarly, Peuquet (1988) states that:

> One mechanism that has been investigated is fuzzy set theory. ... It is based on the assumption that the imprecision in natural languages is probablistic in nature (Robinson *et al.*, 1986) (p. 390).

Both of these writers are actually misquoting their sources, who do not equate fuzziness with probability, but the confusion persists. Indeed, the current author has himself contributed to the literature of misunderstanding, as is discussed below (Fisher, 1992). The misunderstanding is not restricted to geographers, however, and Zadeh (1980) himself has come to the defence of others in attempting to clarify the differences between fuzzy sets and probability. In the spatial database literature, Robinson and Frank (1985) set out the differences between the fuzzy and probability models of uncertainty, but their work has not received the attention it deserves and it pre-dates the work quoted above. Robinson (1988) unfortunately fails to provide a clear distinction, and much other geographical literature and writing on fuzzy sets fail to draw any distinctions. For example, Leung (1988), in an otherwise thorough book on the use of fuzzy sets in spatial analysis, avoids the issue of the distinction completely.

Probability theory has a long tradition dating back to the eighteenth century and before, and includes the work of many luminaries including Laplace, Boole and

Bayes (Gillies, 1973). The theory has been in the geographer's tool box since before the beginnings of the quantitative revolution, and is taught in theoretical and applied statistics courses in many different subjects, including geography, at school and university (e.g. Ebdon, 1977; Silk 1979; Griffith and Amrhein, 1991). As a result, most academics feel that they have a grasp of the foundations of the theory. Fuzzy sets, on the other hand, were first proposed by Professor Zadeh in 1965, and so are a relative newcomer to the field (Zadeh, 1965). Robinson and Strahler (1984) and Burrough (1986, p. 131) were among the first to suggest the use of fuzzy sets in the handling of uncertainty in geographical databases, although Gale (1972), Leung (1979), and Pipkin (1978) advocated the theory earlier as a suitable approach to quantification in behavioural geography.

Perhaps because indices of probability and fuzziness are both measured on a numerical scale from 0 to 1, many GIS users and researchers, are confused about the actual meaning of the terms. Fuzzy sets are often used when subjective probability is indicated by the context, or even when experimentally definable probability can be established. Sometimes the two appear to be synonymous or at least interchangeable. This presumably relates to the previously noted youth of the theory of fuzzy sets, and the relative wealth and availability of texts on probability theory.

This chapter explores three areas where either fuzzy or probable models of spatial regions are appropriate, depending on the conceptual understanding of the issues and the phenomena. The viewshed (the subject of continuing research by the current author) is first examined in Section 6.2, and then general spatial regions and land cover mapping from satellite imagery in Sections 6.3 and 6.4. The issues and the problems in each of these are rather different.

6.2   The viewshed

The viewshed function is common in GIS that are used for analysing surface elevation. From several locations on the surface it is possible to define the area which is visible. The region is usually represented as a binary map, where 1 indicates that a location is visible and 0 that it is not. The decision is based on a simple line-of-sight (LoS) calculation from the viewing point to every other location in the land surface model. If the LoS is intercepted by the surface, then the target location is out of sight; otherwise it is not. Fisher (1991, 1993) has shown that the viewshed is very susceptible both to error in the elevation model and the formulation of the algorithm used. For the same viewing points Fisher showed that it was possible to derive viewsheds varying in area by up to 3100% in the most extreme instance (differences in algorithm give viewsheds that vary between 92 and 2907 pixels).

The viewshed, as it is usually reported, is a binary, deterministic phenomenon. A location is either in-view or out-of-view. Doubt can arise as to whether any particular location on the map is actually in-view or out-of-view, however, but by definition the viewshed is still Boolean, only two alternatives exist, and any location can only be assigned to one or the other. If a method for deriving the probability of this outcome can be defined, then at some level of statistical significance, a location can be assigned to the set of locations which are visible or invisible. Several methods have been suggested to enable derivation of the probabilities which allow realization of alternative probabilities. Fisher (1992, 1993, 1994) has suggested a Monte Carlo simulation approach to this problem, although it was originally and incorrectly identified as a fuzzy viewshed (Fisher, 1992).

The probability, however it is derived, determines the likelihood of a location being visible from the viewing point. There is another way of looking at the uncertainty in the visible area: the degree to which anything can actually be seen. This second approach is the fuzzy set of locations, as opposed to the Boolean set.

If a person is standing at a location, and they look over the landscape, then they may be able to see to a location (the line of sight is not interrupted by the land surface), but there is no guarantee that they can see what is there. Atmospheric attenuation of light, object–background contrast, variable atmospheric conditions, and the eyesight of the individual (among other effects) all contribute to the degree to which an individual can determine and discriminate an object at a location. Thus there is a measure of the degree to which they can see the phenomenon which can be defined by a whole family of mathematical functions, a sample of which is given by Fisher (1994). These formulae are fuzzy membership functions for the fuzzy set of locations which are clearly visible. For any particular version of the Boolean viewshed at some level of probability (confidence) of being visible, it is therefore possible to define a fuzzy membership function for a particular combination of viewer, weather, and object.

In summary, there are two forms of uncertainty or poorly defined viewshed regions (Fisher 1994): *probable viewshed* (visibility) recognizes the uncertainty over whether a location is visible (i.e. there is a direct line of sight between observer and location); and *fuzzy viewshed* (clarity) defines the degree to which any potentially visible target can be distinguished.

6.3 General regions

Much analysis in studying the built environment is based on regions. These can be either real and perceived regions or imposed regions. Imposed regions include all political and legal regions. These are clearly defined and have boundaries which are legally enforced and surveyed with measurable precision and accuracy on the ground. At the other extreme, there are more perceptual regions, the regions we actually live in, the city, the neighbourhood, and the like. Neighbourhoods are not mapped on the ground, and although they may be synonymous with a legally defined region (e.g. the city) they are not spatially synonymous.

Censuses of population throughout the world are arranged in nested hierarchies of regions. In England and Wales, for example, there are enumeration districts at the lowest level of counting, which are agglomerated into wards, into districts, into counties, and finally into regions. Any household can therefore only belong to one enumeration district, and that ED can belong to only one ward, one district, one county, and one region. This is clearly a Boolean regionalization of space. Such a regionalization (in the UK partly similar) exists for government, where generally taxation may be levied on inhabitants from only one set of nested authorities (city or county council and central government in the UK, or city, county, state and federal government in parts of the US). There is usually little doubt as to which region at any level a property and an individual belongs. In exceptional cases houses are built on the borders, but still some decision is made as to how the inhabitants should be counted and taxes assessed. A Boolean decision is made, no degree of belonging is admitted. Much socio-economic analysis is based on such a regionalization of space.

The space we as human beings actually occupy is very different. For example, the inhabitants of a city may have a very precise understanding of the local politics which determine the extent of the city. They may be vaguely aware of the general location of the boundary, but only rarely can they actually trace it. They are commonly even less aware of the political subdivision of the city itself. Those who live further away, though, will undoubtedly have a much more restricted impression of the extent of the city, incorrectly including many politically autonomous suburbs in their definition. How many people consider Westminster to be part of London, for example?

Similarly, socio-economic space is divided into grosser regions: the East Midlands in England, for example, has a clearly defined extent in the nested hierarchy, but where then are the North and South Midlands? If we asked individuals from across the country, the extent of the East Midlands would be very variable, and the North and South Midlands could also be sketched in (they do not exist in the formal regionalization, although the West does).

Attempts have been made to study these perceptual or vernacular regions. Shortridge (1987) and Zelinsky (1980), for example, have used various sources to examine the degree to which people associate names with areas. They use pragmatic measures of degrees of membership, but an appropriate formalism for the perceived extent of New England is the fuzzy set.

At a more behavioural level, the people do not perceive their environment as having hard boundaries. The zone I can drive to in five minutes is a probablistic zone (Ball and Fisher, 1994); there is a probability under any circumstances that I can get to any location within five minutes. Five minutes is a Boolean phenomenon, not admitting to doubt, but in colloquial use of the English language 'five minutes' rarely means that. It probably means less than 15 minutes and rather more than one minute, so the concept is actually much less certain, and is actually a fuzzy concept. The idea of nearness is also fuzzy. What do we mean by near? It will be different for any individual, and may vary by orientation, by a person's usual form of transport, their familiarity, etc. (Fisher and Orf, 1991).

Finally, let us consider a sampling scheme in socio-economic research. The sampling may be for opinion polling, for example. A number of point locations (households) are drawn, within a specific area (a city). But the outline of the city is just one possible area that could have yielded the same point sample. Therefore we can legitimately recognize this as a probable region, it is certainly Boolean, and it is defined by the people included within it, but many areas within the city may not actually be within a real region defined by the city, and so the city outline is only one realization of the boundary defining the sample.

6.4 Land cover mapping

One of the most widespread and important products of satellite imagery is the land cover map (e.g. Campbell, 1987), in which each pixel in the area of the image is assigned to a particular land cover type. Any one pixel can belong to one and only one land cover type and the whole area of that pixel is assigned to that cover type.

In short, the decision is a Boolean assignment of each pixel to a class. More formally, this is defined as

$$LC_{ij} = f(DN_1, DN_2, DN_3, \ldots, DN_n)_{ij},$$ (1)

where LC_{ij} is the land cover type at row i, column j in the image, and DN_1, DN_2, etc., are the digital numbers in each of n bands. The general form of the equation for a textural classifier would include DN values in cells other than just ij.

The normal (supervised) procedure in executing such a classification is to develop a set of training areas which represent each of the land cover types, and then to use summary statistics from those areas as a basis for some numerical procedure to attempt to assign each pixel to a cover type. A number of different methods are in use, but some variant of the maximum likelihood classifier is perhaps the most widely used. This determines the probability of a pixel belonging to all classes and includes a decision rule that saves only the name of the most likely class for the pixel. One variant includes a chi-squared test to determine the confidence with which a pixel can be classified, and this can be used to leave unclassified pixels for which the classification is in doubt (mixed pixels).

It has always been recognized this is an approximation of reality, and that mixed pixels cause major problems. Simply leaving pixels as unclassified is exceedingly unsatisfactory, and the implication that any classified pixel is occupied by a single cover type is also false. In fact, the landscape is not made up of little rectangular plots of uniform land cover which, thankfully, suddenly change their size to match the IFOV of the particular sensor (10 m, 20 m, 25 m, 30 m, 80 m, 500 m, 1 km, etc). At some scale all pixels actually contain a number of different contributing land cover types. The significance and nature of the contribution is clearly dependent on the sensor characteristics, and particularly the IFOV. It would therefore be more correct to say that any pixel has some level of possibility of belonging to each and every cover type identified as being possible in a particular area. This might be written as

$$(\mu_{LC_1}, \mu_{LC_2}, \ldots, \mu_{LC_k})_{ij} = f(DN_1, DN_2, \ldots, DN_n)_{ij},$$ (2)

where μ_{LC} is the degree to which the pixel matches the characteristics of each of k land cover types.

One vehicle to achieving this end is to store the probabilities of the maximum likelihood classifier. A three-dimensional matrix of probabilities is therefore recorded ($i \times j \times k$). This has only rarely been done (Marsh et al., 1980). In this case, the probabilities are actually being treated as a measure of the degree to which a pixel belongs to each land cover type, and not as the basis of justification of a Boolean decision rule. Therefore they have more in common with fuzzy memberships than classic probability, and are indeed being treated as fuzzy memberships. Others have used classification approaches specifically designed to extract the fuzzy memberships either from numerical classifiers (Robinson and Thongs, 1986; Fisher and Pathirana, 1990; Wang, 1990), rule-based systems (Leung and Leung, 1993a,b), neural networks (Civco, 1993), or mixture modelling.

Fisher and Pathirana (1993) have outlined one application of the membership values. They showed that consideration of membership values enabled recognition of pixels where the poorly defined set of pixels in which land cover has changed can be well defined, and the Boolean map of change can be easily refined.

Conceptually, holding membership values for all cover types in each cell is much more satisfactory than holding only the Boolean assignment. The suburban environments studied by Fisher and Pathirana (1990) and Wang (1990) are not characterized by large homogeneous regions, and their similarity to the situations discussed above are not clear. Where large homogeneous areas of vegetation are present, however, the similarity is again apparent. In natural areas abrupt changes in vegetation are very much the exception; intergrades are much more common. Here the fuzzy set approach to regional analysis is much more appropriate than the Boolean, since the transition zone from one vegetation type to another can be mapped, no matter how wide or narrow it is (Foody, 1992). In such a transition zone any location has a degree of belonging to any of the modal vegetation types mapped. It is in fact a direct parallel of the viewshed and general regions, as are the suburban areas.

In summary, remotely sensed land cover types may be recognized as fuzzy sets, due to inherent fuzziness in the land cover types defined, and in the pixel-based data structure employed to image and store the information. The data storage requirements of the approach can become enormous when a matrix $i \times j \times k$ is required. On the other hand, Goodchild *et al.* (1994) indicate that it is rarely necessary to store fuzzy memberships for more than two cover types, although that is not the indication from Fisher and Pathirana (1993). It can easily be argued that conceptually the Boolean model of the classified land cover image is only an approximation, and a rather poor one at that. Essentially, the land cover derived from a satellite image is an inherently poorly defined set of geographic objects with poorly defined boundaries.

6.5 Conclusions

This chapter does not intend to say that fuzzy sets are the only way to define geographic objects with indeterminate boundaries. Other formalisms may also exist and would bear investigation. It is intended to point out that probability is not really appropriate to the problem since the probability is of a Boolean event and in space a Boolean phenomenon or region does not have undetermined or indeterminate boundaries; rather, it has definite determinable boundaries, which may, however, be hard to measure with precision. Some phenomena may be modelled by probability and some by fuzzy sets. Indeed, some discussion which the author has taken part in suggests that all spatial phenomena have both Boolean (with associated probability) and fuzzy versions, although in any particular case one or the other may not be useful or significant. Furthermore, the fuzzy and Boolean interpretations of the set are often actually different phenomena, as in the viewshed example quoted above.

In the extreme case, it can be seen that any location can have fuzzy membership in any number of phenomena. This is a major problem, resulting as it does in the need to store vast amounts of information for every location and every phenomenon ($i \times j \times k$, where k may be regions, viewsheds, or land cover types). It is necessary to examine whether truncation of the amount of data is possible, whether it can be adapted, and whether there are real advantages always to retain all k themes.

The data structure used to model these regions has not been touched on in this chapter, and is a further issue. Much sophisticated database research presupposes the existence of boundaries (van Oosterom, 1994), and it is very unclear whether such structures can accommodate poor definition. Some attempts have been made to implement probability modelling. Any structure that accommodates a surface (TIN or raster) can very clearly be used for the extreme case.

Acknowledgements

The opinions expressed here have developed over a number of years, through discussions with many individuals and collaboration with a number of co-workers. To list them all would be too lengthy, and repetitive, since most are included in the references. I would like to thank them all. I would also like to thank Andrew Frank and Peter Burrough for their invitation to attend the GISDATA meeting.

REFERENCES

BALL, J. and FISHER, P. F. (1994). Visualizing stochastic catchments in geographic networks. *Cartographic Journal*, **31**, 27–32.

BURROUGH, P. A. (1986). *Principles of Geographical Information Systems for Land Resources Assessment*, Oxford: Oxford University Press.

CAMPBELL, J. B. (1987). *Introduction to Remote Sensing*, New York: Guilford Press.

CHANG, L. and BURROUGH, P. A. (1987). Fuzzy reasoning: A new quantitative aid for land evaluation, *Soil Survey and Land Evaluation*, 7(2) , 69–80.

CIVCO, D. L. (1993). Artificial neural networks for land-cover classification and mapping, *Int. J. Geographical Information System*, **7**, 173–185.

EBDON, D. (1977). *Statistics in Geography: A Practical Approach*, Oxford: Blackwell.

FISHER, P. F. (1991). First experiments in viewshed uncertainty: The accuracy of the viewable area, *Photogrammetric Engineering and Remote Sensing*, **57**, 1321–1327.

FISHER, P. F. (1992). First experiments in viewshed uncertainty: Simulating the fuzzy viewshed, *Photogrammetric Engineering and Remote Sensing*, **58**, 345–352.

FISHER, P. F. (1993). Algorithm and implementation uncertainty in the viewshed function, *Int. J. Geographical Information System*, **7**, 331–347.

FISHER, P. F. (1994). Probable and fuzzy models of the viewshed operation, in: Worboys, M. (Ed.), *Innovations in GIS 1*, pp.161–175, London: Taylor and Francis.

FISHER, P. F. and ORF, T. (1991). An investigation of the meaning of near and close on a university campus, *Computers, Environment and Urban System*, **15**, 23–35.

FISHER, P. F. and PATHIRANA, S. (1990). The evaluation of fuzzy membership of land cover classes in the suburban zone, *Remote Sensing of Environment*, **34**, 121–132.

FISHER, P. F. and PATHIRANA, S. (1993). The ordering of multitemporal fuzzy land-cover information derived from Landsat MSS data, *GeoCarto International*.

FOODY, G. M. (1992). A fuzzy sets approach to the representation of vegetation continua from remotely sensed data: An example from lowland heath, *Photogrammetric Engineering and Remote Sensing*, **58**, 221–225.

GALE, S. (1972). Inexactness, fuzzy sets, and the foundation of behavioral geography, *Geographical Analysis*, **4**, 337–349.

GILLIES, D. A. (1973). *An Objective Theory of Probability*, London: Methuen.

GOODCHILD, M. F., CHIH-CHANG, L. and LEUNG, Y. (1994). Visualizing fuzzy maps, in: Hearnshaw. H. M. and Uwin, D. J. (Eds), *Visualization and Geographical Information Systems*, pp.158–167, Chichester: Wiley.

GRIFFITH, D. A. and AMRHEIN, C. (1991). *Statistical Analysis for Geographers*, Englewood Cliffs, NJ: Prentice Hall.

HOWARD, D. C. and BARR, C. J. (1991). Sampling the countryside of Great Britain: GIS for the detection and prediction of rural change, in: Heit, M. and Shortreid, A. (Eds), pp.217–221, *GIS Applications in Natural Resources*, Fort Collins, Co: GIS World.

LEUNG, Y. C. (1979). Locational choice: A fuzzy set approach, *Geographical Bulletin*, **15**, 28–34.

LEUNG, Y. C. (1988). *Spatial Analysis and Planning under Imprecision*, New York: Elsevier.

LEUNG, Y. and LEUNG, K. S. (1993a). An intelligent expert system shell for knowledge-based geographical information systems 1: The tools, *Int. J. Geographical Information Systems*, **7**, 189–199.

LEUNG, Y. and LEUNG, K. S. (1993b). An intelligent expert system shell for knowledge-based geographical information system 2: Some applications, *Int. J. Geographical Information Systems*, **7**, 201–213.

MARSH, S.E., SWITZER, P., KOWALIK, W. S. and LYON, R. J. P. (1980). Resolving the percentage of component terrains within single resolution elements, *Photogrammetric Engineering and Remote Sensing*, **46**(8), 1079–1086.

PIPKIN, J. S. (1978). Fuzzy sets and spatial choice, *Annals of the Association of American Geographers*, **68**, 196–204.

ROBINSON, V. B. (1988). Some implications of fuzzy set theory applied to geographic data-bases, *Computers, Environment and Urban Systems*, **12**, 89–97.

ROBINSON, V. B. and FRANK, A. U. (1985). About different kinds of uncertainty in collections of spatial data, in: *Proc. Auto-Carto 7* (American Congress on Surveying and Mapping, Falls Church, VA) pp. 440–449.

ROBINSON, V. B. and STRAHLER, A. H. (1984). Issues in designing geographic information systems under conditions of inexactness, in: *Proc. 10th Int. Symposium on Machine Processing of Remotely Sensed Data*, Purdue University, Lafayette pp. 198–204.

ROBINSON, V. B. and THONGS, D. (1986). Fuzzy set theory applied to the mixed pixel problem of multispectral landcover databases, in: Opitz, B. (Ed.), *Geographic Information Systems in Government*, pp. 871–885, Hampton, VA: Deerpak.

ROBINSON, V. B., THONGS, D. and BLAZE, M. (1986). Man–machine interaction for acquisition of spatial relations as natural language concepts, in: Opitz, B. (Ed.) *Geographic Information Systems in Government*, pp.443–453, Hampton, VA: Deerpak.

SHORTRIDGE, J. R. (1987). Changing usage of four American regional labels, *Annals of the Association of American Geographers*, **77**(3), 325–336.

SILK, J. (1979). *Statistical Concepts in Geography*, London: George Allen & Unwin.

van OOSTEROM, P. J. M. (1994). *Reactive Data Structures for Geographic Information Systems*, Oxford: Oxford University Press.

WANG, F. (1990). Improving remote sensing image analysis through fuzzy information representation, *Photogrammetric Engineering and Remote Sensing*, **56**, 1163–1169.

WANG, F., HALL, G. B. and SUBARYONO (1990). Fuzzy information representation and processing in conventional GIS software: Database design and application. *Int. J. Geographical Information Systems*, **4**(3), 261–283.

ZADEH, L. A. (1965). Fuzzy sets, *Information and Control*, **8**, 338–353.

ZADEH, L. A. (1980). Fuzzy sets versus, probability, *Proceedings of the IEEE*, **68**(3), 421.

ZELINSKY, W. (1980). North America's vernacular regions, *Annals of the Association of American Geographers*, **70**(1), 1–16.

PART THREE

Languages to Describe Shape

ANDREW U. FRANK

Introduction

The four papers in Part 3 deal with methods of describing geographic objects with uncertain boundaries. Spatial phenomena that are perceived and understood within some context must be represented by some kind of language in order to be communicable. This is independent of whether the object has crisp or undetermined boundaries. A major part of GIS research and the related literature deals with different methods of representing spatial objects with crisp boundaries within a computer system. Very little has been done on the representation of objects with uncertain boundaries.

In Chapter 7 Giacomo Ferrari investigates how natural languages, in particular Italian and English, describe spatial objects and their boundaries. Natural language is one of the most easily observed artifacts reflecting methods of human reasoning. Most languages seem not to have prominent instruments to express the type of boundary of an object (as compared with those used to express the number of objects: singular and plural), but there is at least one American Indian language where boundedness is a grammatical category. Ferrari first considers the bounds on thematic categories and points to differences between different cultures: some languages make further differences than others. Elsewhere, David Mark has compared differences in the concepts of lakes in different languages (Mark, 1993), implicitly asking the question how a count of geographic objects, such as that reported by Sarjakoski in Chapter 6, can be compared between countries: are statements like 'Finland has more lakes than Canada' meaningful? Can they be made meaningful? The question is not trivial, certainly when one is attempting to cross-correlate environmental data across national borders, as is required in the European Union.

Supposedly uniform areas are separated by boundaries: boundary areas are often described in special terms. An interesting case is provided by periodically flooded areas (river floodplain, sea shore), where the boundary area itself is more or less clearly delimited. Cultural objects, as already described by Campari in Chapter 4, sometimes have surprisingly sharp boundaries, even if these are not at all visible to an outsider. Ferrari's chapter describes areas in cities which have sharp boundaries on one side, but completely undetermined ones on the other.

In Chapter 8 Christian Freksa concentrates on the description of properties of objects, and how they can be further subdivided to any useful level of detail. This corresponds with Couclelis' observation that there are very different types of uncertainty in the boundaries of objects: the level of crispness must be adapted for the purpose. Freksa points to the interrelation between concepts: words are only meaningful in opposition to other words – *near* is meaningful when opposed to *far*, but not by itself. These distinctions are not necessarily in a strict hierarchy where one term is divided into others but depend on the purpose for which the distinction is made.

Finally, Freksa points to the difference between imprecise knowledge which relates to well determined places ('either Paris or London'), imprecise knowledge about a boundary (between 5 and 6 m wide), and the uncertainty introduced by the 'indefinite correspondence between concepts and represented world'.

In Chapter 9 Schlieder concentrates on the formal representation of imprecise spatial knowledge. In acquiring knowledge about the world, imprecision is introduced both by the process of recognition and reconstruction. Different sources or sensors may be combined to improve information, but this is not always necessary. Schlieder contrasts methods of image processing and computer vision, which include algorithms to counteract recognition indeterminacy by qualitative spatial reasoning, especially reasoning about shape. Most qualitative reasoning is devoted to reasoning about position in space (i.e. metric relations) or topological relations. Schlieder approaches a representation of shape with an order-based method, which captures information which is neither metric nor topological (for example, convexity). His formalism, based on the order of the connections between three points, captures not only configurations but also the gradual changes between them.

The work by Markus Schneider, described in Chapter 10, was originally motivated by the uncertainty introduced by computer approximations to spatial positions. Geometric objects are imagined to be on a continuous plane, where another point can be inserted between any two points. Computer representation of objects in geographic information systems automatically uses a 'discrete geometry' on a grid, which is represented by finite approximations to real numbers that are imposed by computer hardware. Some of the basic assumptions about geometry can be violated in computer representation; it is not always true that a third point can be introduced on the straight line between two others; nor does the computed point of intersection of two lines always lie on one or other of the two lines. The Realm method overcomes this approximation problem and guarantees the preservation of important geometric properties (e.g. topology) despite approximations. On this basis, Schneider constructs a system to deal with objects with imprecise boundaries. Two different models of imprecise boundaries are described: one models imprecision as a buffer around a line (Perkal's epsilon band), and the other models imprecision, such as a lake with varying water levels. He shows that the second of these models is closed, i.e. the result of any operation on an imprecise geometric object must be an imprecise geometric object.

These four chapters link up with the discussion in Part 2, by contrasting concepts to capture uncertainty in the boundaries, with the representation of knowledge. The purpose and context is not only important for the concepts used, but also determines the representation selected. Natural languages have powerful methods to represent objects with uncertain boundaries. It remains an open question as to which of these methods are universal, i.e. the same in all languages, and which are culturally dependent. The level of distinction in natural language descriptions can vary and some languages make distinctions where others do not, or at least not at the basic concept level (e.g. the Italian language differentiates between *giardino* and *orto*, whereas the English needs to qualify the word *garden* with additional terms).

Part 3 explores formal models for some aspects of spatial phenomena. The discussion in Part 4 specifically concerns topological relations. Together, these formalizations of some aspects of geographic objects form a 'spatial information theory' which must be part of the foundation for the rational development of geographic information systems. Later, in Part 5, Molenaar (Chapter 14) and Hadzilacos (Chapter 16) combine some of the already existing (crisp) theories with aspects of uncertainty, the formal representation of which must capture varying levels of detail. The representation must be able to deal with different levels of uncertainty not only in the feature and its characterization, but also in the representation of its spatial properties. There are uncertainties about location, shape and topological relations (see Part 4) but it seems that information about shape and uncertainties in shape is one of the least well investigated aspects of spatial phenomena.

REFERENCES

MARK, D. M. (1993). Towards a theoretical framework for geographical entity types, in: Frank, A. U. and Campari, I. (Eds), *Spatial Information Theory: Theoretical Basis for GIS*, pp. 270–283, Berlin: Springer.

CHAPTER SEVEN

Boundaries, Concepts, Language

GIACOMO FERRARI

Department of Human Studies, University of Torino at Vercelli, Vercelli, Italy

7.1 Introduction

In linguistics, it is hardly more than a commonplace that the relation between linguistic form, its meaning, and reality is an arbitrary one. Humans analyze reality according to their own categories, which are historically and socially determined, and then translate those categories into linguistic expressions. A consequence of this situation is that the 'lexical distinctions drawn by each language will tend to reflect the culturally important features of objects, institutions and activities in the society in which the language operates' (Lyons, 1969, p. 432). Thus the study of the use of language is a direct way of accessing the categories through which reality is analyzed by people. A classic example of the culture-dependent categorization of reality, and, consequently, of the related lexicon, is the names of colours. It is well known, in traditional linguistics, that although colours are more or less physically well appreciable entities, all languages or linguistic groups have different ways of distributing the names of colours along the continuum between white and black, even dividing such space into different numbers of partitions (Conklin, 1955). Even though the names of colours can be proved to have an objective biological basis (Berlin and Kay, 1969; Kay and McDaniel, 1978), a deeper discussion of the matter leads to the conclusion that 'Color categorization makes use of human biology', but 'Color categories result from the world plus human biology plus a cognitive mechanism that has some of the characteristics of fuzzy set theory plus a culture-specific choice of which basic color categories there are' (Lakoff, 1987, p. 29).

Space is a complex cognitive area, be it the physical space in which we operate, such as rooms, houses, towns, or the geographical space with which we are familiar. It is therefore obvious that the analysis of space plays an important role in the cognitive world of humans, and language reflects this cognitive complexity. If we take reality to be the geographical space, then the words by which geographical objects are called are not directly related to them, but there are meanings (*concepts*) which are connected to referents related to reality. More interestingly for us, we can understand how people analyze their geographical space through the analysis of the linguistic expressions they use.

In this chapter I will discuss the relation between the boundaries, which (more or less) objectively delimit geographical spaces, the concepts related to the description of those spaces, and the linguistic descriptions employed by ordinary people in relation to the same geographic spaces. It appears that this relation does not go always in one direction – from objects to concepts and linguistic expressions – but in many cases follows a winding path, from historical and social tradition, to concepts and the delimitation of spaces.

Some examples, mostly taken from my own personal experience, of linguistic descriptions of (geographical) space, will help to illustrate at least some of the problems raised by a cognitive approach to geographic knowledge. Some requirements for a uniform representation system will also be highlighted.

7.2   Language, concepts and partitions of space

7.2.1   Environment, perception and natural language

Our geographic environment is a complex structure made up of *natural objects*, such as mountains, hills, and valleys or lakes, seas, rivers, creeks, etc., as well as man-made artifacts such as towns, villages, roads, channels, etc. Artifacts are often integrated into natural reality, as in the case of olive groves in a Mediterranean environment, which have modified the slopes they occupy and are often so old that they have created a new 'natural' landscape. Plainly natural objects, on the other hand, are classified, interpreted, 'lived in' by humans, who give names to, make up legends about, and draw borderlines around them, and all this through different temporal layers.

The consequence of this complex integration between humans and their environment is the complexity of the relations between language – and thus conceptual modelling – and environmental reality. This complexity deploys itself along the traditional modalities of human language: a conventional culture-dependent analysis of the environment, and its historical motivations, which often make some conceptualizations completely unnatural and unexpected.

Thus the (linguistic) characterization of geographical spaces is necessarily not universal. As with the case for colours, the correspondence between words, and therefore between concepts, across different languages is quite rare. As sketched above, the general principle is that, given a conceptual space, different cultures provide different partitions of it. Thus, if natural languages provide different analyses of geographical spaces, the criteria by which such spaces are logically delimited differ from culture to culture, so that possibly even the perception on geographic boundaries can be affected by cultural differences.

It is necessary to pause here, to clarify an important assumption underlying this chapter. For humans, objects, and therefore geographic objects, have no objective appearance, shape, boundaries, but their perception depends upon the cognitive categories the culture of a community imposes on its members, and can be studied through the language of the community itself.

7.2.2   Different analyses of the same conceptual area

Linguistics is rich in examples of simple spaces that are analyzed in uncommon ways by exotic cultures, such as Eskimos having a number of names for the simple concept of snow (Whorf, in Carroll, 1956). However, in traditional Europe, where strong pressures towards standardization should have affected the culture, examples of different partitions of the same reality can also be found, leading to the conclusion that it will be quite difficult to 'construct entity typologies that will stand up across individuals and cultures' (Mark, 1993). Mark discussed the universality of geographic terminology on the basis of a comparison of the names of 'water entities' in English, French and Spanish, as presented in geographic glossaries or bilingual dictionaries. Similar results can also be obtained using the most traditional linguistic tool, a monolingual dictionary.

To describe cultivated areas, Italian has *giardino*, an area where flowers and other plants are grown for pleasure; *orto*, where vegetables, but not cereals, are grown on a small scale; and *campo*, where cereals and vegetables are grown on a large scale. English, on the other hand, distinguishes between *garden*, which covers the functions of *giardino* and *orto* (often referred to as a *kitchen garden*), and *field*, which corresponds to *campo*.

An Italian dictionary (Zingarelli) defines *giardino* as 'terreno con colture erbacee e arboree di tipo ornamentale' ((a piece of) land with ornamental herbaceous and arboreal plants), *orto* as 'appezzamento di terreno, di solito cintato, dove si coltivano gli ortaggi' (a piece of land, usually fenced, where vegetables are grown), and *campo* as 'superficie agraria coltivata o coltivabile, compresa entro limiti ben definiti' (an agricultural surface cultivated or cultivable, enclosed within well defined limits). The Oxford English Dictionary (OED) defines *garden* as 'An enclosed piece of ground devoted to the cultivation of flowers, fruit or vegetables', with the additional information that it is 'often used with a defining word, such as flower-, fruit-, kitchen-, etc.'. In a more specific sense, *'field'* is defined as 'land or a piece of land appropriated to pasture or tillage', or 'a piece of ground put to a particular use', or even 'an extent of ground containing some special natural production'.

If dictionaries can be taken as repositories of the common lexical knowledge of a language (Rey-Debove, 1971) the differences between the two languages are quite clear as to the pair *giardino/garden*, where the types of cultivations are explicitly listed both in the Italian and English definitions. In the case of *campo/field*, Italian stresses the agricultural aspect, whereas English keeps to a more generic definition. English has clearly no room for *orto*, which is subsumed by *garden*.

Any speaker of Italian or English will be able to clearly distinguish the three, or two, concepts, and to map them onto her reality when answering the question 'Where are you now, in your {*giardino, orto, campo*}/{*garden, field*}?', thus recognizing the identity of each concept, and also their physical boundaries. For any speaker of Italian the physical boundaries of *giardino* and *orto* are as clear as the concepts themselves, and, probably, as they are diffuse for a speaker of English. Linguistics, as well as logics, have a number of explanations for this human ability to classify concepts and, as a consequence, also real situations. The idea that a set of characteristic properties defines a concept, and therefore an object or a situation, is the most widespread, even if not the only theory, and it has been often criticized.

7.2.3 A uniform representation system

Artificial intelligence has developed, in the past, a powerful formalism to represent objects by means of their properties, the *frames* (Minsky, 1975). A frame consists of a *header*, which contains the name of the concept to be described, and a number of *slots*, each of which contains the specification of a property of the concept. Properties can be specified in terms of a superconcept (in the 'is_a' slot), another concept, a list of concepts, among which one can be selected, or a procedure to establish a new property. The mapping of the dictionary definitions onto a frame formalism, would produce roughly the result shown in Table 7.1.

The differences between the English and Italian characterizations of cultivated areas clearly appear in the formalized version of the definitions. Frame representations have been generated by a simple mapping of definitions onto a quite limited number of properties (slots), thus showing the relative simplicity of the process of building of a cognitive model of a (geographic) domain.

The points we want to make clear in this section are not necessarily related, so it is probably advisable to summarize them.

The first point is that the sole basis for the physical characterization of objects, and therefore of their boundaries, is cognitive. If a language provides a certain analysis of a domain, the recognition of physical objects and boundaries will be a direct function of such a linguistic reality.

The second point is that culturally different characterizations of the same cognitive area can in many cases be reduced to a uniform set of properties with different values. This means that concepts can be compared on a formally homogeneous basis only if homogeneity is realized at a representational (syntactic) level. The consequence of this is that a possible answer to the problems of representation proposed by Mark (1993), is the use of a soundly grounded, uniform representation system, even if this does not exhaust the problem of representing geographical objects, as will be shown later.

7.2.4   Conceptually and physically fuzzy boundaries

In the previous section it was shown that geographic concepts (and any other conceptual spaces described by linguistic means, although they differ across cultures) are quite neatly defined and provide a method of recognizing objects. Natural lan-

Table 7.1

Italian	English

Frame: *giardino*
  is_a: piece_of_land
  type_of_cultivation: ornamental
	Frame: *garden*
	  is_a: piece of_land
	  type_of_cultivation: {flowers, fruit, vegetables}
	  shape: enclosed

Frame: *orto*
  is_a: piece_of_land
  type_of_cultivation: vegetables
  shape: fenced	

Frame: *campo*
  is_a: area
  usage: tillage
  shape: enclosed
	Frame: *field*
	  is_a: piece_of_land
	  usage: {pasture, tillage, ...}

guage tends to define precise boundaries between concepts, even where such boundaries are fuzzy from the physical point of view.

In Italian, *golena* is defined as 'terreno compreso entro gli argini del fiume, invaso dalle acque in periodi di piena' (land enclosed within a river's banks, which is flooded during high water). This is an area that has one definite boundary, but whose property may change, according to the season, from being dry to being flooded. Also *battigia* is defined by the same dictionary as 'parte della spiaggia battuta dalle onde' (portion of the shore, hit by waves), but, very instructively, its synonym *bagnasciuga* is 'zona di una spiaggia di costa bassa ove si rompono le onde e per questo appare ora asciutta e ora bagnata' (area of low shore where the waves break, and therefore appears alternately dry and wet). Thus, areas whose properties swing from one status to its opposite have clear definitions, just as a function of this change. If we adopt the same representation method as before, we obtain Table 7.2.

The use of a uniform representation mechanism again proves adequate to treat this specific typology of cases; the definition of objects such as those described above relies on the very fact that one property (slot) takes a value or its opposite over a definite time period. For these examples, Italian provides a precise terminology, and therefore a precise conceptualization, based on the fact that the value of its characteristic property is swinging.

Properties may, however, change according to different modalities. Transition areas, such as from Mediterranean bush to coastal dune, or from wood to alpine grassland, are not characterized by swinging values, but by decreasing values, such as, in the first example, a decreasing number of shrubs, going towards the sea, or the decreasing amount of sand and dune vegetation, going towards the bush. Although I know of no Italian terms that identify such types of areas, the representational device we have employed here would allow a smooth treatment of such concepts.

7.2.5 Issues in representation

This category of geographic objects, which is characterized by having a clear description except for one property, which takes varying values, is connected with a notion quite common in knowledge and meaning representation, that of *granularity*. The fact that a concept like *bagnasciuga* exists in Italian, does not mean that Italian speakers do not have a precise idea about where the sea finishes and the land starts. In fact, there are two different levels of granularity; if the boundary of the sea is in question, the borderline is clear in the mind of the speaker. If, instead, a more fine-grained knowledge is required, i.e. the focus is on a narrower area, the notion of *bagnasciuga* becomes relevant. The ability to change the scope, and with it the granularity of the representation of the relevant knowledge, is a general property of human cognition.

Table 7.2

Frame: *golena*	Frame: *battigia/bagnasciuga*
is_a: land	is_a: land
bounded_by: river banks	bounded_by: {depends on conditions}
condition: alternating (flooded/dry)	condition: alternating (flooded/dry)
[with time period: high water]	[with time period: waves]

The fact that we have tried to employ a uniform representation formalism in the discussion of the previous examples does not imply that we are suggesting the use of that formalism. Knowledge representation is a complex and ever-growing area of artificial intelligence, and it might certainly be able to offer a more suitable tool for this purpose. However, the following few requisites have appeared during the discussion, which apply to whatever formalism one adopts:

- the most suitable way of representing concepts/objects is by means of hierarchical relations (the 'is_a' link) and sets of properties;
- some concepts/objects are characterized by one property with 'unconstant' values; these objects may have either definite or diffuse boundaries; and
- a granularity relationship similar to a hierarchical one, may hold between some concepts/objects.

7.3   Where properties do not hold

There is a number of objects whose characterization and delimitation cannot be reduced in terms of a finite number of properties. Concepts like *town, village* and *country* can be well reduced in terms of sets of properties, but the boundaries of real towns or villages are often not established on the basis of those properties. A *town* is described as 'an assemblage of buildings, public and private, larger than a village, and having more complete and independent local government; applied not only to a "borough", and a "city", but also to an "urban district", and sometimes also to small inhabited places below the rank of an "urban district"', while a *city* is described as 'a town or other inhabited place' (OED). The definition of *city* is totally dependent upon the definition of *town*, from which it is possible to work out some characteristic properties, such as being 'an assemblage of buildings', having a sort of 'local government', and extending over an area called an 'urban district' (which is not further defined). However, even if these properties allow an observer to roughly classify a generic town, or city, they are not sufficient to allow the identification of the boundaries of a specific town. Towns like Pisa, Rome, or London are, in fact, instances, of the generic concept of town, and a number of social conventions and historical facts contribute to the perception and classification of the environment, beyond generic properties related to the notion of 'town-ness'.

7.3.1   Historically grounded boundaries

As mentioned above, the integration between human culture and the environment also has a historical dimension. Thus, if the borderline between town and non-town can be generally identified in some area, where, for instance, there are fewer buildings, the borderline of a specific town, or other specific (inhabited) areas, is often determined on historical grounds.

7.3.2   Non-perceivable boundaries

Particularly interesting cases are those in which boundaries are not immediately perceived, or even do not exist at all, but are clearly agreed within a community, due to some historically or culturally determined situation.

An important case of non-perceivable, but socially clear boundaries occurs where even the bounded object is historically well known to everyone, even if it is not physically apparent. In Bucharest, for example, there is a place known as *Coada Calului* (Horse's Tail), which is not a square or a crossing, but simply a row of shops standing behind the equestrian statue of Michael the Brave in the direction of the tail of the horse.

In some cases, the historically determined boundaries are as invisible as clear in the minds of the inhabitants of an area. The place where I live, near Pisa, is known as *Badia* (the Abbey), because it is centred around an eleventh-century abbey. Although the boundary of this area, as a whole, is quite diffuse, all the local inhabitants have a precise notion of where some points of the boundary are certainly located, and they are able to identify them precisely. For example, along the main road of the area, Via Vecchia Fiorentina (Old Road to Florence), there is a house whose corner coincides with the border of Badia. There is no visible landmark here, except for a very small gutter behind the row of houses shown in the sketch in Figure 7.1. A larger and deeper gutter a few metres from this point is unanimously attributed to the neighbouring village. All the locals agree that this point is the borderline, and they gave support to their information by the fact that even the priest of the abbey, when leading a procession, stops at that very point and goes back. In fact, one can hypothesize that the boundaries of *Badia* are determined by the area within which the old administration of the Abbey had the right to collect taxes and contributions.

An interesting feature of this situation is what we might call the functional delimitation of space; when asked 'where does Badia finish?', people answer 'at the corner of that house' without taking into account that boundaries should enclose a whole area. This is probably because the only dimension of interest is the one along which one can move and thus cross the boundary, i.e. the road.

In other cases the name, and therefore the extent, of an area has no such definite basis, but evolves from the name of an important landmark. Pisa, as most medieval towns, had walls and therefore gates. In recent times, new quarters have been built outside most of these gates, and have taken the name of the corresponding gate. Thus Pisa has a quarter named *Porta a Lucca* (gate to Lucca), one named *Porta a*

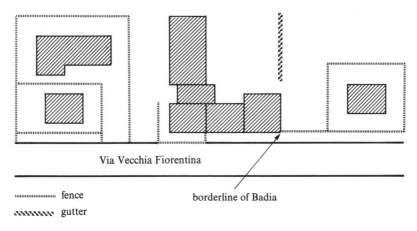

Via Vecchia Fiorentina

‹‹‹‹‹‹‹‹‹‹‹ fence   borderline of Badia
‹‹‹‹‹‹‹‹‹‹‹ gutter

Figure 7.1   Sketch of the border of Badia.

Mare (gate to the sea), another called *Porta Fiorentina* (gate to Florence). All of these quarters stand outside the old gates, which still exist (except Porta Fiorentina), and have extensions that are clear to the local people. In the case of *Porta Fiorentina*, although the gate no longer exists, the limits of the quarter are equally clear to the inhabitants.

The above cases show that most of the boundaries, especially in those areas where human settlements have a long history, need not have a physical or perceptual basis, because they are clearly marked in the minds of the inhabitants, even if the historical origins of such boundaries have been forgotten.

From the point of view of representation, it has been shown that geographic objects like *Badia* or *Porta Fiorentina*, and the corresponding concepts, are instances of generic concepts of a village or a quarter, from which they probably inherit generic properties like having borderlines, being or not being part of a town, etc. Some of the values of these properties can be derived neither from the generic properties nor from direct perception, but are the consequence of historical evolution and, therefore, appear totally unfounded from a synchronic viewpoint.

7.3.3 Localizing places

Most people have a precise idea of their environment, and are able to localize themselves with respect to the question 'where am I now?'. However, the delimitation of an area, and the determination of its borders, is often intermingled with its localization. People will always be able to distinguish whether they are in *Badia* or in one of the neighbouring villages, as well as they will know that living in *Porta a Lucca* implies living outside the wall.

When localizing something, two reference systems are in general available, the absolute one (north, south, east, west), and the relative one (left of, right of, etc), but the identification of relative positions is in many cases the result of the historical evolution of the socio-economic structure of a place, or even of viewpoints whose cognitive foundations may be unclear.

An example of ambiguity of such a culturally determined system of reference is the Boston subway system, which has two directions, *inbound* and *outbound*, which are not totally clear to foreigners. It is intuitively obvious that such expressions have something to do with the centre of Boston, or of some historically relevant area of it. The problem that confuses many visitors is where *in* ceases and gives way to *out*, and if there are areas which are all *in*. In fact, the definition of *in* and *out* are a function of some recursive localization of the boundary of the city centre; nevertheless, there must be a point where it is impossible to go inbound, but the movement is reversed to outbound.

Pisa is divided in two parts, 'di qua d'Arno' (on this side of the Arno) and 'di la' d'Arno' (beyond the Arno), but this is not defined relative to the position of the speaker, but is a sort of absolute discrimination (north/south). Thus, any good Pisan, when located south of the Arno, will refer to his position as 'di la' d'Arno' (beyond the Arno), even if speaking to someone on the same side of the river. Most Pisans, however, would never say that Porta Fiorentina is 'di la' d'Arno', even if it is south of the Arno, as it seems that such a distinction holds only for the areas within the walls.

What I want to stress here, is that the orientation system people use in their own area is in general a relative one, and often depends upon reference systems and landmarks which are largely determined by the historical evolution of the area itself.

7.3.4 Issues in representation

The types of geographic entities and borders discussed above have two character-istics: they are clear in the minds of the local inhabitants, be they perceptually clear or not, and they have a more or less clear historical basis. They do not fall into the same category of geographic concepts discussed in Section 7.2, which are generic concepts and can be described in an abstract representation system. The cases of *Badia* and *Porta Fiorentina* are well localized instances whose properties have been determined by history. The concepts of *village* or *quarter* do not help in the identifi-cation of the borders of *Badia* or *Porta Fiorentina*. The consequences of this situ-ation are as follows:

- When dealing with geographic objects, it is necessary to strengthen the mecha-nism for the representation of the properties of individual entities, taking into account that most of them have a historical motivation which makes their syn-chronic characterization appear totally arbitrary and unfounded.
- An interesting point of research is the identification of the reasons for such his-torical stratifications of geographic description. The examples discussed here are based on my personal experience, supported by people I have interviewed. A more serious method of research, possibly based on formal questionnaires, could be fruitful.

7.4 Conclusions

The first point that has become apparent is that the description, and consequently the identification and delimitation, of geographic objects is more strictly related to the corresponding concept or cognitive model than to perceptive reality; indeed, the latter can be strongly affected by the cognitive background of the observer. The consequence of this is that the description of a geographic object does not coincide with its image alone, but must also be paired with a representation of the underlying concept(s). As a simple example, the representation of a river will consist of an image and the corresponding concept, expressed in some uniform representation formalism.

It has also been shown that the relation between the perceptual appearance of an object, if there is an absolute one, is arbitrary, and the definition of concepts is affected by two aspects of human cognition:

- the analysis of the environment depends upon different culturally determined categories; this leads to cultural divergences in the descriptions of the same con-ceptual space; and
- the historical evolution of human dwellings leaves traces in the analysis of the environment; people often refer to orientation systems which do not correspond to an actual configuration of buildings and landmarks, but to previous situations that may no longer exist.

Geographic data can be described in an absolute way, and referred to an absolute description system, without losing descriptive adequacy. However, knowledge of the mental categories according to which such data are organized and classified in people's minds is the only guide we have to the ways in which people perceive and feel the areas in which they live. Language is the most important repository of such stratified cognitive models of geographic space.

A uniform representation system for the cognitive and conceptual aspects of geographic description can probably be drawn from artificial intelligence and cognitive sciences. This representation may well be based on the notion of concepts having properties, but these properties may behave in unusual ways, such as taking swinging or decreasing values. In addition, in passing from generic concepts, like *mountain*, *sea*, *town* or *village*, to instances like *Monte Bianco* or *Pisa*, a number of historical and traditional elements may add, change or modify the properties.

The important lesson for the design and building of GIS is that, besides the discussions about the objectiveness of geographic boundaries and their cognitive foundations, it is necessary to take into account how local inhabitants perceive their environment, and how this spatial modelling is manifested in the linguistic system. In fact, this is not only a way of studying curious ideas, but may also give an idea of the historical evolution of the environment and of people, who are an integral part of the environment itself. For the future, deeper investigations in the field of local toponymy and of geographic and spatial expression could be undertaken as part of research on areas to be surveyed.

REFERENCES

BERLIN, B. and KAY, P. (1969). *Basic Color Terms: Their Universality and Evolution*, Berkeley, CA: University of California Press.

CARROLL, J. B. (Ed.). (1956). *Language, Thought, and Reality: Selected Writings of Benjamin Lee Whorf*, Cambridge, MA: MIT Press.

CONKLIN, H. C. (1955). Hanunoo color categories, *Southwestern Journal of Anthropology* , **11**, 339–344.

KAY, P. and McDANIEL, C. (1978). The linguistic significance of the meanings of basic color terms, *Language* **54**(3), 610–646.

LAKOFF, G. (1987). *Women, Fire and Dangerous Things*, Chicago: University of Chicago Press.

LYONS, J. (1969). *Introduction to Theoretical Linguistics*, Cambridge: Cambridge University Press.

MARK, D. (1993). Toward a theoretical framework for geographic entity types, in: Frank, A. U. and Campari, I. (Eds), *Spatial Information Theory: Theoretical Basis for GIS*, pp. 270–283, Berlin: Springer.

MINSKY, M. (1975). A framework for representing knowledge, in: Winston, P. (Ed.), *The Psychology of Computer Vision*, pp. 211–277, New York: McGraw-Hill.

REY-DEBOVE, J. (1971). *Étude linguistique et sémiotique des dictionnaires Français contemporains*, Paris: Mouton.

The Shorter Oxford English Dictionary, (OED), (Ed.) Little, W., Oxford: Clarendon Press (since 1993).

ZINGARELLI, N. *Vocabolario della Lingua Italiana*, Bologna: Zanichelli (since 1992).

CHAPTER EIGHT

On the Relations between Spatial Concepts and Geographic Objects

CHRISTIAN FREKSA and THOMAS BARKOWSKY

Fachbereich Informatik, Universität Hamburg, Hamburg, Germany

8.1 Why is it difficult to represent geographic knowledge?

The geographic world surrounding us is extremely complex. When we want to master a given problem in this world, we need to single out particular aspects of current interest from this multifaceted information. So, at any given time, we are only interested in few objects, and concerning these objects again we are regarding only particular properties and/or relations. The capability to isolate the relevant aspects and relating them to each other, results in a unique intellectual efficiency. This efficiency, however, is necessary for successfully operating in the world.

To represent knowledge about the world is to make explicit specific aspects of the world. In making explicit certain aspects we ignore others. In representing knowledge, every single aspect of interest can be represented separately. Alternatively, we can aim at representing different aspects within a single structure; this requires that the different aspects of interest must be compatible, i.e. they must fit into one reference system corresponding to a global view. It is impossible to make all potentially interesting aspects of the world simultaneously explicit within one representation medium, be it a conventional map or a single representation structure in a computer.

8.1.1 The need for different world views

There is an information-structural reason why a unified representation of the geographic world is not sufficient for solving all the tasks that can be solved with that information: we are dealing with geographic objects and with relations between these objects. If we could decide once and for all which entities we should view as objects and which as relations, our task would be simpler; however, an important feature of using world knowledge intelligently is the ability to switch between views; depending on the specific task to be solved, certain entities may be viewed as objects (fixed background entities) and others become the relations we manipulate. For other tasks, these roles may change.

For example, we may consider roads as relevant geographic objects in some context. For solving certain navigation or transport tasks, we may be interested in the intersections between these roads; having roads as objects, intersections naturally can be viewed as (connection) relations between roads. For designing road systems, it may be convenient to view road intersections as the primary objects; the roads then can be viewed as (connection) relations between these intersections. Of course we could say that we want to view both roads and their intersections as objects to obtain a unified representation (after all, a toy train set contains both regular tracks and switches as objects). But this does not really solve the general problem: when we view both the roads and their intersections as (separate) objects, we create a situation in which the roads do not meet the intersecting roads; all the roads meet intersections. To solve navigation tasks, for example, we have to consider relations between roads and intersections; thus we have only shifted the problem to the next level. We encounter similar situations when we consider regions and their boundaries, which may be viewed as objects and relations, respectively, or vice versa.

Therefore, if we want to make all potentially interesting knowledge accessible, we must make it explicit in different structures (maps or computer representations). Creating many different structures for representing knowledge about the same domain becomes expensive, both computationally and in terms of storage, since a lot of implicit information must be carried along to link the knowledge to its domain.

8.1.2   GIS as a mediator between the world and the user

When developing a geographic information system (GIS), we must find an adequate compromise between two extreme possibilities: (1) to acquire and store all knowledge from raw information once and for all before the knowledge is accessed; and (2) to provide unprocessed raw information and to compute specific knowledge on demand. On the one hand, we expect a GIS to contain sufficient data about the geographic world, and on the other, we want to obtain a selective view of the relevant aspects and to discard any other data of less interest. The entities represented in a GIS stand for the real-world objects and their properties. From this point of view, a GIS serves as a mediator between world's reality and the way humans interact with this reality. The user expects the entities represented in the system to show the same properties as the real objects they are standing for.

8.1.3   The human use of spatial knowledge

The human capability to see the world as an inexhaustible source of information may make it desirable to process the knowledge about the world in such a way that it is instantly available to the human user. But if we process this knowledge about the world in advance, we will create many structures which most likely will never be accessed. On the other hand, if we do not provide structured knowledge to the user, great efforts may be required to compute this knowledge when needed.

When we consider the human capability to deal with geographic information we can identify two challenges in the development of an 'intelligent' GIS: (1) how can we model the human ability to focus on relevant information when solving a problem?, and (2) how can we overcome the problem that a GIS can not contain all facts about the real world that might become important in a special context?

8.2   What are spatial concepts?

When we consider real-world objects, we usually are interested in certain properties of these objects, i.e. we regard the objects under certain aspects. For example, when we take a look at a geographic entity, say a lake, we regard it with respect to its horizontal extent, depth, shape, or the like. All these notions that describe spatial aspects of a subset of the world, we call spatial concepts. We will use the term 'concept' in a rather general sense (cf. Church, 1956). Concepts can be anything we have a notion of; thus, 'size' can be a concept and 'big' can be a concept as well. As we predicate aspects of objects using concepts, it is obvious that spatial concepts will play a crucial role in representing knowledge about the geographic world.

8.2.1   Properties of spatial concepts

In the following we present several semantic and structural aspects of spatial concepts which are of particular importance for their representation and for the operations that are to be performed on them. Specifically, we address the relationships between different concepts and the relationships between concepts and real-world entities.

Concepts have meaning in relation to objects

As already pointed out, we use concepts to describe aspects or properties of objects. This means that concepts are related to the aspects of the objects that are described. It is impossible to describe the qualities of objects without using related concepts. Conversely, the meanings of concepts are rooted in their relation to objects, e.g. the concept of a square is related to quadrilateral objects whose sides have equal length and meet at right angles. The mutual relatedness between concepts and objects enforces certain structures upon the concepts; in particular, not every relation between concepts and objects is meaningful.

Concepts have meaning in relation to other concepts

Nevertheless, the meaning of spatial concepts is not only given by their relations to physical objects. The meaning of concepts is given to a large extent by their relations to other concepts (cf. 'lateral thinking'; de Bono, 1969). We illustrate this point using the example of the square: we can view the meaning of the concept 'square' by its relation to all existing physical objects that are square in shape; but we also can view the meaning of 'square' by relating it to imagined or imaginable objects that are square in shape. Rather than viewing the meaning to be made up of an infinite number of relationships with imaginable objects, we can view the meaning to consist of a finite number of relationships between related concepts.

For example, the concept 'square' may be related to the concept 'rectangle' by a 'special_case' relationship, to the concept 'equilateral triangle' and 'rhombus' by a 'near_miss' relationship, to the concept 'shape' by an 'is_a' relationship, etc. In this way, spatial concepts may have meanings without reference to physical instances. In particular, spatial concepts have meanings that are independent of envisioned physical materializations: the shape concept 'square' abstracts from the realization by a section of a checkerboard or a marketplace in a city.

Concepts induce distinctions

Concepts are human-made entities. We do not have any concept that is not related in some way to another concept. We form new concepts in order to make distinctions between features (or sets of features) in the world. Thus, concepts come in pairs or n-tuples, depending on how many cases are to be distinguished (cf. Freud, 1910; Quillian, 1968). The relation between concepts can frequently be viewed as positive, when there are concepts that describe similar aspects (e.g. 'special_case'), or as negative when the description within one concept excludes others, 'square' versus (general) 'rectangle'. Interestingly, we frequently find relations with both positive and negative connotations between a given pair of concepts; (e.g. a square is a special case of a rectangle (positive) and it is not just any rectangle (negative). This fact about human concepts plays a particularly important role in the construction of appropriate models for dealing with such concepts. The property of relatedness between concepts is often important for gaining implicit knowledge about conceptual aspects from the description of other concepts.

Consider a spatial concept like 'big'. In attempting to define the meaning of 'big' people sometimes ask questions like 'how big is "big"?' (Denofsky, 1976). Taking into account the above considerations, we can answer this question on two levels:

(1) When we view the meaning of concepts as given by specific instances, we can provide the answer 'it depends'; specifically, it depends on the type of entity to which the concept 'big' is applied. By applying the notion 'big' to some specific instance(s), we express a distinction from instances which are not 'big'. If everything in the set under consideration was equal in size, it would not make sense to apply the concept.

(2) When we view the meaning of concepts as given by their relations to other concepts, we find a qualification relation to the concept 'size' and a contrast relation to the concept 'small', and possibly to the concept 'very big'.

On level (2), the meaning of the concept 'big' is independent of specific sizes to be distinguished, and is even independent of the types of entities to which it may apply. If we hear about a 'big' something and do not know which something or type of something is being referred to, we still can answer questions about the size of the entity under discussion; specifically, we can say that the size of the something is greater than the size of a corresponding something labelled 'small' and that the size of a 'big' something else of the same category will be of the same order of magnitude as the size of the something. Thus, there is a structural aspect of the concepts of size which is context independent; in essence, it establishes universal conceptual relations that are valid for all situations in which the concept can be used.

To ensure that the concepts are as informative as possible, it is desirable to make available a dynamically adaptable number of concepts, depending on how many objects are to be classified and how significantly they differ from each other (Freksa, 1981). For example, when only a few sizes are to be distinguished, say five, it is sufficient to differentiate five classes by making available appropriate concepts. In natural language, we use contextual adaptivity of conceptual classes all the time: depending on the situation, we implicitly contrast 'big' versus 'small', or 'big' versus 'very big' versus 'rather big' versus 'medium-sized' versus … . Thus, the meaning of spatial concepts is determined by at least three types of relations: (1) between concepts and physical objects, (2) between concepts and related concepts (context of discourse), and (3) between objects and situation context.

Concepts can be substituted for by finer or coarser concepts

One of the most important features in dealing with concepts is that the number of classes they generate can be increased or decreased just as is needed for an adequate description. This means that the spectrum of possible values is reorganized according to the needs of the situation. Of course, the single classes need not to be equal in size, neither with reference to the ranges they cover, nor to the number of actual values they include. Figure 8.1 illustrates this property. It shows that a range associated with a concept, e.g. 'size' is further and further divisible, up to the level of refinement actually needed, or conversely, fine concepts can be merged to form coarser concepts. This property requires a neighbourhood structure for concepts: 'fine' concepts which are in opposition on a level of high resolution are united in harmony on a 'coarser' conceptual level on which fewer distinctions are made (Freksa, 1991). If we assume that new concepts result from the refinement of existing concepts, we will always obtain a neighbourhood structure as depicted in Figure 8.1.

Concepts have no internal structure

We can view concepts as atomic, discrete entities in the sense that we either use or do not use a concept in a given situation. Concepts do not have an internal structure, but can be replaced by a structure of finer concepts or merged to form coarser concepts. It is important to see that if we replace a concept by a structure of finer concepts, the original concept is no longer accessible; therefore we consider concepts as not internally structured.

Concept of Size

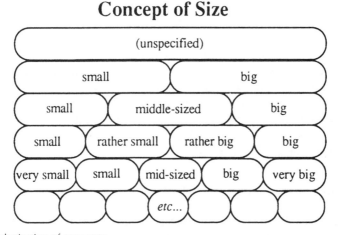

Figure 8.1 Substitution of concepts.

When we replace a concept by a structure of finer concepts, we can observe interesting polarity properties of the competing finer concepts. On the one hand, the finer concepts are *in opposition* to one another, and we introduce the concepts 'big' and 'small' in order to emphasize these opposing qualities; on the other, the new concepts are *neighbours*; they can be merged when this particular distinction is not required. This dualistic relationship between concepts with common roots is extremely useful for generating adequate descriptions in different contexts and for interpreting them in a robust way.

By stating that concepts have no internal structure we imply that it is not meaningful to speak of continuous or discrete concepts. If we assume that concepts are created by refining existing concepts, we do not gain anything by postulating a continuous (as opposed to discrete) internal structure. The assumption of continuous structures results from a world view which is not based on conceptual structures or observed feature values, but on a generalization of low-level features. This world view is heavily influenced by the availability of mathematical models which characterize complex structures in terms of simpler components.

Concepts are crisp in relation to their explicit competitors

An interesting question is whether concepts are *crisp* or *fuzzy* entities, i.e. whether they have sharp or gradual boundaries. To answer this question we must determine what the boundaries of concepts are. If we consider the relation between a given concept and an opposing one, we clearly have a sharp boundary, since opposing concepts result from distinctions, and distinctions are made on the basis of qualitative criteria rather than (arbitrary) gradual values. For example, when we distinguish between 'very big' and 'big' we want to point out a *significant* difference in size.

We can now look at concepts which coexist on different branches of a 'concept hierarchy' as a result of distinctions made in different situations rather than on the basis of an opposition as described above. For example, we may have notions like 'pretty big' and 'rather big' which do not seem to be in opposition but which are not identical either. Do the associated concepts have crisp or fuzzy boundaries to one another? Well, if these concepts are on non-neighbouring branches of our conceptual hierarchy, they may share no boundaries; thus, the question may be ill-posed. 'However', you might argue, 'these concepts address the same feature dimension, they express almost identical ideas, so what is their relationship to one another?'

If we continue to restrict our considerations to the conceptual domain, the relationship between concepts on different branches of the conceptual hierarchy is given by the vertical relationships which connect these concepts; but these relationships may not induce boundaries between the concepts in question. However, if we take into consideration the relationships between spatial concepts and objects in the real or imagined world, the situation changes: when asking the question of applicability of concepts in a concrete situation, the question of fuzzy versus crisp boundaries of concepts appears in a new light.

8.2.2 Relations between concepts and objects

In Section 8.2.1 we discussed spatial concepts and their properties in isolation from the spatial domain to which they correspond. In this section we address aspects pertaining to the correspondence between conceptual structures and real-world entities (cf. Palmer, 1978).

Lack of detail results in fuzziness

Fuzziness is not a property of a single aspect or concept; rather, it is the property of a relation between a concept and an object. When concepts come into existence 'top-down' by refinement of existing concepts rather than 'bottom-up' by definitions in terms of low-level features, then the relation between concepts and specific object properties may be viewed as 'fuzzy', since we may not be able to determine for each low-level value whether the relation holds or not. In a sense, we apply coarse notions to situations of fine resolution; the neighbourhood structure among concepts and domain situations allows for interpolating and extrapolating between central uses of concepts and clearly inapplicable situations. The result is a fuzzy correspondence. Admitting fuzzy correspondence in representations is very useful, as it enables us to place situations into ballparks of closely matching conceptualizations when no precisely matching conceptualization is possible or desirable.

Horizontal and vertical neighbourhood

We can distinguish two dimensions of neighbourhood between concepts: 'horizontal' and 'vertical'. 'Horizontal neighbourhood' refers to competing concepts on the same level of granularity (Freksa and López de Mántaras, 1982), while 'vertical neighbourhood' refers to compatible concepts on different levels of granularity (cf. Figure 8.1) (Zadeh, 1977; Hobbs, 1985). Thus, exploitation of the horizontal dimension allows for the selection of an appropriate description *value*, while exploitation of the vertical dimension allows for the selection of an appropriate description *granularity*, depending on the context of the specific situation.

The combination of the horizontal and vertical degrees of freedom allows for great flexibility in selecting the most appropriate descriptor, in a given situation. For example, suppose coarse granularity is sufficient in a given situation (e.g. 'small', 'medium-sized', 'big'; compare Figure 8.1), but the object to be characterized is equally well described by two competing descriptors (e.g. 'small' and 'medium-sized'), i.e. the situation is a boundary case; then we can go to the next finer level of granularity and typically we will find a concept which very well characterizes the object (e.g. 'rather small'). In this way, the 'fuzziness' of concepts can become instrumental in selecting appropriate descriptors, which then may be in a rather unfuzzy (crisp) relation to the described situation.

Concept hierarchies do not form trees

In refining concept spaces, we do not simply subdivide individual concepts as in a hierarchical tree (cf. Winston, 1975), but we revise the subdivisions according to the neighbourhood structure in such a way that the boundary cases on one level of description become central cases on the next level. In other words, when a concept is needed to refer to a boundary region of two (relatively coarse) concepts, the new (finer) concept typically refers to a region integrating both sides of that boundary (cf.

Figure 8.1). By exploiting the horizontal and vertical dimensions of concept hierarchies, we obtain a powerful tool for tailoring effective and efficient descriptions of
arbitrary situations from arbitrary points of view.

8.3   How do we describe the world with concepts?

In this section we address the problem of structural incompatibilities between the
conceptual domain and the object domain. The structure of the object domain
depends largely on the view taken with respect to this domain: when we view the
spatial domain as *dense* or *continuous*, we implicitly refer to the set of *possible situations* rather than to actual situations. If we take each given situation individually
by itself, we do not obtain a continuous structure. Thus, the structural compatibility
between concepts and objects depends much on the assumed world view.

8.3.1   The properties of the world are assumed to be continuous

When we regard an entity in the world under a certain aspect, say a lake under the
aspect 'area', we assume that *in principle* lakes may have any extent, within a certain
range. This insight is responsible for using a continuous scale, i.e. real numerical
values, to describe areas. In fact, this is a rather sensible approach to data acquisition.

When concerned with special tasks to be performed on these data, the pure
numerical description may not be of major interest. What we actually need is a
classification of the data with respect to the task of current interest. Each individual
task typically induces specific thresholds, i.e. critical values above or below which
there may be a qualitatively different situation. These thresholds may be crisp or
fuzzy.

8.3.2   Concepts are discrete entities

As suggested in Section 8.2, new concepts are formed when new distinctions are
needed; thus concepts – like the symbols denoting them – are inherently discrete
entities. How do we use these discrete entities to describe a world whose objects are
assumed to have continuous-valued properties? To answer this question we will
look at the formation of concepts from classifying real-world entities on the one
hand, and from refining existing concepts on the other.

We have seen that the world of real objects and their properties may appear
quite different from the world of concepts we use for describing the 'real world'.
When we describe certain aspects about the 'real world', we have to match the
(discrete) world of concepts with the (continuously perceived) world of features of
the entities in the real world. To do this, we can (1) go 'bottom-up' from the features
towards the concepts, (2) go 'top-down' from the concepts towards the features, or
(3) combine the two directions. At some point, we will reach a level where the
bottom-up generalization of feature values will reach about the same granularity as
the top-down refinement of the concepts. Nevertheless, we cannot expect a one-to-
one mapping between the top-down conceptualization and the bottom-up gener-

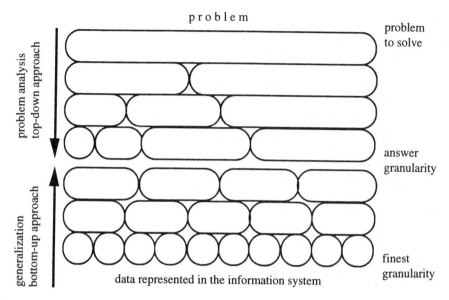

Figure 8.2 Bottom-up versus top-down conceptualization.

alization; rather, there will be incompatibilities. This phenomenon is illustrated in Figure 8.2.

8.3.3 Concepts are adjusted to the situation, not vice versa

As the discussion of the incompatibility between the top-down and the bottom-up conceptualizations shows, it does not make sense to form concepts mechanically from the real-world data alone. This is because there is no absolute measurement for an entity to belong to one class of a concept or to another. The same set of data may require different classifications (even at the same level of granularity) when used within different contexts.

Concepts provide an efficient means for understanding, thinking, planning and communicating. But concepts are not fixed in their reference to real-world entities; rather, they allow to be adapted with some flexibility to the requirements of the specific situation. By using concepts we can see the world as detailed as we need, or, by using concepts we construct our reality of interest.

8.3.4 GIS eventually should employ concept-driven sensors

The epistemological independence between concepts and real-world entities can be used when we are concerned with the level of granularity to be chosen for the representation of data in technical systems. Let us assume that we do not have all the data we need to answer a certain question in a system (this is a rather realistic assumption). Let us further assume that we have implemented an appropriate model of human conceptualization. Then it is quite imaginable to combine a technical system with an automatic information acquisition system (e.g. satellite-based) to obtain the data needed according to the conceptualization wanted directly from the real geographic world.

The task-driven application of conceptualizations to knowledge acquisition would be a very appropriate compromise between the two options described at the outset: (1) to acquire and store all knowledge from raw information once and for all before the knowledge is accessed; and (2) to provide unprocessed raw information and to compute specific knowledge on demand. The compromise consists of pre-processing raw data to a level of conceptually relevant granularity and using this preprocessed data on demand by the conceptualizations formed in specific task contexts.

The suggested approach exhibits a strong correspondence to the human use of geographic maps. A geographic map corresponds to preprocessed raw data: certain features are classified according to criteria from the conceptual level, other features – in particular spatial relations – are processed only with respect to data resolution (granularity). Depending on the task to be performed, the conceptual level determines which relations in the map are further elicited or processed.

8.4 Where is the fuzziness in geographic objects?

In representing knowledge about the geographic world, we must distinguish three kinds of uncertainty: (1) we may not know the precise location of crisply classified geographic entities and thus may be uncertain of their location, or (2) we know the precise locations of the geographic entities including the (possibly gradual) transitions between them, but we are uncertain precisely how to classify them, or (3) we may have a combination of the two possibilities.

8.4.1 Uncertainty due to imprecise knowledge

In the first case, we are confronted with the problem of representing and interpreting incomplete information. Information may be incomplete for a variety of reasons: for example, a piece of information may be associated with a European city and we may be uncertain if it is London or Paris. Thus, we have two *discrete* possibilities with regard to the correct reference, and the information is incomplete with respect to the resolution of this ambiguity. On the other hand, we may be uncertain with regard to the precise location of a geographic object due to measurement errors; here, we obtain a *continuum* of possibilities with regard to the precise location of the object and the information may be considered incomplete with respect to the required precision.

Discrete possibilities are frequently represented by logical disjunctions and interpreted like an 'exclusive or': either London or Paris (but not both). We obtain a special case of the discrete situation when the different possibilities are arranged according to some ordering criterion. Under certain circumstances (which are given in many situations), the possibilities are distributed in such a way that only *neighbourhoods* of possibilities can occur. These possibilities can also be represented by disjunctions: for example, in specifying geographic locations, '38 metres' may mean in more precise terms '37.5 metres' or '37.6 metres' or … '38.4 metres'.

Treating neighbouring possibilities in this way has computational drawbacks (especially computational complexity) and it neglects the spatiality expressed in the neighbourhood of the geographic locations, since it ignores the difference between close and remote locations. Therefore, representing neighbouring possibilities by structures preserving neighbourhood has great advantages. Continuous structures preserve neighbourhood information. Intervals, probability distributions, and possibility distributions are such structures. However, in dealing with real situations, most often we have discrete information to begin with; therefore there are definite advantages in employing discrete models which preserve the ordering information relevant in the domain.

8.4.2   Uncertainty due to indefinite correspondence between concepts and the represented world

In the second case (precise information, unclear classification) we are confronted with a rather different problem: the uncertainty is not associated with the information itself, but with the correspondence between the concepts available for the description of the geographic world and the geographic world itself (cf. Rosch, 1973, 1975). If we use as many concepts as possibilities we want to be able to distinguish and use a one-to-one correspondence, then the problem has vanished. However, if these concepts are not connected appropriately, we again loose the internal structure of the domain and the representation is not very informative. What we usually want is to represent knowledge in such a way that few concepts capture rich situations. In particular, this means that (1) gradual transitions from one concept to another should not be captured by introducing a large number of intermediate concepts, and (2) we need concepts that can be adapted to a multitude of situations if the concepts are to be used universally.

We will address here specifically the representation of transitions. The classical approach to representing gradual transitions is the use of fuzzy sets (Zadeh, 1965). A main utility of the fuzzy set approach is that concepts are split into two levels of consideration: (1) the level on which we talk about concepts (the symbolic level), and (2) the level on which the correspondence to real-world entities is established (the semantic level). In this way, we obtain an interface between the discrete world of concepts and a possibly continuous world of entities to be described (Freksa, 1995). The fuzzy set approach assumes that the concepts associated with the labels of fuzzy sets can be defined precisely. In everyday situations, in particular in non-technical domains, this assumption is usually too strong.

A weaker – and in many situations more realistic – assumption is that we can very well structure the concepts we use and that we can very well structure the entities and features we want to capture, but we have difficulties in establishing a precise correspondence between concepts and real-world entities. The reference of the concepts can then be established by structure matching rather than by correspondence between individual conceptual and real-world entities. As a result of this process, a certain fuzziness in the correspondence between concepts and real-world entities persists: we may have several possibilities in assigning concepts to a given situation. This fuzziness is benign, since the applicable concepts will be very closely related alternatives due to the preservation of the structure.

8.5 Conclusions

We have addressed the problem of representing knowledge about the real world in general, and representing spatial knowledge about the geographic world in particular. The focus of the discussion was on issues related to boundaries in the represented world, in the representing world, and in the relation between these two worlds. We face the challenge of dealing with (1) complex open worlds, i.e. worlds whose dimensions and values cannot be entirely specified; (2) incomplete and imprecise information, i.e. information which cannot be taken at 'face value'; (3) knowledge and concepts of varying granularity causing fuzzy correspondence between concepts and real-world entities; (4) knowledge which is to be used for different tasks requiring different resolution and different conceptualization; and (5) knowledge in different contexts requiring different discrimination capabilities.

We have investigated the role of concepts and conceptual structures in representing knowledge about the real world and emphasized their autonomy with respect to the entities and assumed structures in the represented domain. We have argued that geographic objects are not fuzzy in themselves, but that they are fuzzy with respect to the precision of the underlying knowledge and/or with respect to their classifiability in terms of a given set of concepts. In order to deal with gradual boundaries in a sophisticated way, we propose an approach which takes into account (1) the neighbourhood structure of geographic entities according to a physical model, (2) the vertical and horizontal neighbourhood structure of the spatial concepts, and (3) the correspondence between concepts and geographic entities.

This work is based on studies of representing soft knowledge for the purpose of human–human and human–machine communication about open worlds, and on formal approaches to representing qualitative temporal and spatial knowledge.

REFERENCES

DE BONO, E. (1969). *The Mechanisms of Mind*, New York: Simon & Schuster.

CHURCH, A. (1956). *Introduction to Mathematical Logic*, Vol. 1, Princeton, NJ: Princeton University Press.

DENOFSKY, M. E. (1976). How near is near?, *AI Memo No. 344*, Artificial Intelligence Laboratory, Massachusetts Institute of Technology, Cambridge.

FREKSA, C. (1981). Linguistic pattern characterization and analysis, PhD thesis, Department of Electrical Engineering and Computer Sciences, University of California, Berkeley. Abstract in: *J. of Pragmatics*, **6** (1982), 371–372.

FREKSA, C. (1991). Conceptual neighborhood and its role in temporal and spatial reasoning, in: Singh, M. and Travé-Massuyés, L. (Eds), *Decision Support Systems and Qualitative Reasoning*, pp.181–187, Amsterdam: North-Holland.

FREKSA, C. (1995). Fuzzy systems in AI, in: Kruse, R. *et al.* (Eds) *Fuzzy Systems in Computer Science*, Wiesbaden: Vieweg.

FREKSA, C. and LÓPEZ DE MÁNTARAS, R. (1982). An adaptive computer system for linguistic categorization of 'soft' observations in expert systems and in the social sciences, in: *Proc. 2nd World Conf. on Mathematics at the Service of Man*, Las Palmas.

FREUD, S. (1910). Über den Gegensinn der Urworte, *Jb. psychoanalyt. psychopath. Forsch.*, **2**(1), 179–184.

HOBBS, J. R. (1985). Granularity, *Proc. 9th Int. Joint Conf. on Artificial Intelligence*, pp. 432–435.

PALMER, S. E. (1978). Fundamental aspects of cognitive representation, in: Rosch E. and Lloyd B. (Eds), *Cognition and Categorization*, Hillsdale: Erlbaum.

QUILLIAN, M. R. (1968). Semantic memory, in: Minsky, M. (Ed.), *Semantic Information Processing*, Cambridge, MA: MIT Press.

ROSCH, E. (1973). On the internal structure of perceptual and semantic categories, in: Moore, T. E. (Ed.), *Cognitive Development and the Acquisition of Language*, New York: Academic Press.

ROSCH, E. (1975). Cognitive representations of semantic categories, *Journal of Experimental Psychology: General*, **104**, 192–233.

WINSTON, P. H. (1975). Learning structural descriptions from examples, in: Winston, P. H. (Ed.), *The Psychology of Computer Vision*, New York: McGraw-Hill.

ZADEH, L. A. (1965). Fuzzy sets, *Information and Control*, **8**, 338–353.

ZADEH, L. A. (1977). Fuzzy sets and their application to pattern classification and clustering analysis, in: van Ryzin, J. (Ed.), *Classification and Clustering*, New York: Academic Press.

ZADEH, L. A. (1979). Fuzzy sets and information granularity, in: Gupta, M. M., Ragade, R. K. and Yager, R. R. (Eds), *Advances in Fuzzy Set Theory and Applications*, Amsterdam: North-Holland.

CHAPTER NINE

Qualitative Shape Representation

CHRISTOPH SCHLIEDER

Institute for Computer Science and Social Research, University of Freiburg, Germany

9.1 Introduction

In this chapter we discuss the problems that arise with undetermined boundaries from a spatial reasoning perspective. Spatial reasoning is a subfield of AI research on knowledge representation which studies formalisms for encoding spatial information (Kak, 1988; Chen, 1990; McDermott, 1992). One major concern in this field is the comparison of different representational formalisms in terms of their expressive power; another is the analysis of the computational costs of the underlying inference mechanisms. Such questions seem similar to those raised by algorithmic disciplines outside AI which also deal with spatial information, e.g. computational geometry. Although a number of techniques can be found in both areas, there is a clear difference in emphasis: the algorithmic disciplines consider well-posed problems and try to find optimal solutions for them, whereas spatial reasoning seeks to draw inferences even from incomplete spatial information, that is, from ill-posed problems. Suboptimal results may well be acceptable in spatial reasoning if the efficiency of the inference process can be increased that way. For example, a strategy solving a spatial constraint satisfaction problem could be considered useful if it works for some frequently encountered cases, even though it may fail to produce a solution for every solvable problem.

The approach to shape representation advocated in this chapter applies the idea of qualitative abstraction which is implicitly used in many spatial reasoning formalisms to a special kind of spatial information, namely, ordering information. Section 9.2 focuses on the indeterminacy of spatial information in general. Sections 9.3 and 9.4 describe the procedure of qualitative abstraction and the concept of ordering information, Section 9.5 then develops the ordering information approach to qualitative shape description.

9.2 Boundary and shape in qualitative spatial reasoning

The ability to cope with a certain indeterminacy of spatial data and the trade-off between efficiency and optimality make spatial reasoning techniques attractive to a number of application domains such as computer vision, robot navigation, image information systems and GIS. In contrast to many special purpose solutions that have been developed in these field, spatial reasoning focuses on general inference techniques and conceptual issues. Inferences and concepts can be handled naturally within the framework of mathematical logic. This logic-oriented approach amounts to representing spatial information by the formulas of some logical language. Syntactic rules operating on formulas describe the valid inferences, and axioms implicitly define the concepts (constants, functions, relations, etc.) that appear in the language. Even though it is not always practicable to give a spatial representation formalism this strict logical form, the idea of axiomatization is generally thought of as the ideal to achieve because it is a prerequisite for any further analysis of the formalism's properties.

Approaches that adhere to the axiomatic ideal often characterize themselves as *qualitative spatial reasoning* (Freksa, 1992; Cohn *et al.*, 1993). We leave aside the question of whether there are other approaches that could equally be considered as 'qualitative', and simply adopt the attribute as a distinctive label for the purpose of the following discussion. It will be shown that there is more to qualitative spatial reasoning than just axiomatization. The importance of axiomatization was stressed in the very beginning of spatial reasoning research associated with naive physics. Hayes (1985) proposed to capture our commonsense reasoning about the physical world in a first-order logic axiomatization. A primary goal of his naive physics consists in identifying conceptual clusters, i.e. sets of concepts that are highly inter-related and possess only few logical connections to concepts outside the cluster. Hayes refers to 'a fairly autonomous theory of shape' as a typical example of what could turn out to be a conceptual cluster.

It should be pointed out that the attribute 'naive' can be understood in two quite different ways: an axiomatization can be intended to model some human cognitive competence, or it can be conceived as an approximate, qualitative description of the physical word itself. The latter position is frequently found in connection with engineering applications where the efficiency of computation has a high priority but similarity to human reasoning is of lesser importance. Interestingly, the two research orientations are not always clearly separated. Often, a spatial reasoning formalism claims cognitive adequacy for only some of its assumptions (i.e. some axioms or inference rules), the other being motivated by engineering considerations.

9.2.1   Recognition indeterminacy and reconstruction indeterminacy

Before going into problems of representing undetermined boundaries it seems useful to state clearly what kind of indeterminacy is usually handled by qualitative approaches to spatial reasoning. Indeterminacy of spatial information may arise in at least two different ways. First, there is what could be called *recognition indeterminacy*, that is, indeterminacy arising from competing interpretations of a given set of spatial data. Spatial information gathered through visual sensors will nearly always be subject to this type of indeterminacy. The schematic drawing in Figure 9.1 illustrates this fact by the pixel pattern shown on the left. Assume that an image analysis program recognizes the pattern as being with equal probability the picture of a

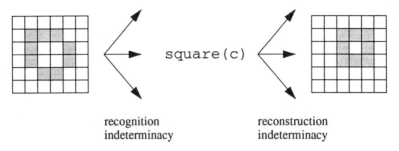

recognition
indeterminacy

reconstruction
indeterminacy

Figure 9.1 Two kinds of indeterminacy of spatial information.

square or the picture of a circle; we then have an example of recognition indeterminacy relating to shape information. However, the recognition task is generally considered to lie outside the scope of qualitative spatial reasoning. Consequently, certain questions that stand in close connection with undetermined boundaries such as noise cleaning, edge detection or ridge–ravine labelling are basic techniques of computer vision rather than spatial reasoning (Haralick and Shapiro, 1992).

Since spatial reasoning is not primarily concerned with building a qualitative description from sensor information most of the research has been devoted to another type of indeterminacy, which we will designate as *reconstruction indeterminacy*. This indeterminacy arises from the fact that the qualitative description of a spatial configuration or shape generally allows for different geometrical realizations. If, for example, a shape description consisted just of the information that *square(c)* holds for some object *c*, then a square of any size and position will realize the description (Figure 9.1). Note that the distinction between recognition and reconstruction indeterminacy can be made only with respect to the system of spatial concepts of the reasoning formalism in use.

This is how undetermined boundaries come into qualitative spatial reasoning. Reasoning starts with qualitative shape information, ideally specified by a set of formulas of some logical language. A high-level computer vision system or a natural language processing system may be thought of as producing this input. By combining the spatial information from different sources the shape of the object becomes more and more constrained as does the boundary. Note, that our use of the term shape information does not imply any commitment to the type of spatial information that is used – it may be topological ('the shape is connected') or metrical ('the shape is a disk') or other. Take the following well-known geometrical theorem as an example for a shape inference rule. The intersection of convex sets is convex. Knowing that a shape arises from the intersection of two convex shapes will allow us to infer its convexity. As even this simple example demonstrates, it is not at all necessary to completely determine the shape in order to draw inferences. We will soon see that qualitative spatial reasoning follows a principle of least commitment in the sense that inferences are made on a minimal information base.

To sum up, because of recognition indeterminacy different boundaries may correspond to the same shape concept and because of reconstruction indeterminacy a shape concept may be realized by different boundaries. Thus, a correspondence between boundary and shape will have to be conceived as many-to-many mapping. Qualitative spatial reasoning has focused on recognition indeterminacy but has also developed a way to deal with reconstruction indeterminacy, namely, the conceptual neighbourhood structures which will be described next.

9.3 Two levels of qualitative abstraction

From the point of view of naive physics spatial reasoning appears as a field dealing with the conceptual supercluster 'space', which comprises at least two subclusters. 'position' and 'shape'. The distinction between representation problems related to object position and to object shape is common not only in naive physics but also in other areas which motivated spatial reasoning research such as computer vision and robot navigation. Also, in cognitive modelling a different representation of position and shape was proposed because neuropsychological evidence suggests that the two kinds of spatial information are processed along different cortical visual pathways (Kosslyn *et al.*, 1990). We certainly cannot conclude from such results that object position and object shape are to be treated independently of each other. But it seems as if in many spatial reasoning applications shape concepts are more closely related to each other than to position concepts. Note that the conceptual cluster metaphor does not exclude the existence of a few critical relations between the clusters.

Much more attention has been devoted to reasoning about object position than to reasoning about shape – this is at least the impression that results from analyzing the literature on qualitative approaches (for an overview, see Hernandez, 1994; Freksa and Röhrig, 1993). In particular, the case of reasoning about one-dimensional position has been extensively studied, the most comprehensive theory from a mathematical point of view being that of Ligozat (1990). At first glance, the approaches appear quite different in the choices they make of relation concepts and inference rules. Nevertheless, it is possible to identify a kind of standard procedure which most of them follow in dealing with the indeterminacy of spatial information. The procedure consists in distinguishing two *levels of qualitative abstraction*, which allow us to consider first reconstruction indeterminacy and only thereafter recognition indeterminacy. We will describe the levels of abstraction by making reference to the simplest spatial reasoning task, namely inferences about the relative position of two intervals on a directed line.

Developing a spatial reasoning formalism may be viewed as a three-step process, including a preparatory step and two steps of qualitative abstraction. First, the reasoning task has to be fixed by specifying a configuration space. Next, the reasoning formalism is axiomatized, that is, a set of qualitative relations is described together with appropriate inference rules. Finally, a conceptual neighbourhood structure is defined on the qualitative relations. It will soon become clear what exactly is meant by configuration space, qualitative relations and conceptual neighbourhood structure. It should merely be noted that the qualitative relations constitute the first level of qualitative abstraction and the conceptual neighbourhood structure the second level. We will now describe these three steps.

9.3.1 Qualitative relations

Consider the example of reasoning about the relative position of two closed intervals on the real line. Let the first interval be $[x, y]$ and the second $[u, v]$; then every point (x, y, u, v) of R^4 will denote a specific position for each of the two intervals on the real line. The four-dimensional space defined by all possible interval positions is called the *configuration space* of the reasoning task. Similarly, for other task a configuration space may be specified as the space of all possible positions or shapes one wants to reason about. The central idea behind qualitative spatial reasoning is to abstract from configuration space, a goal that is usually achieved by partitioning the configuration space into equivalence classes. Figure 9.2 depicts the configuration space; for ease of visualization the interval $[u, v]$ is kept fixed to $[-1, 1]$, which reduces the space to two dimensions, x and y. In fact, only half of the configuration space is shown: the shaded areas in the Figure 9.2 correspond to the points that satisfy the additional constraint $x > y$.

Let us look at the first step of qualitative abstraction. The left diagram in Figure 9.2 shows equivalence classes which are the 0-, one- and two-dimensional cells of a cell decomposition of the configuration space. The cells are labelled by symbols: $=$ is a 0-cell, m, s, f, mi, si and fi are 1-cells, and $<$, o, d, $>$, oi and di are 2-cells. To state explicitly the equivalence relation, we introduce an auxiliary function $range(x) := r_1$ iff $x < -1$, r_2 iff $x = -1$, r_3 iff $-1 < x < 1$, r_4 iff $x = 1$, r_5 iff $x > 1$, and define

$$(x, y) \equiv (x', y') \quad \text{iff} \quad range(x) = range(x') \quad \text{and} \quad range(y) = range(y').$$

Every equivalence class with respect to this relation consists of points in R^2 that describe relative positions of the two intervals which are not distinguished on this level of qualitative abstraction. In other words, every equivalence class defines a *qualitative relation* that holds between the interval $I_1 = [x, y]$ and the interval $I_2 = [-1, 1]$. The qualitative relations obtained are the well-known 13 interval relations of Allen (1983), which are usually written in infix notation: $I_1 m I_2$, 'I_1 meets I_2', is thus just another way of expressing that (x, y) lies within the cell labelled m.

It is important not to confuse the equivalence relation with the qualitative relations. The latter are called relations only because of the special nature of the configuration space: A point of the space describes the relative position of two intervals, that is, a binary relation between spatial objects. Generally, the points of the configuration space of a spatial reasoning task describe an n-ary relation. In some cases, however, they just describe the position or the shape of a single spatial object. Often

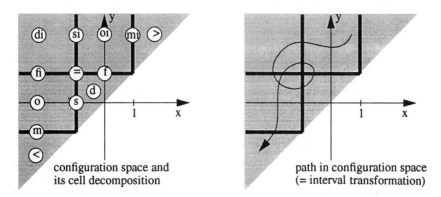

configuration space and
its cell decomposition

path in configuration space
(= interval transformation)

Figure 9.2   Cell decomposition of the configuration space.

such unary spatial relations are also referred to as qualitative predicates or qualitative concepts. Defining the set of qualitative relations constitutes the crucial conceptual issue in developing a qualitative spatial reasoning formalism. Usually this is
done without explicit reference to an equivalence relation on the configuration
space. One simply takes care that every possible position (or shape) is described by
some relation and that no position (or shape) is described by more than one relation. Both criteria are captured by the concept of equivalence class partitioning of
the configuration space.

The typical spatial inference follows the relational composition paradigm. Given
the (binary) qualitative relation between spatial objects X and Y, and between Y
and Z, what constraints are there on the qualitative relation between X and Z?
Usually, such inference rules are described in the form of a composition table such
as the one Allen (1983) established for the interval relations. Although compositions
of binary relations have mostly been studied in qualitative spatial reasoning, inference rules can also be formulated for more than two premises and can generally
involve n-ary relations. Interestingly, the efficiency of qualitative inferences relies on
reconstruction indeterminacy. The introduction of qualitative relations permits one
to reason about configurations without completely determining them, that is,
without numeric computations in the configuration space. We already mentioned
this as the principle of least commitment with reference to the set of qualitative
relations that is chosen.

9.3.2 Conceptual neighbourhood

Figure 9.2 shows a special type of equivalence relation, namely, one that gives rise
to a cell decomposition of the configuration space. This is no accident. Not every
equivalence relation provides an adequate qualitative abstraction – there is a continuity principle that we want to be satisfied. Any continuous transformation of the
interval $[x, y]$, e.g. a rigid movement or a distortion describes a path in the configuration space, as depicted in the right diagram of Figure 9.2. Knowing the actual
interval relation allows one to predict which relation will occur next; oi, for
instance, may only be followed by f, si, mi or $=$. This is what Freksa (1992) calls the
conceptual neighbourhood structure. In this sense, oi and f are said to be conceptual
neighbours. Using the topological notion of cell decomposition we can give a more
general definition of conceptual neighbourhood, one that applies to any spatial
reasoning task.

Definition 1
Let S be the n-dimensional configuration space of a spatial reasoning task and
let D be a cell decomposition of S into 0-, 1-, ..., n-cells. Two qualitative
relations are *conceptual neighbours* iff the corresponding cells are incident.

Characterizing conceptual neighbourhood is the second step of qualitative
abstraction. It permits dealing with a certain kind of recognition indeterminacy. As
an example, consider a visual sensor that provides information about the relative
position of two intervals. The intervals on the left in Figure 9.3 seem to meet.
However, sensor data are noisy, so the intervals could also be disjointed or overlap
a bit. This recognition indeterminacy can be expressed by describing the situation
by a set of conceptually neighboured qualitative relations, namely $\{<, m, o\}$.

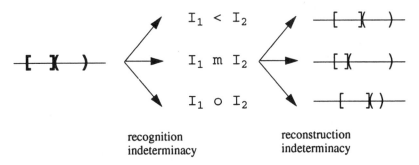

$I_1 < I_2$

$I_1 \, m \, I_2$

$I_1 \, o \, I_2$

recognition
indeterminacy

reconstruction
indeterminacy

Figure 9.3 Modelling recognition indeterminacy by a conceptual neighbourhood.

Working with sets (i.e. disjunctions) of neighboured qualitative relations has shown to be an alternative to approaches based on fuzziness – the conceptual neighbourhood structure even helps to speed up the inference process, as was shown by Freksa (1992). This combination of qualitative relations with conceptual neighbourhoods in order to deal with recognition and reconstruction indeterminacy can be seen as the essential idea of qualitative spatial reasoning (Figure 9.3).

Stating that only those equivalence relations which give rise to a cell decomposition of configuration space are acceptable for the qualitative abstraction, of course leaves us only with a necessary criterion. It does not tell us how to choose the qualitative relations so as to meet the requirements of the application we have in mind; this still belongs to the art of developing a spatial reasoning formalism. Let us summarize the qualitative approach to reasoning about shape as it will be developed in Section 9.5. First, a configuration space that adequately captures shape will have to be specified; we then look for an appropriate equivalence relation, and finally we describe the conceptual neighbourhood structure.

9.4 The gap between topological and geometric information

Spatial information is subject to recognition and reconstruction indeterminacy, as discussed in Section 9.3. This is a functional difference expressing how the information affects spatial reasoning. Usually there is also a difference in quality, which concerns the degree of determination: spatial information is distinguished according to the invariants it describes. Special consideration has been given to two transformation groups in this context: the group of homeomorphism, which defines the topological invariants, and the group of isometries, which defines the geometrical invariants. The functional and qualitative types of indeterminacy combine. In other words, recognition indeterminacy may be found for topological information as for geometric information: the same also holds for recognition indeterminacy.

From the very beginning of spatial reasoning research the need for specialized representational formalisms for topological and geometrical information was clearly seen. A separate topological representation level proved to be particularly useful in reasoning tasks where spatial information is acquired incrementally. Such is the case for a mobile robot which explores a terrain providing the system with successive pieces of partial information (e.g. McDermott and Davis, 1984; Kuipers and Levitt, 1988). Finding that a room is connected to another room constitutes a typical piece of topological information. If a topological representation level is missing, this information can only be accumulated, that is, stored without relating it to other information. Obviously, integration is preferable to accumulation because it provides the base for specific inferences such as topological path planning. Whenever it shows that some information can not be integrated into any of the representational levels provided by a spatial reasoning formalism, there is reason to ask whether introducing a new level will solve the problem. This is how we will proceed.

An example of information about object shape that can not be integrated into either a topological or a geometric representation level is convexity. Knowing that a polygon is convex does not specify any metric properties such as the length of a side. On the other hand, convexity can not be considered a topological property since it is not preserved by topological transformations. In fact, there is a gap between the topological and geometric degree of determination. The two-dimensional ordering information falls into this gap, and so does the ordering information approach to shape representation proposed in Section 9.5. Although we outline the theoretical background and do not go into representational details (e.g. efficient data structures), the following discussion of ordering information implicitly pleads for an additional representational level intermediate between the topological and the geometric levels.

9.4.1 Two-dimensional ordering information

A characteristic of spatial relations such as 'X is seen to the right of Y' or 'X lies north of Y' consists in the fact that they make reference to a system of axes, i.e. some collection of directed lines. These axes may be locally defined, as happens to be the case for the right–left distinction, or globally, as for the cardinal directions. We will call a spatial relation which is defined with reference to a system of axes an *orientation relation*. Judging from the overview article of Freksa and Röhring (1993) orientation relations are prominent in qualitative spatial reasoning: 8 of the 10 approaches reviewed make use of systems of axes in some form or another. Orientation relations are based on the distinction between the left and right sides of a line in the plane. Similarly, the one-dimensional interval relations depend on distinguishing two 'sides' of a point on a line. Although one may prefer to speak in the one-dimensional case about points being ordered instead of points having sides, there is a general connection between orientation and ordering that holds in any dimension. Analogues of the concept of linear order can be defined for points in the plane and in higher dimensions.

Different notions of higher-dimensional order have been studied in discrete and computational geometry from a combinatorial and an algorithmic perspective. General introductions into these fields are given by Edelsbrunner (1987) and Pach (1993). More specifically related to our problem is a series of works by Jacob Goodman and Richard Pollack concerned with the ordering of points in two dimensions. In a recent survey article their results are summarized and related to other topics in discrete geometry (Goodman and Pollack, 1993). Although many of their results extend to higher dimensions, the following account is restricted to two-dimensional order since we discuss only two-dimensional shape.

The *orientation of three points* lying in an oriented plane is said to be positive if visiting the points in the order p_i, p_j and p_k requires a counterclockwise turn. In that case we write $[p_1 p_2 p_3] = +$. If these visits require a clockwise turn, the orientation is said to be negative, $[p_1 p_2 p_3] = -$. For collinear points the orientation is defined to be zero, $[p_1 p_2 p_3] = 0$, which includes all situations in which incidences occur, e.g. $p_1 = p_2 = p_3$. Note that given the orientation $[p_1 p_2 p_3]$, the orientation for any permutation of the points is determined:

$$[p_1 p_2 p_3] = + \Leftrightarrow [p_1 p_3 p_2] = - \Leftrightarrow [p_2 p_1 p_3] = - \Leftrightarrow$$
$$[p_2 p_3 p_1] = + \Leftrightarrow [p_3 p_2 p_1] = - \Leftrightarrow [p_3 p_1 p_2] = +.$$

A set $P = \{p_1, \ldots, p_n\}$ of points in the plane is called a *configuration*. The *triangles* of the configuration are all its subsets constituted of three points. There are thus $\binom{n}{3}$ triangles for a configuration of n points. The simplest type of planar ordering is defined in terms of the triangles: this is the only one we need to consider.

Definition 2
Let $P = \{p_1, \ldots, p_n\}$ be a configuration then the *triangle orientations* of P are given by $\{[p_i p_j p_k] \; p_i p_j p_k \text{ is a triangle of } P\}$.

Figure 9.4 shows an example of how triangle orientations describe a configuration. The configuration depicted on the left consists of four points and has four triangles for which the following orientations are found

$$[1\ 2\ 3] = + \qquad [1\ 2\ 4] = - \qquad [1\ 3\ 4] = + \qquad [2\ 3\ 4] = +.$$

Looking at the lines defined by the points provides a simple way of visualizing how the triangle orientations constrain the relative positions of a point, $[1\ 2\ 4] = -$ can be interpreted as stating that 4 lies in the half-plane on the right side of the directed line from 1 to 2. Point 4 also appears in two other triangles, $[1\ 3\ 4] = +$ and $[2\ 3\ 4] = +$. The shaded area in the diagram depicted in Figure 9.4 corresponds to the intersection of the half-planes determined by these three triangle orientations. Moving the point 4 within the marked area will not change the triangle orientations.

The set of triangle orientations defines what Goodman and Pollack call the *order type* of the configuration. Both configurations shown in Figure 9.4 are said to be of

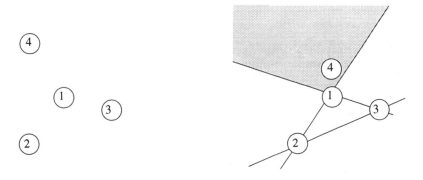

Figure 9.4   Constraints expressed by triangle orientations.

the same order type in the sense that there is a one-to-one correspondence between the points of the configurations, which preserves the orientation of the triangles. Several geometric definitions exist for order types and it is not always trivial to see their equivalence; the interested reader is referred to Goodman and Pollack (1983).

9.4.2   An example: qualitative robot navigation

Reasoning with two-dimensional ordering relations played an important role in the robot navigation task formulated by Levitt and Lawton (1990). We present this task as an example because it involves reasoning with regions whose shapes are determined on the level of ordering information, while remaining completely undetermined on the geometric level. The scenario concerns a mobile robot equipped with a visual sensor by which it identifies surrounding landmarks. It is assumed that the order in which the landmarks appear from left to right can always be recovered even if, due to measurement errors, the visual angle under which the landmarks appear can not be determined. Levitt and Lawton (1990) describe a special representational layer of their QUALNAV system, which organizes information about the visual order of landmarks. A qualitative navigation strategy uses the information to plan a path from the robot's present position to a given goal position. It should be noted that this qualitative navigation is not intended to be used as a stand-alone strategy. In fact, the QUALNAV system uses it as supplementary strategy which is more reliable, though much less accurate, than geometrical navigation.

Landmarks are considered to be points at the qualitative navigation level. In addition, it is assumed that every landmark can be seen and identified from any observer position. Two classes of positions with respect to two landmarks 1 and 2 are distinguishable: positions from which 1 is seen to the left of 2, and positions from which 1 is seen to the right of 2. If the observer's position is denoted by 0, the two classes may also be described by triangle orientations, $[0\ 1\ 2] = -$ and $[0\ 1\ 2] = +$. Increasing the number of landmarks will increase the number of qualitative positions that can be distinguished by triangle orientations. Figure 9.5 shows a configuration of landmarks, together with the lines defined by them. The lines decompose the plane into regions which correspond to the distinguishable qualitative observer positions.

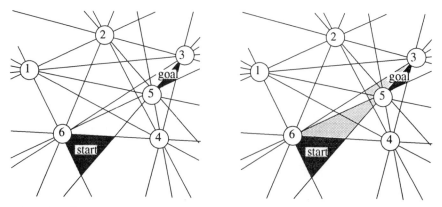

Figure 9.5   A qualitative navigation task and one of its solutions.

The qualitative navigation task consists in planning a path between a start and goal region, both specified in a qualitative map that encodes the configuration of landmarks in terms of ordering information. Note that Figure 9.5 only shows one of infinitely many realizations of the same qualitative map. As our interest is in reasoning about shape we will not go into the details of path planning. One crucial point is worth noting, however. Successful navigation depends on finding the right characterization of the qualitative observer positions in terms of stable visual properties. For this purpose, the QUALNAV system simply uses the circular order in which the landmarks appear, e.g. $4 > 5 > 6 > 1 > 2 > 3$ for the goal region in Figure 9.5. As detailed in Schlieder (1993), this encoding fails to capture the full ordering information. It is only by analyzing the ordering information involved in the problem that the connection to the theory of order types becomes obvious and an optimal pathfinding algorithm can be formulated.

Shape appears in the navigation task insofar as the qualitative map determines the regions of equivalent observer positions. What constrains these regions can be expressed in terms of two-dimensional ordering relations. Figure 9.6 shows two different realizations of the qualitative map for a configuration of four landmarks. The dark shaded regions will be triangular in all realizations; the constraints imposed by the triangle orientations are even stronger, since they allow us to describe how the region's shape will alter when the points of the configuration move. However, no metric property of the triangle can be inferred. We thus find that observer positions which can be distinguished with respect to the visual order of some landmarks constitute examples of shapes that are determined on the ordering information level. Such regions appear naturally in all kinds of problems related to questions of the visibility of objects and may therefore be relevant not only to navigation but also to fields such as architecture. An example from regional planning is provided by Zewe and Koglin (1994), who address the problem of the visibility of power lines in a landscape on a geometric level.

The indeterminacy we encounter in the navigation task is reconstruction indeterminacy of the very same kind as that found in the interval relation example. We may think of the map's regions as being qualitative relations describing an observer's position relative to some landmarks. Abstracting from configuration space, that is, from the geometry of the problem, provides once more the basis for efficient

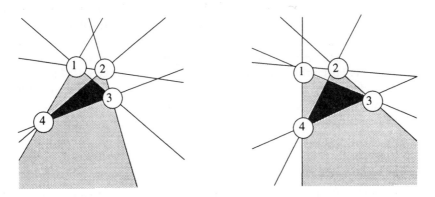

Figure 9.6 Indeterminacy of two-dimensional ordering information.

inference processes. The path-planning algorithm described in Schlieder (1993) may serve as a demonstration for this claim. Remembering that the qualitative relations are the map's regions it is not difficult to visualize the conceptual neighbourhood structure. It simply consists in the incidence relations between regions. The light shaded regions in Figure 9.6 are the conceptual neighbours of the central triangular region. We will now see how more complex, i.e. less constrained shapes can be described by ordering information.

9.5 The ordering information approach to shape representation

The navigation example demonstrates that there are shapes whose boundaries are undetermined without being fuzzy. Such shapes are the ideal candidates for a qualitative shape description which applies the process of qualitative abstraction described in Section 9.3 to inferences with two-dimensional ordering relations. The qualitative observer positions that were identified in the qualitative navigation example constitute a rather constrained class of shapes. They all are convex, for instance. In a more realistic navigation scenario, which allows extended objects to appear as landmarks or obstacles, descriptions of more complex shapes become a necessity.

9.5.1 Qualitative shape relations

The two-dimensional ordering relations are defined for points. As a consequence, shape description will have to make use of some reference points. The reference points are to be thought of as parameters which completely specify the boundary, that is, they constitute the inputs to some curve-fitting technique which produces the boundary. Not knowing the exact location of the reference points will thus cause the boundary to be undetermined in a way that depends on the interpolation algorithm used. For polygonal boundaries it is natural to choose the vertices as reference points. Since our interest is not in curve fitting, we restrict qualitative shape description to this case. Boundaries will always be polygons linking the reference points in the order in which they are numbered.

The central idea of the *ordering information approach to reasoning about shape* consists in constraining the location of the reference points by specifying triangle orientations. It is not necessary to completely determine the order type of the configuration formed by the reference points. Often, for the purpose of qualitative inferences it suffices to fix some of the triangle orientations. See Figure 9.7 for an example of how a shape is more and more constrained by adding ordering information until it finally falls into a class of shapes that may be characterized as Z-shapes.

Each of the following triangles consists of successive vertices of a hexagon. The triangle orientations may be interpreted as describing the concavities and convexities of the boundary.

$$[1\ 2\ 3] = -\qquad [2\ 3\ 4] = +\qquad [3\ 4\ 5] = -$$
$$[4\ 5\ 6] = -\qquad [5\ 6\ 1] = +\qquad [6\ 1\ 2] = -$$

A positive orientation $[i\ j\ k] = +$ corresponds to a concave vertex j, and a negative orientation to a convex vertex (assuming a clockwise vertex numbering). Many

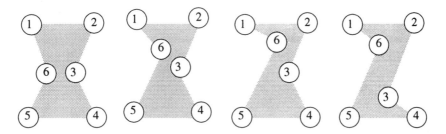

Figure 9.7 Successive specifications of a shape.

shapes satisfy these triangle orientations; for instance, all of the polygons shown in Figure 9.7. Taking into account the triangle formed by the points 1, 4 and 6 allows us to distinguish the first polygon from the others, for which [1 4 6] = +. Similarly, adding successively [1 3 6] = + and [1 4 6] = − leads to more constrained qualitative shape descriptions for which the third and fourth polygons provide realizations. The whole set of nine triangle orientations may be considered to represent qualitatively the class of Z-shaped hexagons.

9.5.2 A system of qualitative shape relations

Having presented the central idea behind shape description at the ordering information level, we will now combine it with the framework of qualitative abstraction in order to reason about shape. All three steps – the specification of the configurations space, the definition of the qualitative relations and the characterization of the conceptual neighbourhood structure – will be formulated explicitly. In particular, we want to give a complete overview of the qualitative relations and therefore choose a small and simple shape reasoning task: changes in the shape of quadrilaterals.

The reference points are the quadrilateral's four vertices. The position of a reference point is specified by its x- and y-coordinates, and we find that each possible configuration C belongs to an eight-dimensional *configuration space* S:

$$C = (x_1, y_1, \ldots, x_4, y_4) \in S = R^8.$$

The orientation of a triangle formed by points i, j and k may easily be expressed in terms of coordinates:

$$[i\ j\ k] = \text{sgn} \begin{vmatrix} x_i & x_j & x_k \\ y_i & y_j & y_k \\ 1 & 1 & 1 \end{vmatrix}.$$

There are four triangles that can be formed with the reference points. We will arrange the triangle orientations associated with a configuration $C \in S$ in form of a vector

$$t(C) = ([1\ 2\ 3], [1\ 2\ 4], [1\ 3\ 4], [2\ 3\ 4]).$$

To qualitatively abstract from the configuration space S, the following equivalence relation on S is introduced:

$C \equiv D$ iff $t(C) = t(D)$,

that is, two configurations are equivalent iff they have the same triangle orientations. Note that this equivalence relation does not allow a relabelling of the points as does the definition of order types mentioned in Section 9.4.

Definition 3
The qualitative relations of the quadrilateral shape reasoning task are the equivalence classes r_i of S with reference to $\equiv$. For $C \in r_i$ we also write $r_i(C)$.

Having introduced the equivalence relation only as a means to define the qualitative relations we will prefer the notation $r_i(C)$, which emphasizes that r_i is viewed as an unitary shape relation (or shape predicate, shape concept). What interests us next is to give an overview of the qualitative relations. How many of them are there?

This question may be answered by enumerating the equivalence classes, which amounts to finding all different triangle orientation vectors. However, not all of the 3^4 sign patterns that arise from combining $-$, 0 and $+$ orientations for the four triangles are possible. The triangle orientation vectors $v = (- + - +)$ for instance, can not be realized. Starting with any triangle for which $[1\ 2\ 3] = -$, one will find that the half-planes that correspond to possible locations of point 4 according to the other three triangle orientations have an empty intersection – the point can thus not be placed. Obviously, choosing an orientation for a triangle puts constraints on the other triangle's orientations. Axiomatizing this constraint leads to a characterization of all possible sign patterns in terms of an algebraic structure called an oriented matroid (Björner *et al.*, 1993). We do not need to go into this theory for the purpose of our example, as we would if we were treating configurations of more than four points.

In enumerating the equivalence classes we will make a further restriction, namely, that to non-zero orientations. In other words, we are considering only quadrilaterals with vertices in general positions (no three points on a line). The only reason for this restriction is to reduce further the number of qualitative relations in order to yield a number in which they may be all presented in a table (Figure 9.8).

From the 16 combinatorially possible triangle orientation vectors, two are impossible to realize. The qualitative relations corresponding to the remaining 14 vectors have been numbered in Figure 9.8 by interpreting the sign pattern as a binary number, e.g. $(- + + -) = 0110 = 6$. In Figure 9.8 there are four non-simple polygonal shapes whose sides intersect: 3, 6, 9, 12. Two of the quadrilaterals, 0 and 15, are convex, the remaining eight are concave. Each shape has a mirror image. Since a reflection causes all triangle orientations to be switched, the mirror image of shape i will be the shape $15 - i$.

9.5.3 Reasoning about changes in shape

The restriction of non-zero orientations signifies that only the eight-dimensional cells of the partitioning of the configuration space are considered. We will define the conceptual neighbourhood for qualitative relations accordingly. This is to say, 8-cells will be neighboured, although in fact they are separated by lower-

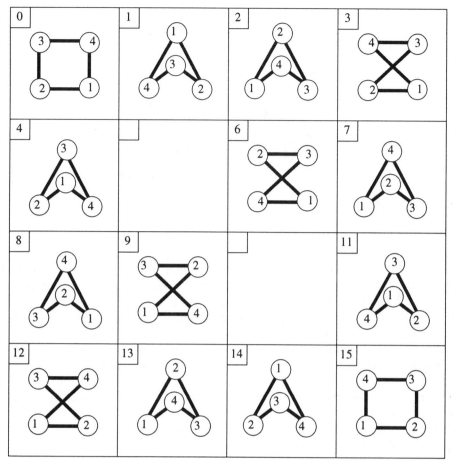

Figure 9.8 Qualitative relations for quadrilateral shapes.

dimensional cells. To each qualitative relation r_i there is an associated triangle orientation vector, denoted t_i. Using the Hamming distance $d(t_i, t_j)$ between two vectors t_i and t_j, which simply gives the number of components in which they differ, we will define the conceptual neighbourhood structure.

Definition 4

Two qualitative relations r_i and r_j of the quadrilateral shape reasoning task are said to be conceptual neighbours $n(r_i, r_j)$ iff $d(t_i, t_j) = 1$.

The conceptual neighbourhood structure resulting from this definition is depicted in Figure 9.9. Also shown are the two impossible shapes that would correspond to the numbers 5 and 10 in Figure 9.8. All edges denote Hamming distances of 1 between two triangle orientation vectors, but only the heavy edges indicate conceptual neighbourhood. Thus the conceptual neighbourhood structure appears to be embedded in the hypercube formed by the distances between all combinatorially possible vectors.

Reasoning about changes in shape will rely on the conceptual neighbourhood structure. Let us consider an example. For a quadrilateral whose vertices form a

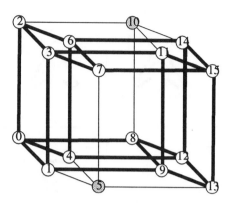

Figure 9.9   Conceptual neighbourhood structure.

configuration C, it is known that $r_1(C)$ holds. The quadrilateral thus has a concave vertex 3. A transformation of the shape occurs, after which $r_8(C)$ holds. That is, the shape now has its concavity at a different vertex, namely 2. As the quadrilateral in this case happens to be a solid shape, not just a wire frame, we want to know whether this transformation is physically plausible, i.e. whether it can occur without a non-simple shape appearing as an intermediate result. This question can be answered by inspecting the conceptual neighbourhood structure: 1–0–8 is a transformation which in this sense is physically plausible. Even more general inferences can be reduced to this type of ordering information reasoning about shape. The conceptual neighbourhood structure reveals, for instance, that in order to transform a shape into its mirror image one has to pass through a physically implausible shape. In other words, performing a reflection requires that the shape is twisted around itself (or moved outside the plane).

9.6   Conclusions

The quadrilateral relations described in this chapter constitute just a simple example of qualitative reasoning about shape. More interesting than the inference formalism itself is the method used to state the formalism, i.e. in the process of qualitative abstraction. Although qualitative approaches have been applied successfully to several spatial reasoning problems, the underlying process of qualitative abstraction has never been stated explicitly. Consequently, we have provided a detailed description of qualitative abstraction. We described qualitative abstraction as a two-step process – the first step consists in defining an appropriate set of qualitative relations, and the second step in deriving the conceptual neighbourhood structure.

Qualitative approaches have been used mostly in the context of reasoning about object positions. Making the steps of qualitative abstraction explicit allows us to apply the same method to reasoning about shape. We have shown how qualitative abstraction can be applied to a special type of spatial information, which we call ordering information, and provided a simple example of this type of inference: reasoning about quadrilateral shape. The interest in ordering information for shape description lies in the fact that it is less constrained than metric information but is more constrained than topological information.

By combining qualitative abstraction and ordering information a whole class of formalisms for reasoning about shape may be obtained. The most prominent characteristic of these formalisms is that they deal with boundaries in a quite different way than fuzzy models do. Qualitative formalisms are best suited for crisp boundaries with unknown locations. Such boundaries arise in connection with man-made geographical entities found in application areas such as town planning or architecture. Genuine fuzzy boundaries, such as boundaries between soil types, seem to be less easy to handle by qualitative abstraction. Fuzzy models, on the other hand, have difficulties with crisp but unknown boundaries. We should thus consider fuzzy modelling and qualitative abstraction as complementary rather than competing methods for describing the boundaries of geographic entities.

REFERENCES

ALLEN, J. (1983). Maintaining knowledge about temporal intervals, *Communications of the ACM*, **26**, 832–843.

BJÖRNER, A., LAS VERGNAS, M., STURMFELS, B., WHITE, N. and ZIEGLER, G. (1993). *Oriented Matroids*, Cambridge: Cambridge University Press.

CHEN, S. (Ed.) (1990). *Advances in Spatial Reasoning*, Norwood, NJ: Ablex.

COHN, A., RANDELL, D. and CUI, Z. (1993). A taxonomy of logically defined qualitative spatial relations, in: Guarino, N. and Poli, R. (Eds), *Proc. Int. Workshop on Formal Ontology*, Padova.

EDELSBRUNNER, H. (1987). *Algorithms in Combinatorial Geometry*, Berlin: Springer.

FREKSA, C. (1992). Temporal reasoning based on semi-intervals, *Artificial Intelligence*, **54**, 199–227.

FREKSA, C. and RÖHRIG, R. (1993). Dimensions of qualitative spatial reasoning, in: Piera Carreté, N. and Singh, M. (Eds), *Qualitative Reasoning and Decision Technologies*, pp. 483–492, Barcelona: CIMNE.

GOODMAN, J. and POLLACK, R. (1983). Multidimensional sorting, *SIAM J. Comput.*, **12**, 484–507.

GOODMAN, J. and POLLACK, R. (1993). Allowable sequences and order types in discrete and computational geometry, in: Pach, J. (Ed.), *New Trends in Discrete and Computational Geometry*, pp. 103–134, Berlin: Springer.

HARALICK, R. and SHAPIRO, L. (1992). *Computer and Robot Vision*. Reading, MA: Addison-Wesley.

HAYES P. (1985). The second naive physics manifesto, in: Hobbs, J. and Moore, R. (Eds) *Formal Theories of the Commonsense World*, pp. 1–36, Norwood, NJ: Ablex.

HERNANDEZ, D. (1994). *Qualitative Representation of Spatial Knowledge*, Berlin: Springer.

KAK, N. (Eds) (1988). Spatial reasoning (special issue). *AI Magazine*, **9**.

KOSSLYN, S., FLYNN, R., AMSTERDAM, J. and WANG, G. (1990). Components of high-level vision: A cognitive neuroscience analysis and accounts of neurological syndromes, *Cognition*, **34**, 203–277.

KUIPERS, B. and LEVITT, T. (1988). Navigation and mapping in large-scale space, *AI Magazine*, **9**, 25–43.

LEVITT, T. and LAWTON, D. (1990). Qualitative navigation for mobile robots, *Artificial Intelligence*, **44**, 305–360.

LIGOZAT, G. (1990). Weak representations of interval algebras, in: Dietrich, T. and Swartout, W. (Eds), *Proc. AAAI-90*, pp. 715–720.

MCDERMOTT, D. (1992). Reasoning, spatial, in: Shapiro, S. (Eds), *Encyclopedia of Artificial Intelligence*, 2nd edn., pp. 1322–1334, New York: Wiley.

McDermott, D. and Davis, E. (1984). Planning routes through uncertain territory, *Artificial Intelligence*, **22**, 107–156.

Pach, J. (Ed.) (1993). *New Trends in Discrete and Computational Geometry*, Berlin: Springer.

Schlieder, C. (1993). Representing visible locations for qualitative navigation, in: Piera Carrete, N. and Singh, M. (Eds), *Qualitative Reasoning and Decision Technologies*, pp. 523–532, Barcelona: CIMNE.

Zewe, R. and Koglin, H. (1994). 3DOG: Ein Verfahren zur Beurteilung der Sichtbarkeit von Hochspannungsfreileitungen, in: Englert, G. (Ed.), *Proc. 1st Workshop on Visual Computing*, pp. 251–272, Darmstadt: Fraunhofer Institut für Graphische Datenverarbeitung.

CHAPTER TEN

Modelling Spatial Objects with Undetermined Boundaries using the Realm/ROSE Approach

MARKUS SCHNEIDER

Praktische Informatik IV, FernUniversität Hagen, Hagen, Germany

10.1 Introduction

The diversity of geometric applications has led to many proposals both for the modelling of spatial data and for the design of new data models and query languages integrating traditional alphanumeric data as well as geometric data. In the literature the general opinion prevails that special data types are necessary to model geometry and to enable geometric data to be efficiently represented in database systems. These data types are commonly denoted as *spatial* or *geometric data types* (SDTs), such as, for example, *point, line* or *region*. We speak of *spatial objects* as occurrences of spatial data types. Thus, we take an entity-oriented view of spatial phenomena. The definition of spatial data types and operations expressing the spatial semantics visible at the user level and the mechanisms for providing them to the user are to a high degree responsible for the design of a spatial data model and the performance of a spatial database system being the basis of GIS and have a great influence on the expressiveness of spatial query languages. This is true regardless whether we consider spatial objects with sharp or undetermined boundaries and whether a DBMS uses a relational, complex object, object-oriented, or some other data model. Hence, the definition and implementation of spatial data types is probably the most fundamental issue in the development of spatial database systems.

Spatial data types modelling objects with sharp boundaries are routinely used in the descriptions of spatial query languages (e.g. Egenhofer, 1989; Güting, 1988; Joseph and Cardenas, 1988; Lipeck and Neumann, 1987; Svensson and Huang, 1991) and have been implemented in some prototype systems (e.g. Güting, 1989; Orenstein and Manola, 1988; Roussopoulos *et al.*, 1988), even if only a few formal definitions have been given for them (Güting, 1988; Güting and Schneider, 1993, 1995; Gargano *et al.*, 1991; Scholl and Voisard, 1989). For spatial objects with undetermined boundaries analogous approaches, especially formal ones, are unknown to the author.

The treatment of spatial objects with undetermined, vague or blurred boundaries is especially problematic for computer scientists confronted with the difficulties of how to model such objects in their database system, how to finitely represent them in a computer format, how to develop spatial index structures for them, and how to draw them. They are accustomed to the abstraction process of simplifying spatial phenomena of the real world to simply structured, manageable, and sharply bounded objects of Euclidian geometry such as points, lines, and regions. On the other hand, this abstraction process itself, mapping reality onto a mathematical model, introduces a certain kind of vagueness and imprecision.

Spatial objects with undetermined boundaries are difficult to model and so far have rarely been supported in spatial database systems. Two categories of vagueness and indeterminacy concerning spatial objects can be distinguished. *Uncertainty* relates either to a lack of knowledge about the position and shape of an object with an existing, real border, or to the inability to measure such an object precisely, *Fuzziness* describes the vagueness of objects which certainly have an extent, but which inherently do not have a precisely definable border.

At least three alternatives are conceivable as general design methods for the modelling of spatial objects with undetermined boundaries: (a) fuzzy models (Banai, 1993; Dutta, 1991; Heuvelink and Burrough, 1993; Leung *et al.*, 1992; Leung and Leung, 1994), (b) probabilistic models (Finn, 1993), and (c) transfer and extension of data models, methods and concepts for spatial objects with sharp boundaries to spatial objects without clear boundaries. In this chapter we pursue the third approach and extend the Realm/ROSE model (Güting and Schneider, 1993, 1995) as an algebraic model for handling spatial objects with sharp boundaries to a model for spatial objects with undetermined boundaries which contemporaneously obeys general criteria for the design of spatial data types and which preserves the properties of the Realm/ROSE model. The idea is to consider determined zones surrounding the undetermined borders of the object and expressing its minimal and maximal extent. The zones serve as a description and separation of the space that certainly belongs to the region object, and the space that is certainly outside. Similar approaches have been pursued by Clementini and Di Felice, and Cohn and Gotts (Chapters 11 and 12, this volume) who presuppose some kind of zone concept for the modelling of vague spatial objects. But, in contrast to this chapter, they are mainly interested in classifications of topological relationships between vague spatial objects and not in a precise formal modelling of the objects themselves.

Section 10.2 introduces general criteria for the design of spatial data types regardless of the determinacy or indeterminacy of its objects. Section 10.3 sketches the Realm/ROSE model, and Section 10.4 shows how this concept can be used to formally model general region objects with undetermined boundaries having good closure properties.

10.2 General criteria for the design of spatial data types

General design criteria for spatial data types are stated which are considered to be relevant for the modelling of spatial objects and which are valid regardless of whether we consider objects with sharp or undetermined boundaries. The current modelling approaches for spatial objects with determined boundaries only partially follow these criteria. Within the framework of the Realm/ROSE model we have attempted to take all these criteria into account and to offer a satisfactory solution in a single model. The design criteria are in detail:

- *Generality.* It should be feasible to model spatial objects as occurrences of SDTs as general as possible. A line object should be able to model the ramifications of the Nile delta, for example. A region object should be able to represent a collection of disjoint areas, each of which may have holes. This allows, for instance, the German state of Niedersachsen including the state of Bremen and the offshore islands in the North Sea to be modelled as one object.

- *Closure properties.* The domains of spatial data types like *point*, *line* and *region* must be closed under union, intersection, and difference of their underlying point sets. This allows the definition of powerful data type operations with good closure properties. By observing this criterion we can avoid geometric anomalies which can occur, for instance, when conventional operations in set theory and point set topology are carried out, although for this case the problem has been solved by regularized operations (Tilove, 1980).

- *Rigorous definition.* The semantics of SDTs, that is, the possible values for the types and the functions associated with the operations, require *formal*, clear and unique definitions in order to avoid ambiguities for both the user and the implementor.

- *Finite resolution, numerical robustness and topological correctness.* The formal definitions must take into account the *finite representations available in computers*. This has so far been neglected in definitions of SDTs. It is left to the programmer to close the gap between theory and practice, which leads rather inevitably not only to numerical but also to topological errors.

- *Geometric consistency.* Distinct spatial objects may be related through geometric consistency constraints (e.g. adjacent regions have a common boundary, or two lines meet at a point). The definition of SDTs must offer facilities to enforce such consistency.

- *Extensibility.* Even though the designer of a spatial database system may provide a good collection of spatial data types and operations, there will always be applications requiring further operations on existing types, or requiring new types with new operations. A type system should therefore be extensible for new data types.

- *Data model independence.* Spatial data types as such are rather useless; they need to be integrated into a DBMS data model and query language. However, a definition of an SDT should be valid regardless of a particular DBMS data model and therefore should not depend on it. Instead, the SDT definition should be based on a general abstract interface to the DBMS data model.

These design criteria have to be transferred to and realized at the implementation level when constructing spatial database systems.

10.3 The Realm/ROSE model: An informal overview

In this section we present a short, intuitive and informal overview of the realm concept and the ROSE (RObust Spatial Extension) algebra, both of which support an entity-oriented view of spatial reality and which were originally only planned for spatial objects with sharp boundaries. Formal definitions are omitted here, and the reader interested in details is referred to Güting and Schneider (1993, 1995).

A *realm* used as a basis for spatial data types is essentially a finite set of points and *non-intersecting* line segments over a discrete domain (Figure 10.1) and from a graph-theoretical point of view can be viewed as a planar graph over a finite resolution grid. Intuitively, it describes the complete underlying geometry of an application. All spatial objects such as points, lines and regions can be defined in terms of points and line segments present in the realm. In fact, in such a database spatial objects are never created directly, but only by selecting some suitable realm objects and composing them to spatial objects. They are never updated directly, but updates are performed on the realm and from there propagated to the dependent spatial objects. Hence, all spatial objects occurring in a database are *realm-based*.

Figure 10.2 shows some spatial objects defined over the realm of Figure 10.1. The realm-based spatial data types are called **points, lines** and **regions**, and are the sorts (types) of the ROSE algebra. Hence, A and B represent **regions** objects, C is a **lines** object, and D a **points** object. One can imagine A and B as belonging to two adjacent countries, C as representing a river, and D a city.

The underlying grid of a realm arises simply from the fact that numbers have a finite representation in computer memory. In practice, these representations will be of fixed length and correspond to INTEGER or REAL data types available in pro-

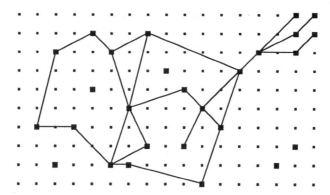

Figure 10.1   An example of a realm.

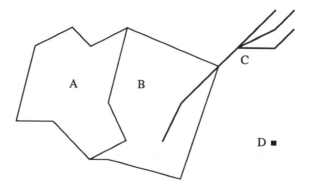

Figure 10.2   Spatial objects defined over the realm of Figure 10.1.

gramming languages. Of course, the resolution selected for a concrete application will be much finer than can be shown in Figure 10.1.

The Realm concept as a basis of spatial data types serves the following purposes:

- It guarantees good *closure properties* for the computation with spatial data types above the realm. The algebraic operations for the spatial data types are defined in a way to construct only spatial objects that are realm-based as well. For example, the intersection of region B with line C (the part of river C lying within country B) is also a realm-based **lines** object. So the spatial algebra is closed with respect to a given realm. This means in particular that no two objects of spatial data types occurring in a geometric computation have 'proper' intersections of line segments. Instead, two initially intersecting segments have already been split at the intersection point when they were entered into the realm. One could say that any two intersecting SDT objects have already 'become acquainted' when they were entered into the realm. This is a crucial property for the correct and efficient implementation of geometric operations.

- It shields geometric computation in query processing from numeric correctness and robustness problems. This is because such problems arise essentially from the computation of intersection points of line segments which normally do not lie on the grid (see Note 1). With realm-based SDTs, *no new intersection points have to be computed* in query processing. Instead, the numeric problems are treated *below* and *within* the realm level, i.e. whenever updates are made to a realm.

- It provides the programmer with a precise specification on all levels of the model that directly lends itself to a correct implementation. In particular, this means that the spatial algebra obeys algebraic laws precisely in theory as well as in practice.

- It enforces *geometric consistency* of related spatial objects. For example, the common part of the borders between countries A and B is exactly the same for both objects.

Certain structures can be constructed in a realm that serve as a basis for the definition of SDTs. Let us view a realm as a planar graph. Then an *R-cycle* is a cycle of this graph. An *R-face* is an *R*-cycle possibly enclosing some other disjoint *R*-cycles corresponding to a region with holes. An *R-unit* is a minimal *R*-face. These three notions support the definition of a **regions** data type. An *R-block* is a maximal connected component of the realm graph; it supports the definition of a **lines** data type. For all of these *realm-based structures* predicates are defined to describe their possible topological relationships.

The ROSE algebra contains very general data types **points**, **lines** and **regions** (Figure 10.3). Let *R* be a realm. Then a **points** object is a set of *R*-points. There are two alternative views of **lines** and **regions**. The first 'flat' view is that a **lines** object is a set of *R*-segments and a **regions** object is a set of *R*-units. The other 'structured' view is equivalent but 'semantically richer'. A **lines** object is a set of disjoint *R*-blocks and a **regions** object a set of (edge-) disjoint *R*-faces. For example, it is now possible to represent the whole area of a state, including islands or separate land areas, in a single **regions** object, or a complete highway network in a single **lines** object. Especially, the modelling of 'regions with holes' is now possible.

The definition of these data types guarantees very good closure properties. They are closed under the geometric operations *union, intersection* and *difference* with

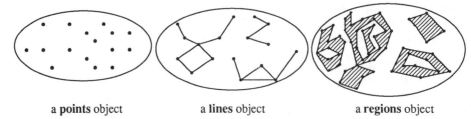

a **points** object   a **lines** object   a **regions** object

Figure 10.3   Examples of spatial data type objects of the ROSE algebra.

regard to the same realm. That is, the result of such an operation is a realm-based object as well and corresponds to the definitions of the spatial data types informally given above. The validity of the closure properties is based on the reduction of the geometric operations to the corresponding set-theoretic ones.

The spatial operations of the ROSE algebra (Güting and Schneider, 1995) are divided into four classes. Note that the last group of operations manipulates not only spatial objects but also the geographical objects with which they are associated. The classes are:

- spatial predicates expressing topological relationships (e.g. **inside, adjacent, disjoint**),
- operations returning atomic spatial objects (e.g. **intersection, contour, plus, minus**),
- operations returning numbers (e.g. **length, distance, diameter, area**),
- operations on sets of geographical objects (e.g. **overlay, fusion, closest, decompose**).

10.4 Using the Realm/ROSE approach for modelling spatial objects with undetermined boundaries

An extension of the Realm/ROSE model to spatial objects with undetermined boundaries leads to very general data types **vpoints**, **vlines** and **vregions**, where the prefix 'v' stands for the term 'vague', which unifies the two categories of uncertain and fuzzy spatial objects. These vague objects are to be defined by 'sharp' means using some concepts and definitions of the Realm/ROSE model. Within the framework of this chapter we confine ourselves to the formal treatment of general *regions with undetermined boundaries* or *vague regions* with possibly existing *vague holes* and to the treatment of their closure properties.

The central idea is to approximate each of the undetermined boundaries of a region object, that is, its outer boundary line and the boundary lines of each of its possibly existing holes, by *zones* modelling, a kind of 'irregular spatial interval' which we call *border zone* and *hole zone*, respectively (Figure 10.4). A border zone is modelled by two or more simple cycles, one representing its *outer border* and one or more representing its *inner border(s)*. A hole zone is modelled by two simple cycles representing its *inner border* and its *outer border*. Matching inner and outer borders surround the undetermined borders of the outer boundary line and the holes on both sides, so that for a vague region border zone and hole zones express the vagueness of the real, undetermined boundary lines which lie somewhere between the outer border and the inner border(s) of the zones. For a zone the area of its

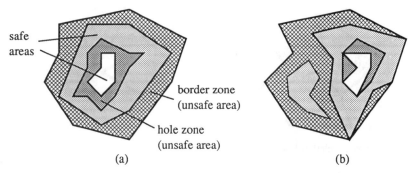

safe areas

border zone
(unsafe area)

hole zone
(unsafe area)

(a)   (b)

Figure 10.4   (a) A vague region with a vague hole and corresponding zones; (b) note that the inner border of a border or hole zone may have common segments and common points with its outer border and that there can be more than one inner border for border zones.

inner border(s) (the 'safe' area) is always contained in the area of its outer border. All the zones together serve as a description and separation of the space that certainly belongs to the vague region and the space that is certainly outside. Hence, the maximal extent of a vague region is given by the outer border of its outer boundary line and the inner borders of its holes. The minimal extent is given by the inner border(s) of its outer boundary line and the outer borders of its holes.

As an example, we can consider a lake which has a minimal water level in dry periods, and a maximal water level in rainy periods. In dry periods there can be puddles where there is no water. Small islands in the lake which are less flooded in dry periods and more flooded in rainy periods can then be modelled as vague holes. Even in rainy periods an island is never flooded completely. The example shows that this modelling approach is suitable for describing vague regions, and that it corresponds with the user's conceptual view and intuition of spatial vagueness.

We now take a more formal view of vague regions and use concepts of the Realm/ROSE model. Let $N = \{0, \ldots, n-1\} \subseteq \mathbb{N}$. An N-point is defined as a pair $(x, y) \in N \times N$. An N-segment is a pair of distinct N-points (p, q). P_N denotes the set of all N-points and S_N the set of all N-segments. Two N-segments *meet* if they have exactly one end point in common.

An *R-cycle* is a cycle in the graph interpretation of a realm, defined by a set of R-segments $S(c) = \{s_0, \ldots, s_{m-1}\}$, such that

(i) $\forall i \in \{0, \ldots, m-1\} : s_i$ *meets* $s_{(i+1) \bmod m}$.
(ii) No more than two *segments* from $S(c)$ meet in any point P.

Cycle c partitions the set P_N into three subsets, $P_{in}(c)$, $P_{on}(c)$, and $P_{out}(c)$, or R-points lying inside, on and outside c. Let $P(c) := P_{on}(c) \cup P_{in}(c)$. Cycles are interesting because they are the basic entities over realms for the definition of objects with a spatial extent. The relationships that may be distinguished between two R-cycles c_1 and c_2 are shown in Figure 10.5.

The following terminology is used for these configurations: c_2 is *(area-)inside* (i, ii, iii), *edge-inside* (ii, iii), or *vertex-inside* (iii) c_2. c_1 and c_2 are *area-disjoint* (iv, v, vi), *edge-disjoint* (v, vi), or *(vertex-)disjoint* (vi).

The meaning is that (i) c_2 is (with reference to *area*) inside c_1; (ii) in addition, it has no common edges with c_1; and (iii) it does not even have common vertices with

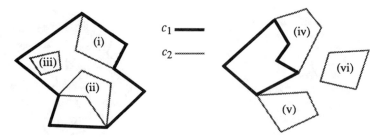

Figure 10.5   Possible relationships between two R-cycles.

c_1. Similarly, (iv) c_2 is *disjoint* with reference to *area* with c_1, (v) in addition, it has no common edges with c_1, (vi) and does not even have common vertices with c_1. *area-inside* is the standard interpretation of the term *inside*, and *vertex-disjoint* is the standard interpretation of the term *disjoint*. Formally, these predicates are defined as follows:

c_1 *(area-)inside* c_2 $: \Leftrightarrow P(c_1) \subseteq P(c_2)$

c_1 *edge-inside* c_2 $: \Leftrightarrow c_1$ *area-inside* $c_2 \land S(c_1) \cap S(c_2) = \varnothing$

c_1 *vertex-inside* c_2 $: \Leftrightarrow c_1$ *edge-inside* $c_2 \land P_{on}(c_1) \cap P_{on}(c_2) = \varnothing$

c_1 and c_2 are *area-disjoint* $: \Leftrightarrow P_{in}(c_1) \cap P(c_2) = \varnothing \land P_{in}(c_2) \cap P(c_1) = \varnothing$

c_1 and c_2 are *edge-disjoint* $: \Leftrightarrow c_1$ and c_2 are *area-disjoint* $\land S(c_1) \cap S(c_2) = \varnothing$

c_1 and c_2 are *(vertex-)disjoint* $: \Leftrightarrow c_1$ and c_2 are *edge-disjoint* $\land P_{on}(c_1) \cap P_{on}(c_2) = \varnothing$

Based on the concept of R-cycles, the notions *R-face* and *R-unit* are introduced to describe regions from two different perspectives and which are used equivalently. Both of them essentially define polygonal regions with holes. An R-unit is a 'minimal' R-face in the sense that any R-face within the R-unit is equal to the R-unit. Hence R-units are the smallest regional entities that exist over a realm.

An *R-face* f is a pair (c, H), where c is an R-cycle and $H = \{h_1, \ldots, h_m\}$ is a (possibly empty) set of R-cycles such that the following conditions hold (let $S(f)$ denote the set of segments of all cycles of f).

(i) $\forall i \in \{1, \ldots, m\} : h_i$ *edge-inside* c.
(ii) $\forall i, j \in \{1, \ldots, m\}, i \neq j : h_i$ and h_j are *edge-disjoint*.
(iii) Each cycle in $S(f)$ is either equal to c or to one of the cycles in H (no other cycle can be formed from the segments of f).

The last condition ensures uniqueness of representation, that is, there are no two different interpretations of a set of segments as sets of faces. The grid points belonging to an R-face f are defined as

$$P(f) := P(c) \backslash \bigcup_{i=1}^{m} P_{in}(h_i).$$

Let $F(R)$ denote the set of all possible R-faces, and let $U(R)$ denote the set of all R-units for a realm R. The equivalence of two representations of a region over a realm is formally established (Güting and Schneider, 1993) as a set of (pairwise) edge-disjoint R-faces, and as a set of area-disjoint R-units. Operations called *faces* and *units* are defined to convert between the two formal representations. Hence the equivalence can be expressed as: $\forall F \subseteq F(R) : faces(units(F)) = F$. The operation *units* is defined as $units(F) := \{u \in U(R) | \exists f \in F : u \ area\text{-}inside \ f \}$. The operation *faces* basically works as follows: From a given set of area-disjoint R-units, its multiset of boundary segments is formed. Then, all segments that occur twice are removed. The remaining set of segments defines uniquely a set of edge-disjoint R-faces.

Now we are able to formally define vague regions. Two equivalent definitions are conceivable, each of which expresses a slightly different conceptual view. The first definition supports the zone concept, whereas the second definition emphasizes the maximal and minimal extent of a vague region.

Let $C = (c^{out}, C^{in})$ denote a pair of a single R-cycle c^{out} and a non-empty set of R-cycles $C^{in} = \{c_1^{in}, \ldots, c_n^{in}\}$, and let $H = (H^{out}, H^{in})$ denote a pair of (possibly empty) sets of R-cycles $H^{out} = \{h_1^{out}, \ldots, h_m^{out}\}$ and $H^{in} = \{h_1^{in}, \ldots, h_m^{in}\}$. Then a **vague region** vr is a pair (C, H) so that the following conditions are satisfied:

(i) $\forall i \in \{1, \ldots, n\}: c_i^{in}$ *area-inside* c^{out}.
(ii) $\forall k, l \in \{1, \ldots, n\}, k \neq l: c_k^{in}$ *edge-disjoint* c_l^{in}.
(iii) $\forall k \in \{1, \ldots, m\} \ \exists l \in \{1, \ldots, n\}: h_k^{out}$ *edge-inside* c_l^{in}.
(iv) $\forall k, l \in \{1, \ldots, m\}, k \neq l: h_k^{out}$ *edge-disjoint* h_l^{out}.
(v) There exist two bijective functions $f: \{1, \ldots, m\} \to H^{out}$ and $g: \{1, \ldots, m\} \to H^{in}$ such that

$$\forall i \in \{1, \ldots, m\}: g(i) \ area\text{-}inside \ f(i).$$

These conditions reflect the informal zone model of a vague region presented above. C models the unsafe border zone consisting of the outer border c^{out} and a non-empty set C^{in} of inner borders. H models the unsafe hole zones consisting of a set H^{out} of outer borders and a corresponding set H^{in} of inner borders. The conditions describe the inclusion relationships between the four kinds of cycles that occur in unsafe border and hole zones; the disjointedness relationship between each pair of inner borders of the unsafe border zone, and the disjointedness relationship between each pair of outer borders of the unsafe hole zones. Condition (v) requires that exactly one inner border of H^{in} lies inside exactly one outer border of H^{out}. Note that f and g are total functions, since domain and co-domain of each function have the same cardinality. All conditions together prevent a proper intersection of borders (cycles).

Based on these five conditions, an equivalent definition for vr as a pair of **regions** objects (R^{out}, R^{in}) can be given. The first component of this pair is defined as $R^{out} := \{(c^{out}, H^{in})\}$ which describes a set with a single R-face. R^{out} is a **regions** object, since all inner borders h_i^{in} of unsafe hole zones are edge-inside to the outer border c^{out} of the unsafe border zone (follows from conditions (v), (iii), and (i)), and since they are pairwise edge-disjoint (follows from conditions (v) and (iv)). R^{in} represents a set of edge-disjoint R-faces and is defined as $R^{in} := \{(c_i^{in}, H_i^{out}) | i \in \{1, \ldots, n\}\}$ with $H_i^{out} := \{h \in H^{out} | h \ edge\text{-}inside \ c_i^{in}\}$. R^{in} is a **regions** object, since all inner borders of the unsafe border zone are pairwise edge-disjoint (follows from condition (ii)), and since for an inner border c_i^{in} all cycles of H_i^{out} are pairwise *edge-disjoint* (follows from condition (iv)). It is obvious that $\bigcup_{i=1}^{n} H_i^{out} = H^{out}$ and that $\forall k, l \in \{1, \ldots, n\}, k \neq l$: $H_k^{out} \cap H_l^{out} = \varnothing$.

Intuitively, R^{out} represents the maximal and R^{in} the minimal extent of vr. We will assume this latter definition as the formal definition of a vague region. Let us now define what it means that two vague regions $vr_1 = (R_1^{out}, R_1^{in})$ and $vr_2 = (R_2^{out}, R_2^{in})$ are edge-disjoint (for the definition of the predicate *edge-disjoint* between **regions** objects/R-faces see Güting and Schneider, 1993):

$$vr_1 \ edge\text{-}disjoint \ vr_2 :\Leftrightarrow R_1^{out} \ edge\text{-}disjoint \ R_2^{out}.$$

The realm-based structure of a vague region forms the basis for the definition of the spatial data type **vregions**.

> *For a given realm R, a value of type **vregions** is a set of pairwise*
>
> *edge-disjoint vague regions.*

We now have to show the closure properties of the data type **vregions**, that is, it must be closed under the geometric operations union, intersection and difference with regard to the same realm. Let w.l.o.g. (without loss of generality) $VR_1 = \{(R_1^{out}, R_1^{in})\}$ and $VR_2 = \{(R_2^{out}, R_2^{in})\}$ be two (one-component) **vregions** objects. Then

$$\textbf{\textit{union}} \ (VR_1, VR_2) := decompose(\textbf{\textit{union}}(R_1^{out}, R_2^{out}), \textbf{\textit{union}}(R_1^{in}, R_2^{in})).$$

For **intersection** and **difference** the definitions are analogous. Since each of the three geometric operations when applied to the **regions** objects R_1^{out} and R_2^{out} normally leads to a **regions** object, say R^{out}, with more than one edge-disjoint R-face, R^{out} must be decomposed into its R-faces, and the uniquely matching set of edge-disjoint R-faces from the same geometric operation applied to R_1^{in} and R_2^{in} must be assigned to each such R-face in order to form a set of edge-disjoint vague regions. This is the task of the operation *decompose*, whose formal definition is omitted here. The definitions of the geometric operations can be simply generalized to many-component **vregions** objects. Due to the underlying realms, these operations both in theory *and* in practice obey the usual algebraic laws, for instance, commutative, associative and distributive laws.

10.5 Conclusions

The first part of the paper enumerated relevant design criteria for the modelling of spatial data types that are valid regardless of whether we consider objects with sharp or undetermined boundaries. The second part showed how general vague region objects with good closure properties can be defined on the basis of the Realm/ROSE approach, an algebraic model for constructing sharply bounded spatial objects. Using and extending a 'sharp' model can lead to success and meet the user's conceptual view and intuition of spatial vagueness.

Future work will have to relate to the formal definition of the data types **vpoints** and **vlines** and, of course, to the formal definition of vague spatial operations like vague topological relationships.

Acknowledgement

This work was supported by the DFG (Deutsche Forschungsgemeinschaft) under grant Gu 293/1-7.

Note

[1] Methods for the treatment of numeric correctness problems below and within the realm level and especially the important problem of mapping an application's set of intersecting line segments into a realm's set of non-intersecting line segments are interesting and complex, but beyond the scope of this chapter (see Güting and Schneider, 1993).

REFERENCES

BANAI, R. (1993). Fuzziness in geographical information systems: Contributions from the analytic hierarchy process. *Int. J. Geographical Information Systems*, **7**(4), pp. 315–329.

DUTTA, S. (1991). Topological constraints: A representational framework for approximate spatial and temporal reasoning. *Symp. on Advances in Spatial Databases*, pp. 161–180.

EGENHOFER, M. J. (1989). *Spatial SQL: A Spatial Query Language*, Report 103, Dept. of Surveying Engineering, University of Maine.

FINN, J. T. (1993). Use of the average mutual information index in evaluating classification error and consistency. *Int. J. Geographical Information Systems*, **7**(4), 349–366.

GARGANO, M., NARDELLI, E. and TALAMO, M. (1991). Abstract data types of the logical modeling of complex data, *Information Systems*, **16**(5).

GÜTING, R. H. (1988). Geo-relational algebra: A model and query language for geometric database systems, *Proc. Int. Conf. on Extending Database Technology*, pp. 506–527.

GÜTING, R. H. (1989). Gral: An Extensible Relational Database System for Geometric Applications. *Proc. 15th Int. Conf. on Very Large Databases*, pp. 33–44.

GÜTING, R. H. and SCHNEIDER, M. (1993). Realms: A foundation for spatial data types in database systems. *Proc. 3rd Int. Symposium on Large Spatial Databases*, pp. 14–35.

GÜTING, R. H. and SCHNEIDER, M. (1995), Realm-based spatial data types: The ROSE algebra, *VLDB Journal*, **4**, 100–143.

HEUVELINK G. B. M. and BURROUGH, P. A. (1993). Error propagation in cartographic modelling using Boolean logic and continuous classification, *Int. J. Geographical Information Systems*, **7**(3), 231–246.

JOSEPH, T. and CARDENAS, A. (1988). PICQUERY: A high level query language for pictorial database management, *IEEE Trans. Software Engineering*, **14**, 630–638.

LEUNG, Y. and LEUNG, K. S. (1993). An intelligent expert system shell for knowledge-based geographical information systems: 1 The Tools. *Int. J. Geographical Information Systems*, **7**(3), 189–199.

LEUNG, Y., GOODCHILD, M. and LIN, C.-C. (1992). Visualization of fuzzy scenes and probability fields. *Proc. 5th Int. Symp. on Spatial Data Handling*, pp. 480–490.

LIPECK U. and NEUMANN, K. (1987). Modelling and manipulating objects in geoscientific databases. *Proc. 5th Int. Conf. on the Entity-Relationship Approach*, pp. 67–86.

ORENSTEIN, J and MANOLA, F. (1988). PROBE spatial data modeling and query processing in an image database application, *IEEE Trans. Software Engineering*, **14**, 611–629.

ROUSSOPOULOS, N., FALOUTSOS, C. and SELLIS, T. (1988). An efficient pictorial database system for PSQL, *IEEE Trans. Software Engineering*, **14**, 639–650.

SCHOLL, M. and VOISARD, A. (1989). Thematic map modelling, *Proc. 1st Int. Symposium on Large Spatial Databases*, pp. 167–190.

SVENSSON, P. and HUANG, Z. (1991), Geo-SAL: A query-language for spatial data analysis, *Proc. 2nd Int. Symposium on Large Spatial Databases*, pp. 119–140.

TILOVE, R. B. (1980). Set membership classification: A unified approach to geometric intersection problems, *IEEE Trans. Computers*, **C-29**, 874–883.

PART FOUR

Qualitative Topological Relations and Indeterminate Boundaries

ANDREW U. FRANK

Introduction

The two chapters in Part 4 treat an important and well defined problem, namely topological relations and spatial reasoning applied to objects with uncertain boundaries. Despite the fact that objects like Scandinavia or London do not have crisp boundaries, we readily conclude that London is outside Scandinavia. It seems that uncertain boundaries need not affect topological reasoning.

The two independent approaches to the essentially same question begin from two different starting points and somewhat different perspectives. Both chapters embed the problem in a formal frame, applying formal logic, axiomatization and topological reasoning to clarify the issues. Their results can be compared and confirm each other. But there are significant differences.

Elsewhere, topological relations between extended objects have been classified by Egenhofer and Herring (1991) using a nine-point-intersection schema. Objects are divided into an interior, a boundary and an exterior. Relations between extended 2D objects in a 2D space can be classified considering the intersection between these three parts (interior, boundary, exterior) of the two objects. It is sufficient to observe whether an intersection between the interior, exterior or boundary of the two objects is empty or not, in order to clearly label eight relations. These relations are, given their definitions, invariant under topological transformations, i.e. they are topological relations.

In Chapter 11 Clementini and Di Felice start with connected crisp regions and then coarsen the boundary. They then apply the methods of Egenhofer and Herring (1991) to construct all the relations differentiated by the 9-intersection. This results in 44 topological relations which can be identified between objects with broad boundaries. If one assumes that the boundaries are relatively small compared to the regions, then four relations become impossible and only 40 remain. These can be arranged in conceptual neighbourhoods in the same way as the relations with sharp boundaries (i.e. Hamming distance in the 9-intersection code); indeed, the clusters for objects with coarse boundaries are a superset of the clusters for objects with sharp boundaries.

In Chapter 12 Cohn and Gott start with a detailed discussion of what it means to say that an object has a crisp or a broad boundary. They introduce a relation between two versions of an object A and A', which states that A' is a crisper representation than A, which is defined as asymmetric, irreflexive and transitive. Their definition of conceptual neighbourhood is based on gradual changes ('crisping' or 'coarsening') of the objects involved. This development leads to a few more relations than those differentiated by the Clementini and Di Felice approach (but which is similar to the nine-point-intersection method applied to regions with holes).

The formalization of the relation between two versions of the same object, one being crisper than the other, is a substantive contribution. It fits within the schema of differentiation of eight topological relations between objects (disjoint, meet, overlap, covered by, covers, inside, contains, and equal) independently of how it is formalized. More relations can be differentiated if broad boundaries are considered, which form clusters closely associated with the eight previously listed. Clementini and Di Felice qualify them with the term 'nearly x'. This fits with results from experimental observations of people describing topological relations and the geometric situation associated with them.

The methods discussed here, especially the concept of varying levels of crispness, will eventually be useful and will be built into geographic information systems. They can help to solve the problems listed at the end of Chapter 17 (Brändli), which describes methods for extracting geomorphological structure from digital terrain models and indicates problems in the practical programming due to uncertainty in both the concepts and the data.

REFERENCES

EGENHOFER, M. J. and HERRING, J. (1991). *Categorizing Binary Topological Relationships between Regions, Lines and Points in Geographic Databases*, Technical Report, University of Maine, Department of Surveying Engineering.

CHAPTER ELEVEN

An Algebraic Model for Spatial Objects with Indeterminate Boundaries

ELISEO CLEMENTINI and PAOLINO DI FELICE

Università di L'Aquila, Dipartimento di Ingegneria Elettrica, Poggio di Roio, L'Aquila, Italy

11.1 Introduction

Conventional modelling in geographic information systems (GIS) made us familiar with geographic objects whose spatial extents are clear and universally recognized. When we think of geographic objects with indeterminate boundaries, we immediately face the difficulty of drawing them and representing them in a computer format. The difficulty comes from a well established approach in GIS modelling which tries to simplify each geographic entity and represent its shape by using simple objects of Euclidean geometry, like points, lines and regions. The boundaries of such objects are fixed by geometric definitions. This simplification process can make it difficult to represent the real nature of geographic entities when the application requires more detail. Burrough (1992) thinks that it is necessary to use basic geographic entities that take into account the complexity and uncertainty in position, topology and attributes.

Objects with indeterminate boundaries can be partitioned in two broad categories: (a) objects which have sharp boundaries whose position and shape are unknown or cannot be measured exactly; and (b) objects which do not have well-defined boundaries or for which it is useless to fix a boundary.

The first kind of indeterminate boundary is what is usually called positional uncertainty and involves all geographic data, due to errors in measurements and the finite representation of computer formats. Positional uncertainty affects GIS reliability especially in certain thematic maps, such as soil maps, land use maps, and geological maps. Different models have been developed for the representation of positional uncertainties, such as fuzzy models (Dutta, 1991; Leung *et al.*, 1992; Banai, 1993; Heuvelink and Burrough, 1993; Leung and Leung, 1993) and probabilistic models (Blakemore, 1984; Goodchild and Dubuc, 1987; Finn, 1993; Shibasaki, 1993).

The second kind of indeterminate boundary is very different since the uncertainty is not due to the limitations of current technology, but intrinsically belongs to the nature of the object. Goodchild (1992) says that many human geographical constructs are implicitly uncertain, including spatial objects like 'the Indian Ocean' and 'Europe'. This kind of geographic object has received little attention in the GIS literature, despite its potential interest in several applications. We can think of regions which surely have an extent, resulting from our daily experience, but whose boundaries have never been fixed, or are impossible to define, like mountains and plains, the city and the country, seas and oceans. No one knows the exact boundaries of such regions, but everyone fully understands their meaning and can reason about them. L'Aquila could be described as a town in the mountains, while Rome is in the plains. 'I live in the country' is opposed to 'I live in the city', but there are intermediate situations, such as 'I live on the outskirts of the city'. Weather forecasts can report bad conditions on the Ligurian Sea, moving towards the Thyrrenian Sea, but where does the Ligurian Sea end and the Thyrrenian Sea begin? Geographers usually model these objects as fields (Couclelis, 1992); we will pursue the 'object' view, which is preferable when people need to perform spatial reasoning and treat spatial relations.

In this chapter we propose an algebraic model for handling regions with indeterminate boundaries. Although we restrict our attention to regions, the results could be extended to lines and points. The model is an extension of an existing one for representing 'simple' regions with sharp boundaries and their topological relations. In many applications, the indeterminacy in the boundary can be restricted to a well-defined area which acts as a separation between the space that surely belongs to the region and the space that is surely outside. The model for spatial objects with indeterminate boundaries proposed here describes the uncertainty in the boundary as a two-dimensional exact zone surrounding the object. This assumption is very simple and cannot take into account all kinds of boundary fuzziness, but it has the advantage that standard topology can be used and that known approaches for objects with well-defined boundaries can be extended to handle indeterminate boundaries. The 9-intersection model (Egenhofer and Herring, 1991) will be used as a basis for describing the topological relations involving such objects. Expected results are similar to topological relations between regions in Z^2 (Egenhofer and Sharma, 1993), since rasters have boundaries with an extent of unit 1. Boundaries with an extent, or 'broad' boundaries, were also introduced by Al-Taha (1992), even if all topological relations between such regions were not found exhaustively.

Section 11.2 introduces the spatial data model for objects with broad boundaries. Section 11.3 recalls the 9-intersection as a model for topological relations between simple objects, Section 11.4 applies it to objects with broad boundaries. Section 11.5 proposes names for topological relations by grouping them using conceptual neighbourhood criteria, and Section 11.6 presents a final discussion.

11.2 The spatial data model for objects with broad boundaries

Simple regions, lines and points have been widely discussed in the literature (Egenhofer and Franzosa, 1991; Egenhofer and Herring, 1991; Clementini et al., 1993; Clementini and Di Felice, 1995). A simple region is defined as a closed, homogeneously two-dimensional, simply connected subset of $\mathbb{R}^2$. A simple line is a one-dimensional variety embedded in $\mathbb{R}^2$ with exactly two end-points and no self-intersections. A simple point is a connected 0-dimensional subset of $\mathbb{R}^2$. For such objects, interior (°), boundary (∂) and exterior (⁻) are defined with the standard topological meaning (Munkres, 1975).

In this section we introduce *regions with broad boundaries*. These objects differ from simple ones with regard to their boundary definition. We can define an *inner boundary* and an *outer boundary* for the region. Both are closed curves and the area surrounded by the inner boundary is contained in the area surrounded by the outer boundary (Figure 11.1). These two boundaries represent the indeterminacy of the region, expressing the minimum and the maximum extent of the region itself. Any of the closed curves that are between the two boundaries could actually act as the boundary of the region.

Definition 1. A *region with a broad boundary* A is made up of two simple regions, A_1 and A_2, with $A_1 \subseteq A_2$, where ∂A_1 is the *inner boundary* of A and ∂A_2 is the *outer boundary* of A.

Definition 2. The *broad boundary* ΔA of a region with a broad boundary A is a closed connected subset of $\mathbb{R}^2$ with a hole. ΔA comprises the area between the inner boundary and the outer boundary of A, such that $\Delta A = \overline{A_2 - A_1}$, or equivalently $\Delta A = A_2 - A_1^{\circ}$. If $A_1 \subset A_2$, then ΔA is two-dimensional; in the limit case $A_1 = A_2$, ΔA is a one-dimensional circle. If $\partial A_1 \cap \partial A_2 \neq \emptyset$, then ΔA is not homogeneously two-dimensional and may present one-dimensional parts and separations in its interior.

Definition 3. The interior of a region with a broad boundary A is defined as:

$A^{\circ} = A_2 - \Delta A$.

Definition 4. The exterior of a region with a broad boundary A is defined as:

$A^- = \mathbb{R}^2 - A_2$.

Definition 5. The closure of a region with a broad boundary A is defined as:

$\bar{A} = A^{\circ} \cup \Delta A$.

Following from these definitions, the interior and exterior of a region with a broad boundary are open sets, while the broad boundary is a closed set. Simple regions

Figure 11.1   A region A with broad boundary.

can be seen as special cases of regions with broad boundaries in which the inner and outer boundaries coincide, and therefore $\Delta A = \partial A$.

In most applications, the extent of the broad boundary of a region can be considered to be significantly smaller than its interior. To facilitate the discussion about topological relations, we will later assume that 'the extent of the broad boundary of a region is much smaller than its interior: $\Delta A \ll A^\circ$. This assumption is quasi-topological, since it describes an aspect of the object that remains invariant with respect to some common topological transformations such as rotation and scaling.

The broad boundary has a variable extent and can be also one-dimensional in some parts in order to model geographic entities that fit this behaviour. In the case of the Ligurian Sea on a large scale, the corresponding region has a sharp boundary where the sea meets the Italian coast and a broad boundary where it meets the Thyrrenian Sea and the rest of the Mediterranean.

11.3   The 9-intersection model for objects with sharp boundaries

Topological relations are spatial relations that are preserved under such transformations as rotation, scaling and rubber sheeting. Binary topological relations between two objects, A and B, in $\mathbb{R}^2$ can be classified according to the intersection of A's interior, boundary, and exterior with B's interior, boundary, and exterior. The nine intersections between the six object parts describe a topological relation and can be concisely represented by the following 3×3 matrix, called the *9-intersection* matrix

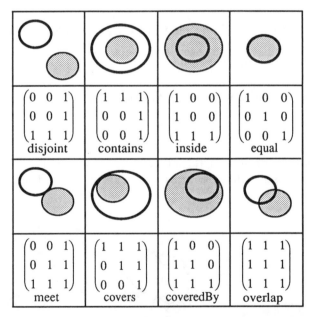

Figure 11.2   The eight topological relations between two simple regions and their corresponding 9-intersection matrices.

(Egenhofer and Herring, 1991):

$$\begin{pmatrix} A°\cap B° & A°\cap \partial B & A°\cap B^- \\ \partial A\cap B° & \partial A\cap \partial B & \partial A\cap B^- \\ A^-\cap B° & A^-\cap \partial B & A^-\cap B^- \end{pmatrix}.$$

By considering the values empty (0) and non-empty (1), we can distinguish between $2^9 = 512$ binary topological relations. For two simple regions with well-defined boundaries, only eight of them can be realized; for two simple lines there are 36 different relations in $\mathbb{R}^2$; and between a simple line and a region there are 19 different topological relations. Each set of relations provides a complete coverage and is mutually exclusive (Egenhofer and Franzosa, 1991). Figure 11.2 shows prototypical examples of the eight relations that can be realized between two regions. The 9-intersection model has been applied not only to simple objects; for instance, Egenhofer *et al.* (1994) extended such a model to encompass regions with holes, that is, regions with disconnected exteriors.

11.4   The 9-intersection for objects with broad boundaries

In this section, we determine all the realizable 9-intersection matrices for two regions with broad boundaries. The nine intersection sets that we consider have a broad boundary in place of the ordinary boundary of the region. Therefore, we redefine the 9-intersection for regions with broad boundaries as follows:

$$\begin{pmatrix} A°\cap B° & A°\cap \Delta B & A°\cap B^- \\ \Delta A\cap B° & \Delta A\cap \Delta B & \Delta A\cap B^- \\ A^-\cap B° & A^-\cap \Delta B & A^-\cap B^- \end{pmatrix}.$$

Egenhofer and Herring (1991) present geometric conditions for simple regions that allow the 2^9 different matrices to be restricted to eight matrices corresponding to eight different topological relations. The same can be done for regions with broad boundaries. We start by observing that the geometric conditions for such regions are less restrictive, since the boundary is two-dimensional. The following geometric conditions are valid for simple regions with sharp boundaries, and can be extended to regions with broad boundaries simply by changing the occurrences of boundary with broad boundary:

(1) the exteriors of two regions intersect with each other;

(2) A's boundary intersects with at least one part of B, and vice versa;

(3) if A's interior intersects with B's interior and exterior, then it must also intersect with B's boundary, and vice versa;

(4) if both boundaries do not intersect with each other, then at least one boundary must intersect with its opposite exterior;

(5) if both boundaries intersect with the opposite interiors, then the boundaries must also intersect with each other;

(6) if A's interior intersects with B's exterior, then A's boundary must also intersect with B's exterior, and vice versa.

On the other hand, we can observe that the following geometric conditions are valid for simple regions with sharp boundaries, but are no longer valid for regions with broad boundaries:

- if both interiors are disjoint, then A's interior intersects with B's exterior, and vice versa;
- if A's interior is a subset of B's closure, then A's boundary must be a subset of B's closure as well, and vice versa;
- if A's interior intersects with B's boundary, then it must also intersect with B's exterior, and vice versa;
- if both interiors are disjoint, then A's boundary cannot intersect with B's interior, and vice versa;
- if the interiors do not intersect with each other, then A's boundary must intersect with B's exterior, and vice versa;
- if both interiors do not intersect with each other, then at least one boundary must intersect with its opposite exterior.

These geometric conditions need to be replaced by a set of less restrictive geometric conditions that hold for regions with broad boundaries. The conditions to be added to the six already given are as follows:

(7) if both interiors are disjoint and A's broad boundary intersects with B's interior, then the two broad boundaries must intersect with each other, and vice versa;

(8) if A's interior is a subset of the B's closure, then A's broad boundary must intersect with B's closure, and vice versa;

(9) if both interiors are disjoint, then A's interior intersects either with B's broad boundary or with B's exterior, and vice versa;

(10) if A's interior does not intersect with B's closure, then A's broad boundary must intersect with B's exterior, and vice versa;

(11) if A's broad boundary intersects with B's interior and exterior, then it must also intersect with B's broad boundary, and vice versa;

(12) if A's closure is a subset of B's interior, then A's exterior must intersect with B's interior, and vice versa.

By applying these 12 geometric conditions, it is possible to reduce the 2^9 matrices to 44 matrices, for which a geometric realization for regions with broad boundaries exists. To facilitate this task, we adopt the following strategy: we start from the valid 4-intersection matrices (the upper left part of the 9-intersection matrices), which involve interiors and broad boundaries alone, and then for each of them find out the realizable intersections involving exteriors.

The 4-intersection matrices with 0 and 1 values are 16. The 11 realizable 4-intersection matrices for regions with broad boundaries are as follows:

$$\begin{pmatrix} 0 & 0 \\ 0 & 0 \end{pmatrix}, \begin{pmatrix} 0 & 0 \\ 0 & 1 \end{pmatrix}, \begin{pmatrix} 0 & 1 \\ 0 & 1 \end{pmatrix}, \begin{pmatrix} 0 & 0 \\ 1 & 1 \end{pmatrix}, \begin{pmatrix} 0 & 1 \\ 1 & 1 \end{pmatrix}, \begin{pmatrix} 1 & 1 \\ 1 & 1 \end{pmatrix}, \begin{pmatrix} 1 & 0 \\ 1 & 1 \end{pmatrix}, \begin{pmatrix} 1 & 1 \\ 0 & 1 \end{pmatrix},$$

$$\begin{pmatrix} 1 & 0 \\ 1 & 0 \end{pmatrix}, \begin{pmatrix} 1 & 1 \\ 0 & 0 \end{pmatrix}, \begin{pmatrix} 1 & 0 \\ 0 & 1 \end{pmatrix}.$$

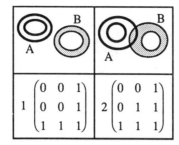

Figure 11.3 'Disjoint' and 'meet' between regions with broad boundaries.

These matrices corresponds to the eight relationships that also hold for regions with sharp boundaries (disjoint, meet, overlap, coveredBy, covers, inside, contains, equal), plus three other matrices.

The five 4-intersection matrices that are not realizable are:

$$\begin{pmatrix} 1 & 0 \\ 0 & 0 \end{pmatrix}, \begin{pmatrix} 0 & 1 \\ 0 & 0 \end{pmatrix}, \begin{pmatrix} 0 & 0 \\ 1 & 0 \end{pmatrix}, \begin{pmatrix} 0 & 1 \\ 1 & 0 \end{pmatrix}, \begin{pmatrix} 1 & 1 \\ 1 & 0 \end{pmatrix}.$$

The first matrix above is impossible from conditions (3) and (8). The second and third matrices are impossible from condition (7). The last two matrices are impossible from condition (5).

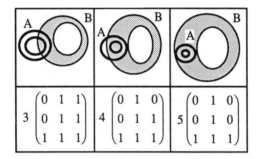

Figure 11.4 The three relations derived from the 4-intersection $\begin{pmatrix} 0 & 1 \\ 0 & 1 \end{pmatrix}$.

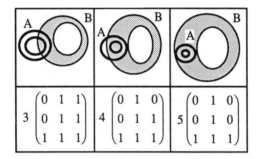

Figure 11.5 The three relations derived from the 4-intersection $\begin{pmatrix} 0 & 0 \\ 1 & 1 \end{pmatrix}$.

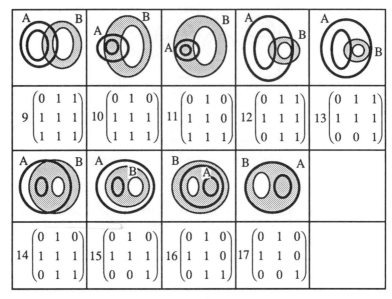

Figure 11.6 The nine relations derived from the 4-intersection $\left(\begin{smallmatrix} 0 & 1 \\ 1 & 1 \end{smallmatrix}\right)$.

Now, let us find all the realizable 9-intersection matrices that share the 11 possible 4-intersection matrices. From the 'disjoint' 4-intersection matrix, by applying conditions (1), (2) and (9), we obtain only one possible 9-intersection matrix (case 1, Figure 11.3). From the 'meet' 4-intersection matrix, by applying conditions (1), (9) and (10), we obtain case 2 (Figure 11.3).

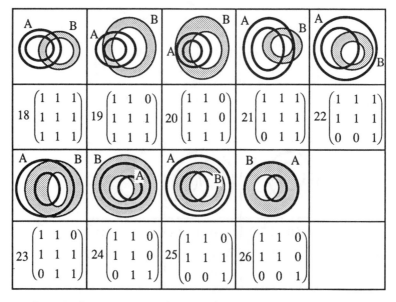

Figure 11.7 The nine relations originating from 'overlap'.

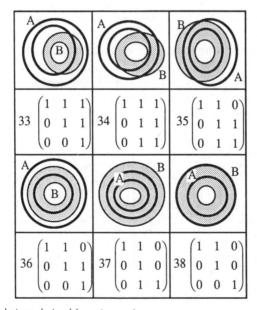

Figure 11.8   The six relations derived from 'coveredBy'.

From conditions (1), (6), (9) and (10), the 4-intersection matrix $\begin{pmatrix} 0 & 1 \\ 0 & 1 \end{pmatrix}$ gives the three relations of Figure 11.4 (cases 3–5), while its symmetric matrix $\begin{pmatrix} 0 & 0 \\ 1 & 1 \end{pmatrix}$ gives three other relations (cases 6–8, Figure 11.5). From conditions (1) and (6), the matrix $\begin{pmatrix} 0 & 1 \\ 1 & 1 \end{pmatrix}$ gives nine different 9-intersection matrices (cases 9–17, Figure 11.6).

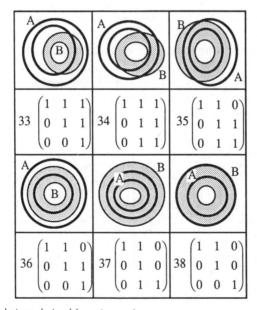

Figure 11.9   The six relations derived from 'covers'.

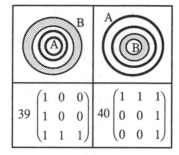

$$39 \begin{pmatrix} 1 & 0 & 0 \\ 1 & 0 & 0 \\ 1 & 1 & 1 \end{pmatrix} \quad 40 \begin{pmatrix} 1 & 1 & 1 \\ 0 & 0 & 1 \\ 0 & 0 & 1 \end{pmatrix}$$

Figure 11.10 'Inside' and 'contains' between regions with broad boundaries.

From conditions (1) and (6), the 4-intersection matrix corresponding to the 'overlap' relation can be expanded to form the nine different relations of Figure 11.7 (cases 18–26). By applying conditions (1), (3) and (6), the six relations of Figure 11.8 (cases 27–32) can be derived from the 'coveredBy' relation matrix and, symmetrically, the six relations of Figure 11.9 by the 'covers' relation matrix (cases 33–38). The 'inside' and 'contains' 4-intersection matrices each have one corresponding 9-intersection matrix, which are obtained by applying conditions (1), (3), (4), (11) and (12) (cases 39–40, Figure 11.10). The 'equal' intersection matrix splits into four different relations (cases 41–44, Figure 11.11), which are obtained from the application of conditions (1) and (3).

11.5 Conceptual neighbourhoods

The 44 different relations we have defined can be organized in a graph having a node for each relation and an arc for each pair of matrices differing for exactly one intersection value. In this way, we build the conceptual neighbourhood graph (Egenhofer and Al-Taha, 1992) for regions with broad boundaries (Figure 11.12). An arc in that graph expresses a gradual transition from one topological relation to another. To simplify the graph, we have excluded from it the four relations (cases 14–17) that are not possible by adding the assumption that the broad boundary of a region is much smaller than its interior (*Small boundaries assumption*). In fact, in these four cases we need very thick boundaries and small interiors in order to find geometric interpretations for them.

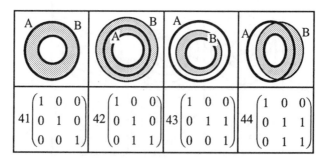

$$41 \begin{pmatrix} 1 & 0 & 0 \\ 0 & 1 & 0 \\ 0 & 0 & 1 \end{pmatrix} \quad 42 \begin{pmatrix} 1 & 0 & 0 \\ 0 & 1 & 0 \\ 0 & 1 & 1 \end{pmatrix} \quad 43 \begin{pmatrix} 1 & 0 & 0 \\ 0 & 1 & 1 \\ 0 & 0 & 1 \end{pmatrix} \quad 44 \begin{pmatrix} 1 & 0 & 0 \\ 0 & 1 & 1 \\ 0 & 1 & 1 \end{pmatrix}$$

Figure 11.11 The four relations derived from 'equal'.

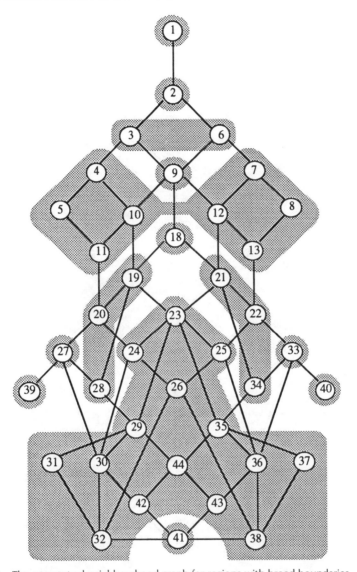

Figure 11.12   The conceptual neighbourhood graph for regions with broad boundaries.

By observing the geometric interpretations given for the remaining 40 relations, we can group them into several clusters that share similar geometric properties. We maintain the same names for the eight relations that also hold in the case of crisp boundaries, and, in addition, each cluster is given a relation name (Figure 11.13), which is reasonable to indicate the relation, even though we do not exclude other possible names or a different clustering for the graph. The current relation names ('close', 'nearly meet', etc.) are influenced by the small boundaries assumption. As an example, consider case 11 in Figure 11.6. This case is labelled 'close' in Figure 11.13, but if we allowed the interior to be very small with respect to the broad boundary, then it would not make much sense to say that A is close to B.

The conceptual neighbourhood graph is symmetric. A geometric interpretation of this aspect is that the configurations on the left part of the graph have region A

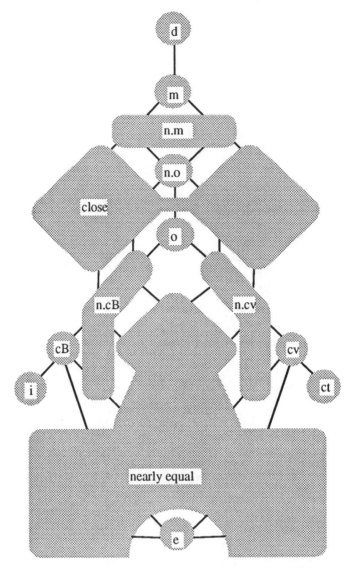

Figure 11.13　The clustering of the conceptual neighbourhood graph; d = disjoint, m = meet, n.m = nearly meet, n.o = nearly overlap, o = overlap, n.cB = nearly coveredBy, n.cv = nearly covers, cB = coveredBy, cv = covers, i = inside, ct = contains, e = equal.

smaller than region *B* and vice versa for those on the right. The configurations lying on the centre line are related to configurations in which *A* and *B* have comparable extents or about which nothing can be said *a priori*. This qualitative interpretation is due in most cases to the quasi-topological small boundaries assumption.

　　So far, the discussion has been related to the general case in which the boundary of both regions and the topological relation between them are indeterminate. We can think of situations in which there is less indeterminacy. In some cases, there is enough evidence of the topological relation between two regions even if both have indeterminate boundaries. This could be the case of the topological relation between

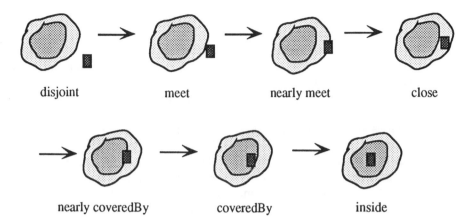

| disjoint | meet | nearly meet | close |

| nearly coveredBy | coveredBy | inside |

Figure 11.14   The topological relations between an object with broad boundary (a forest) and a smaller simple object (a house).

the Ligurian and Thyrrenian Seas, which reasonably is a 'meet', even if it is not important to specify the exact position of the boundary between the two.

It is also worthwhile to consider the topological relations between a region with a broad boundary and a region with a crisp boundary. Some of the relation names we adopted can be illustrated with an example. Let us consider a simple region with an exact boundary representing a small object like a house, and a region with broad boundary representing a forest. Figure 11.14 shows the various relative positions of the two objects, from 'disjoint' to 'inside', through several intermediate clusters. The second step is a 'meet', then there is a 'nearly meet'. If the house lies entirely in the broad boundary, we say that the house is 'close' to the forest. This is reasonable, since we can not say for sure whether the house intersects the forest or not, but we can qualitatively assert that the house is on the outskirts of the forest. Moving toward the interior of the forest, the house becomes 'nearly coveredBy' the forest. Eventually the house is definitely 'coveredBy' and 'inside' the forest.

11.6   Discussion

In this chapter we have extended an existing model for topological relations between objects with sharp boundaries (the 9-intersection) to model topological relations between objects with broad boundaries. This extension maintains all the properties of the 9-intersection model: among others, it gives a mutually exclusive set of relations and is an algebraic basis for spatial reasoning. The model is simple and allows us to describe the indeterminacy in boundaries using the same language adopted for exact boundaries. The clustering of similar configurations gives a set of names that provide qualitative descriptions of topological scenes involving objects with broad boundaries. This set of names is a superset of the names for simple objects. The 9-intersection model can also be successfully used to describe the relations between simple lines and regions with broad boundaries. The model can be easily implemented in an existing GIS once two polygons representing the inner and outer boundaries of each indeterminate region are fixed.

Cohn and Gotts (Chapter 12, this volume) have independently found a set of 46 relations for their egg–yolk theory. A first difference between the two approaches is that in Cohn and Gotts' approach topological relations are logically defined, while in our approach they are geometrically defined. In more detail, Cohn and Gotts classify different configurations according to the relations (out of five base relations) holding between the pairs egg–egg, yolk–yolk, and yolk–egg. In essence, Cohn and Gotts' approach reuses the same relations holding between simple regions with sharp boundaries to describe topological configurations between egg–yolk pairs, whereas our approach is an extension of the 9-intersection leading to the consideration of a new set of base relations defined in terms of geometric criteria (empty or non-empty intersections between interiors, broad boundaries, and exteriors).

The approach of Cohn and Gotts may be considered to be similar to that of Egenhofer *et al.* (1994), where regions with holes were considered. Establishing a correspondence between an egg–yolk pair and a region with a hole, we have almost the same analysis. In the case of Egenhofer *et al.* (1994), however, the eight base relations deriving from the 4-intersection were considered instead of the five base relations of Cohn and Gotts.

Another difference between the two approaches is in the classification of some limit configurations involving inner and outer boundaries. This may be illustrated with an example. Since in Cohn and Gotts' approach the case in which two regions are disjoint, and the case in which two regions are externally tangent, are grouped together in the same base relation, the configuration in which two egg–yolk pairs touch along their outer boundaries falls into the 'disjoint' category, whereas in our approach the same configuration falls into the 'meet' category, since we define broad boundaries as closed point sets and therefore their intersection is non-empty if they touch along their outer boundaries. This difference also leads us to have 44 cases versus 46. In fact, both cases 30 and 31 of Cohn and Gotts' approach fall into case 36 of the present approach. Analogously, cases 37 and 38 fall into our case 30.

Acknowledgements

This work was partially supported by the Italian National Council of Research under grant No. 93.01595.PF69. Max Egenhofer's comments on an earlier draft are gratefully acknowledged. Many improvements were made after discussions with Tony Cohn and the comparison with his contribution to this book. Furthermore, we would like to thank two reviewers among the participants of the GISDATA Specialist Meeting for providing useful insights and comments.

REFERENCES

AL-TAHA, K. (1992). *Deformation Analysis and Geometric Interpretations of Imprecise Binary Topological Relations*, Technical Report, University of Maine, Department of Surveying Engineering.

BANAI, R. (1993). Fuzziness in geographical information systems: Contributions from the analytic hierarchy process, *Int. J. Geographical Information Systems* 7(4), 315–329.

BLAKEMORE, M. (1984). Generalization and error in spatial databases, *Cartographica*, **21**, 131.

BURROUGH, P. A. (1992). Development of intelligent geographical information systems. *Int. J. Geographical Information Systems*, **6**(1), 1–11.

CLEMENTINI, E. and DI FELICE, P. (1995). A comparison of methods for representing topological relationships, *Information Sciences* **3**, 149–178.

CLEMENTINI, E., DI FELICE, P. and VAN OOSTEROM, P. (1993). A small set of formal topological relationships for end-user interaction, in: Abel, D. and Ooi, B. C. (Eds), *Advances in Spatial Database*, 3rd *Int. Symposium, SSD '93*, LNCS 692, pp. 277–295, Singapore: Springer.

COUCLELIS, H. (1992). People manipulate objects (but cultivate fields): Beyond the raster-vector debate in GIS, in: Frank, A. U., Campari, I. and Formentini, U. (Eds), *Theories and Methods of Spatio-Temporal Reasoning in Geographic Space*, LNCS 639, Pisa, Italy, pp. 65–77, Berlin: Springer.

DUTTA, S. (1991). Topological constraints: A representational framework for approximate spatial and temporal reasoning, *Advances in Spatial Databases, 2nd Int. Symposium, SSD '91*, Zürich, Switzerland, pp. 161–180.

EGENHOFER, M. J. and AL-TAHA, K. (1992). Reasoning about gradual changes of topological relationships, *Theories and Models of Spatio-Temporal Reasoning in Geographic Space*, Pisa, Italy, pp.196–219, Berlin: Springer.

EGENHOFER, M. J., CLEMENTINI, E. and DI FELICE, P. (1994). Topological relations between regions with holes, *Int. J. Geographical Information Systems*, **8**(2), 129–144.

EGENHOFER, M. J. and FRANZOSA, R. (1991). Point-set topological spatial relations, *Int. J. Geographic Information Systems*, **5**(2), 161–174.

EGENHOFER, M. J. and HERRING, J. (1991). *Categorizing Binary Topological Relationships Between Regions, Lines, and Points in Geographic Databases*, Technical Report, University of Maine, Department of Surveying Engineering.

EGENHOFER, M. J. and SHARMA, J. (1993) Topological relations between regions in R^2 and Z^2, in: Abel, D. and Ooi, B. C. (Eds), *Advances in Spatial Database, 3rd Int. Symposium, SSD '93*, Singapore, LNCS 692, pp.316–336, Berlin: Springer.

FINN, J. T. (1993). Use of the average mutual information index in evaluating classification error and consistency, *Int. J. Geographical Information Systems* **7**(4), 349–366.

GOODCHILD, M. (1992). Geographical information science, *Int. J. Geographical Information Systems*, **6**(1), 31–45.

GOODCHILD, M. F. and DUBUC, O. (1987). A model of error for choropleth maps, with applications to geographic information systems, *Proc. Auto-Carto 8*, p. 165.

HEUVELINK, G. B. M. and BURROUGH, P. A. (1993). Error propagation in cartographic modelling using Boolean logic and continuous classification. *Int. J. Geographical Information Systems* **7**(3), 231–246.

LEUNG, Y., GOODCHILD M. and LIN, C.-C. (1992). Visualization of fuzzy scenes and probability fields, *5th Int. Symposium on Spatial Data Handling*, Charleston, SC, pp. 480–490.

LEUNG, Y. and LEUNG, K. S. (1993). An intelligent expert system shell for knowledge-based geographical information systems 1: The tools, *Int. J. Geographical Information Systems*, **7**(3), 189–199.

MUNKRES, J. R. (1975). *Topology: A first Course*, Englewood Cliffs, NJ: Prentice Hall.

SHIBASAKI, R. (1993). A framework for handling geometric data with positional uncertainty in a GIS environment, in: Lu, H. and Ooi, B. C. (Eds), *GIS: Technology and Applications*, pp. 21–35, Singapore: World Scientific.

CHAPTER TWELVE

The 'Egg–Yolk' Representation of Regions with Indeterminate Boundaries

A. G. COHN and N. M. GOTTS

AI Division, School of Computer Studies, University of Leeds, Leeds, UK

12.1 Introduction

The topic of this chapter is how best to deal with vagueness in spatial representation and reasoning, particularly within the framework of 'RCC theory' (Randell *et al.*, 1992b; Cohn *et al.*, 1995), which provides a representation of topological properties and relations in which regions rather than points are taken as primitive. We are concerned here with regions having vague or indeterminate *boundaries* but with a known location, not with vaguely *located* entities.

Many of the spatial regions we consider in everyday contexts do not have precise boundaries: consider urban areas, clouds of gas, galaxies, and habitats of particular plants. Such 'regions' will be called 'vague' or 'noncrisp' here, in contrast to 'crisp' regions: those with precisely defined boundaries. RCC theory as originally devised has no means of representing and reasoning about noncrisp regions.

What properties do we require of a treatment of spatial vagueness? First, it should be logically consistent. Second, it should as far as possible respect our intuitions: the more we are able to express the kinds of things we want to say about vague regions, the better. Third, it should be computationally tractable, both in itself, and when combined with spatial information expressed in precise terms. Finally, if it can be linked to an existing treatment of precise spatial information in a straightforward fashion, so much the better. We present here an attempt to build a system for representing and reasoning about vague regions on the basis of an existing representation for crisp ones. This approach is referred to as the 'egg–yolk' representation, for reasons which will become clear. We do not claim to have produced a complete solution to the problems of representing regions with undetermined boundaries, but believe we have made considerable progress toward one.

12.2 The RCC theory of spatial regions

The focus of research on spatial representation and reasoning at Leeds has been to evaluate, extend and implement a theory[1] of space and time based upon Clarke's (1981, 1985) 'calculus of individuals based on connection', and expressed in the many-sorted logic LLAMA (Cohn, 1987).[2] Our revised and extended theory, now known as 'RCC theory' has been developed in a series of papers (Randell and Cohn, 1989, 1992; Randell, 1991; Randell *et al.*, 1992b; Cohn *et al.*, 1995; Bennett, 1994; Gotts, 1994). The most distinctive feature of Clarke's 'calculus of individuals', and of our work, is that extended *regions* rather than points are taken as fundamental, and it is partly this feature which makes RCC theory a promising basis for a treatment of vague regions. Our formal theory supports regions having either a spatial or a temporal interpretation (temporal 'regions' are periods of time). Informally, these regions may be thought of as infinite in number, and 'connection' may be any relation from external contact (touching without overlapping) to spatial or temporal identity. Spatial regions may have one, two, three, or even more than three dimensions, but in any particular model of the formal theory, all regions are of the same dimensionality. Thus, if we are concerned with a two-dimensional model, such as one in which regions are areas of land, the boundary lines and points at which these regions meet are not themselves considered regions. Indeed, they cannot be referred to *directly* within RCC theory, in which they are all assigned to sort NULL, which is disjoint from sort REGION. However, as Gotts (1994) shows, a great deal can be specified about the properties and relations of such lower-dimensional boundary entities while referring explicitly only to relations between the higher-dimensional regions they bound.

   The basic part of the formal theory assumes a primitive dyadic relation: C (x, y), read as 'x connects with y' (where x and y are regions). Two axioms are used to specify that C is reflexive and symmetric. C can be given a topological interpretation in terms of points incident in regions. In this interpretation, C (x, y) holds when the topological *closures* of regions x and y share at least one point.

   Using the relation C, further dyadic relations are defined. In particular, a set of eight 'base relations' can be defined, of which one and only one will hold between a given pair of regions (these relations thus form a jointly exhaustive and pairwise disjoint or JEPD set). These are illustrated in the upper part of Figure 12.1. The eight are: DC (DisConnected), EC (Externally Connected), PO (Partially Overlapping), TPP (Tangential Proper Part), NTPP (Non-Tangential Proper Part), EQ or = (Equal), TPPI (Tangential Proper Part Inverse), NTPPI (Non-Tangential Proper Part Inverse).

   This set of eight base relations, referred to as the 'RCC-8' relations, is used in

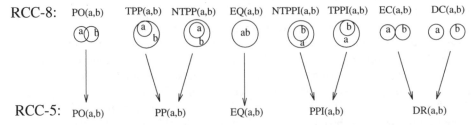

Figure 12.1 The RCC-8 and RCC-5 sets of JEPD relations between region pairs.

much of our work on qualitative spatial relations. In this paper, a smaller JEPD set of relations will be used: RCC-5, whose relation to RCC-8 is illustrated in Figure 12.1.[3] RCC-5 consists of the relations {PO ,PP ,EQ ,PPI ,DR} (Partially Overlapping, Proper Part, Equal, Proper Part Inverse, and Distinct Regions). Considered relative to RCC-8, RCC-5 lumps together TPP and NTPP as PP, TPPI and NTPPI as PPI, and EC and DC as DR. Within RCC-5, C is no longer available as a primitive: we can use PP (among other possibilities) as a primitive from which to define all the other RCC-5 relations.

12.3   Describing and reasoning about vague regions

In this chapter we attempt to extend our connection-based approach to qualitative spatial representation and reasoning to regions with indeterminate boundaries: vague or 'noncrisp' regions. What kinds of things do we want to be able to say about vague regions? How do we want vague regions to relate to 'crisp' regions, and to each other? What formalism will most efficiently and intelligibly express these relations?

We want to be able to say at least some of the same sorts of things about vague regions as about crisp ones: that one contains another (southern England contains London, even if both are thought of as vague regions), that two overlap (the Sahara desert and West Africa), or that two are disjoint (the Sahara and Gobi deserts). In these cases, the two vague regions represent the space occupied by distinct entities, and we are interested in defining a vague area corresponding to the space occupied by either of the two, by both at once, or by one but not the other.

We may also want to say that one vague region is a 'crisper' version of another. For example, we might have an initial (vague) idea of the area inundated in a flood, and then receive information from a systematic survey, reducing the imprecision in our knowledge. In this case, the vagueness of the vague regions is a matter of our ignorance; in others, it appears intrinsic: consider an informal geographical term like 'southern England'. Here, the uncertainty about whether particular places (somewhere to the north of London but to the south of Birmingham) are included cannot be resolved by evidence, unless we simply decide to take the majority view across a particular sample of opinion as decisive. There is a degree of arbitrariness about any particular choice of an exact boundary, and for most purposes, none is required. But *if* we decide to define a more precise version of such an informal term for some purpose, the choice of a precise definition is by no means *wholly* arbitrary: generally, we can distinguish between more and less 'reasonable' choices of more precise description. We need to keep the distinction between ignorance-based and intrinsic vagueness in mind, but many of the problems of representation and reasoning are the same in both situations, and the two are often found together: consider the area inhabited by a particular bird species or subspecies. Here, our knowledge may indeed be limited, but in addition, just where the limits lie may depend on whether we count areas sometimes strayed into on the fringes of the bird's range, or where we decide to say that one species or subspecies ends and another begins.

We can begin by considering what we may want to say, and what we want to hold true, when we consider relations between alternative vague-region versions of an entity's spatial extent. We will need a name for the relation between two estimates of which one is a refinement or 'crisping' of the other, which we could express CR(X, Y), read: 'Vague region X is a crisping of vague region Y' (we will use upper-

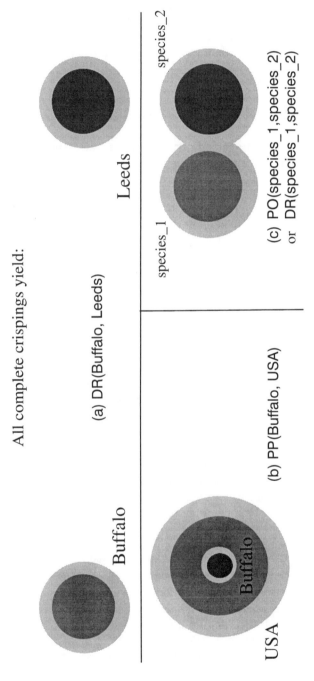

Figure 12.2   What relationships are there between complete crispings of these regions?

case letters for variables ranging over vague regions, in contrast to the lower-case letters used for variables ranging over the normal, crisp regions of RCC theory). On the basis of CR, we can define further relations: CRI, the inverse of CR, the relation of 'mutual approximation' (MA) holding between two vague regions which have a common 'crisping', and so forth. What sort of properties do we want CR to have? First, it should be asymmetric (if one vague region is a crisper version of a second, then the second cannot simultaneously be a crisper version of the first), irreflexive (a vague region is not a crisping of itself) and transitive (if vague region A is a crisping of vague region B, and B of C, then A is a crisping of C). We would also want there to be alternative, and incompatible ways of crisping a given vague region (if there are no such alternatives, what can be meant by calling a region 'vague'?).

We will also find it useful to be able to describe relations between a vague region and a completely crisp one (i.e. a normal region of the type dealt with in RCC theory); in particular, the relation between a crisp and a vague region where the former can be regarded as a '*complete* crisping' (CCR) of the latter. In fact, we can use the possible relations between complete crispings of two vague regions representing *different* spatial entities in classifying the possible relations between such pairs of vague regions, as we will now demonstrate.

Obviously, depending on the initial configuration of the two vague regions, various possibilities might arise. For example if the two noncrisp regions are sufficiently far apart – for example if we consider both Leeds (England) and Buffalo (New York State) to be noncrisp – then it seems reasonable to insist that for such a configuration the relationship between any complete crisping of both regions would be DR (see Figure 12.2a). Similarly, it would seem reasonable that however we made Buffalo and the USA crisp, then Buffalo would always be a proper part of the USA (Figure 12.2b). In other cases things might not be so clear-cut: consider two species with core habitats close to but distinct from each other, but which may encounter each other in an intermediate zone marginal for both (Figure 12.2c). Here, the relationship between a pair of complete crispings of the two regions might be either DR or PO. In general, there will be some *set* of (one or more) RCC-5 relations each of which may hold between complete crispings of a pair of vague regions. This set of possible relations can itself be regarded as a relation holding between the two noncrisp regions. Which sets of RCC-5 base relations can 'hold' in this way? It seems reasonable to assume that each such set will be a *conceptual neighbourhood* (Freksa, 1992; Cohn *et al.*, 1995) of the RCC-5 set of relations. Given a set of JEPD binary

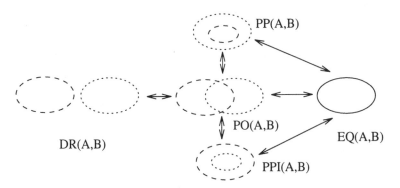

Figure 12.3 Immediate conceptual neighbours among the RCC-5 relations.

relations, such as RCC-5 or RCC-8, we can call any pair of them *immediate conceptual neighbours* if each can be transformed into the other by a process of gradual, continuous change which does not involve passage through any third relation. In the case of RCC-5 (Figure 12.3), all the other four relations are immediate conceptual neighbours of PO; and PP and PPI are immediate conceptual neighbours of EQ; but there are no other pairs of immediate conceptual neighbours.

Any subset of such a set of JEPD relations is a *conceptual neighbourhood* if any pair belonging to the subset are themselves immediate conceptual neighbours, or can form the ends of a 'chain' of relations in which each adjacent pair are immediate conceptual neighbours. There are exactly 23 possible conceptual neighbourhoods in the RCC-5 set of relations: five of rank 1, six of rank 2, seven of rank 3, four of rank 4, and one of rank 5:

{{DR }, {PO }, {PP }, {PPI }, {EQ };

{DR ,PO },{EQ ,PP }, {EQ ,PPI }, {PO ,PP }, {PO ,PPI }, {EQ ,PO };

{DR ,PO ,PP}, {DR ,PO ,PPI}, {PO ,PP ,EQ}, {PO ,PPI ,EQ}, {PP ,PPI ,EQ},

  {DR ,PO ,EQ}, {PO ,PPI ,PP};

{DR ,PO ,PP ,PPI }, {EQ ,PO ,PP ,PPI }, {DR ,PO ,PPI ,EQ }, {DR ,PO ,PP ,EQ };

{DR ,EQ ,PO ,PP ,PPI }}.

Can all of these arise as sets of possible relations between the complete crispings of pairs of noncrisp regions? We will argue the contrary. First, consider {EQ }: if this were the only possible relationship between complete crispings of two regions, the regions must surely have been crisp (and EQ) in the first place. We believe all the other rank 1 conceptual neighbourhoods are possible: we have already effectively argued for DR above when discussing Leeds and Buffalo (Figure 12.2a); for PP, and PPI just imagine one very small noncrisp region right in the centre of a very much larger one (Figure 12.2b), while for PO consider two large regions, each considerably overlapping the other, but still with a significant subregion not part of the other (like the Sahara and West Africa).

Now consider rank 2 conceptual neighbourhoods; the only one we would want to argue is not possible is {EQ ,PO }: given that the two regions are similar enough to crisp to EQ and dissimilar enough to crisp to PO then why should they not crisp to PP or PPI? Consider a pair of noncrisp regions X and Y, which have at least one pair of complete crispings that are EQ, and another pair that are PO. Consider a pair of these complete crispings of X and Y that are EQ. To reach a pair of crispings that are PO, either both of the members of the EQ pair must be enlarged (with their areas of enlargement not completely coinciding), or both must be shrunk (with their areas of shrinkage not completely coinciding). In the first case, enlarging only the complete crisping of X will give a PPI pair, enlarging only the complete crisping of Y will give a PP pair. If the PO pair is reached by shrinking the members of the EQ pair, shrinking only the complete crisping of X will give a PP pair, and shrinking only the complete crisping of Y a PPI pair. So if complete crispings can be chosen independently for any two noncrisp regions, {EQ ,PO } is not among those sets of possible relations that can occur between regions.[4] All the other rank 2 conceptual neighbourhoods seem reasonable: {DR ,PO } is possible if two noncrisp regions slightly overlap each other as in Figure 12.2(c) – the crisping might retain the overlap or result in discrete regions; for {EQ ,PP }, consider an island, with vague boundaries owing to tides or erosion, and some part of the island where a particular plant could live. The latter region might sometimes be equal to and sometimes a proper part of the former, but could never be in a PPI or PO relation to it. The duality of PP and PPI means that {EQ ,PPI } is also reasonable. We want to allow {PO ,PP } in order to cope with a small region which is somewhere around the edge of a very much larger region – the former region might be a part or might partially overlap once the two regions are completely crisped; again duality forces us to allow {PO ,PPI } as well.

Similar arguments lead us to accept just two rank 3 conceptual neighbourhoods: {DR ,PO ,PP } and {DR ,PO ,PPI }; one rank 4 neighbourhood: {EQ ,PO ,PP ,PPI } and the sole rank 5 neighbourhood {DR ,EQ ,PO ,PP ,PPI } (it seems clear that this should be allowed since if we had two very vague regions superimposed then surely any relationship might result depending on the crisping of each region).

This gives 13 out of the 23 possible conceptual neighbourhoods of RCC-5 relations as possible sets of relations between the complete crispings of a pair of non-crisp regions:

{{DR }, {PO }, {PP }, {PPI },

{DR ,PO }, {EQ ,PP }, {EQ ,PPI }, {PO ,PP }, {PO ,PPI },

{DR ,PO ,PP }, {DR ,PO ,PPI },

{EQ ,PO ,PP ,PPI },

{DR ,EQ ,PO ,PP ,PPI }}.

This constitutes a set of 13 JEPD possible relations between vague regions (compared to five in RCC-5). Can we refine this classification any further? We believe so. The crucial observation is that if one of a pair of vague regions is made completely crisp, then, in general, not all of the relationships that were originally possible will remain so. However, rather than explore this in detail now, we will return to this analysis after presenting the 'egg–yolk' interpretation of vague regions, and perform the analysis within that interpretation.[5]

12.4   The egg–yolk theory

Lehmann and Cohn (1994) suggest using two (or possibly more) concentric sub-regions, indicating degrees of 'membership' in a vague region. (In the simplest, two-subregion case, the inner subregion is referred to as the 'yolk', the outer as the 'white', and the inner and outer subregions together as the 'egg'.) This egg–yolk approach is proposed in the context of the problem of integrating heterogeneous databases, where the notions of 'regions' and 'spatial relations' are used metaphorically to represent sets of domain entities, and relations between these sets. Is it a satisfactory approach to spatial vagueness in a literal sense? How does it relate to the crisp/vague distinction outlined above?

The egg–yolk formalism developed by Lehmann and Cohn (1994) allows for just five base relations (corresponding to the RCC-5 set discussed in Section 12.2: DR, PO, PP, PPI and EQ) between any egg–egg or yolk–yolk pair, or any egg and the yolk belonging to another egg (a yolk is always a PP of its own egg). This choice of a set of base relations is not an intrinsic feature of the approach, but corresponds to the demands of the database interpretation, where the concept of two sets of entities being 'externally connected' (EC), and the distinction between tangential and non-tangential proper parts are not useful. The RCC-5 set produces 46 possible relations between a pair of egg–yolk pairs,[6] as shown in Figure 12.4. Can these 46 possibilities be identified with the possible relations between complete crispings of two noncrisp regions? We believe they can.

At first glance, there is one apparent problem with the egg–yolk approach: the most obvious interpretation is that it simply replaces the precise dichotomous division of space into 'in the region' and 'outside the region' of the basic RCC theory by an equally precise trichotomous division into 'yolk', 'white' and 'outside' – and this appears contrary to our intuitions about how vagueness 'works'. What we appear to need if we want to reflect the way people represent and reason with vague regions is something different: not only is there a 'doubtful' or 'borderline' zone around the edges of a vague region, but that zone itself does not have precise boundaries. Is there a way of using the egg–yolk formalism that is consistent with this? What is the relationship between the intuitive idea of a vague region, and an egg–yolk pair of nested crisp regions?

We are currently working on a self-contained axiomatization of reasoning about relations between vague regions, which can then be modelled or interpreted within the egg–yolk formalism. Here, we can only give an informal characterization of what this approach involves. The 'egg' and 'yolk' of an egg–yolk pair are taken to represent conservatively defined limits on the possible 'complete crispings' or precise versions of a vague region: any acceptable complete crisping must lie between the inner and outer limits defined by yolk and egg, but we do *not* assert that any crisp region meeting this constraint should be considered an acceptable complete crisping of the vague region. We leave undefined what additional conditions, if any, must be met by such a complete crisping. Thus, in this use of the egg–yolk formalism we do not claim that the borders of egg and yolk represent precise limits of a noncrisp region's penumbra of vagueness, but rather that the entire penumbra (and perhaps more) lies between these limits. This gives us the kind of indefiniteness in the *extent* of vagueness, or 'higher-order vagueness', that intuition demands. Consider the vague region 'beside my desk'. There are some precisely defined regions, such as a cube 10 cm on a side, 5 cm from the right-hand end of my desk, and 50 cm from the floor, that are undoubtedly contained within any reasonable complete crisping of this noncrisp region. There are others, such as a cube 50 m on a side centred at the front, top right-hand corner of the desk, that contain no such reasonable complete crisping. These two could correspond to the 'yolk' and 'egg' of an egg–yolk pair corresponding to the vague region 'beside my desk'; but there are precisely defined regions including the smaller and lying within the larger of the pair that would *not* be reasonable complete crispings of 'beside my desk'. We do not and need not specify *exactly* where the limits of acceptability lie.

The set of possible egg/yolk configurations is shown in Figure 12.4.[7] Configuration 1, given our interpretation of vague regions in terms of egg–yolk pairs of RCC regions, shows a pair of noncrisp regions such that *any* pair of complete crispings of the two are DR (as in Figure 12.2a). Configuration 2 (compare Figure 12.2c) can be interpreted as a pair of noncrisp regions X (dashed) and Y (dotted), such that it may be possible to choose complete crispings of the two to be DR or PO, and for any complete crisping of X or Y, a DR complete crisping of the other

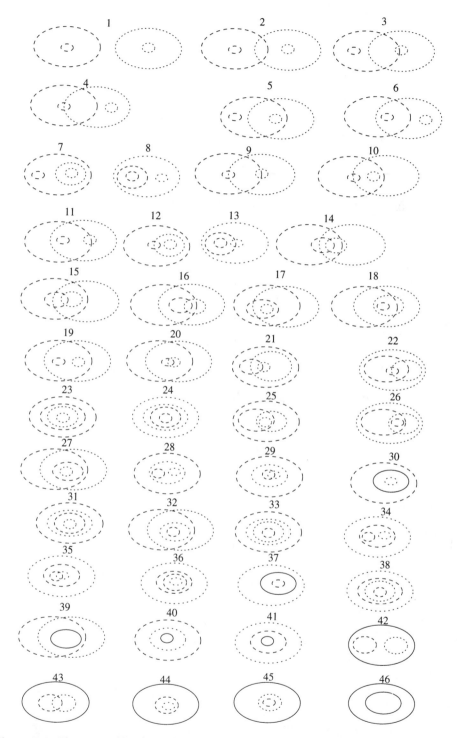

Figure 12.4   The 46 possible relationships between two egg–yolk pairs.

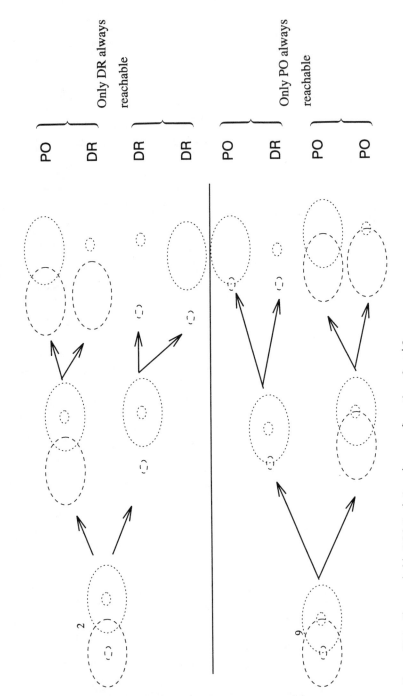

Figure 12.5 The reachable RCC-5 relations from configurations 2 and 9.

may be available, but there are some complete crispings of each for which no PO complete crisping of the other can be found. Similarly, although configuration 9 has the same set of possible RCC-5 relations between any complete crispings of the vague regions involved, only PO is always reachable. These two configurations are illustrated in Figure 12.5. A complete analysis of the possible sets of alternative complete crispings of egg–yolk pairs yields the result that the set of possibilities for a given pair of egg–yolk pairs always coincides with one of the 13 conceptual neigh-bourhoods referred to in Section 12.3. Depending on how it is done, as noted above, completely crisping one region may reduce the range of possibilities that can *then* be created by completely crisping the other.

An analysis taking account of the sets of possibilities that remain *however* either the dotted or the dashed egg–yolk pair is crisped, as well as on the full set of pos-sibilities available before either is crisped, distinguishes all 46 region pairs except for the pair 39–43. We are still investigating how this pair can best be distinguished. One approach is to list not just those RCC-5 relations that can always be reached however X is completely crisped, or however Y is completely crisped, but all *minimal* sets of RCC-5 relations of which at least *one* can always be reached under the same circumstances. In the case of configurations 39 and 43, no single RCC-5 relation between complete crispings always remains possible, once either the dotted or the dashed vague region is completely crisped. In configuration 39, however, any complete crisping of the dashed region will leave at least one from each of the following sets of relations reachable: {EQ ,PO }, {EQ ,PPI }, {PP ,PO }, {PP ,PPI }. In the case of configuration 43, the corresponding sets of relations are: {EQ ,PO }, {EQ ,PP }, {PPI ,PO }, {PP ,PPI } (the second and third sets differ, with PP replacing PPI, and vice versa). If we consider crisping the dotted region first, we find that the list of minimal sets of relations for configuration 39 is the same as that found for configuration 43 when crisping the dashed region first; and conversely, the list of minimal sets of relations for configuration 43 is the same as that found for configu-ration 39 when crisping the dashed region first.[8] The situation is illustrated in Figure 12.6.

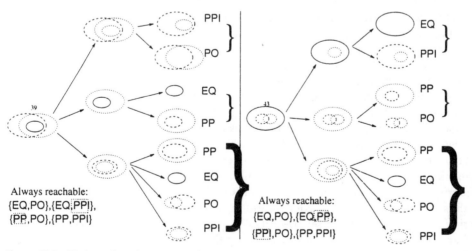

Figure 12.6 Distinguishing between configurations 39 and 43.

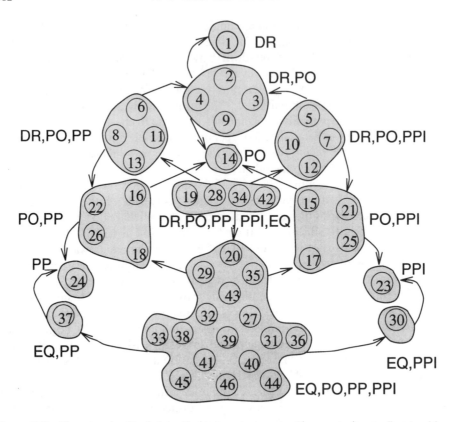

Figure 12.7   Clustering the 46 relations. Each group represents either a set of mutually crispable relations, or configurations whose complete crispings have the same RCC-5 relations. The arrows represent the crisping relationships or a subset relationship between a set of complete crispings.

With as many as 46 separate configurations or relations, it may often be helpful to cluster them. There are two conceptually different, but as it turns out, effectively identical ways of structuring the data. The first is to use the idea of possible RCC-5 relations between complete crispings of the configurations, as outlined earlier. This clusters the relations into 13 groups (see Figure 12.7). The arrows between the groups represent the fact that one set of complete crispings is a subset of another (thus the singleton sets have no outward arrows). Alternatively, one can compute the sets of configurations which are mutual *partial* crispings of each other, i.e. for any two elements in such a set, each configuration can be partially crisped into any of the others, by expanding one or both yolks and/or shrinking one or both egg, but leaving each egg with a distinct yolk and white. It turns out that one obtains exactly the same set of 13 sets of configurations; in this case one can interpret the arrows of Figure 12.7 as indicating that each configuration in the set the arrow stems from can be crisped to any relation in the set it points to. It is also fairly easy to find linguistic labels for the 13 clusters of configurations, based on the set of possible complete crispings. For example, the set {DR ,PO ,PPI }, as it applies to two egg–yolk regions X and Y, could be summed up by asserting that X is *not* a part of Y, and the set {DR ,PO } as 'neither region is part of the other'.

12.5 Related work

Probably the closest piece of work to ours is that of Clementini and Di Felice (see Chapter 11, this volume), which was written simultaneously and entirely independently in response to the call for contributions to the GISDATA workshop on regions with indeterminate boundaries which prompted our work. However, whereas our work is based on our previous logical formulation of spatial representation and reasoning, theirs is based on the point set theoretic approach inspired by Egenhofer and co-workers (Egenhofer and Franzosa, 1991; Egenhofer, 1991; Egenhofer and Herring, 1991; Egenhofer and Al-Taha, 1992). They have generalized the '9-intersection' approach of Egenhofer and Herring (1991), which classifies relations between pairs of 2D regions according to whether or not each of the interior, boundary and exterior of one region intersect (share any points with) each of the interior, boundary and exterior of the other, giving nine possible intersections to consider in all.

If a pair of 2D regions with normal 1D boundaries are considered within this approach, then just eight relations are obtained (Egenhofer, 1991), corresponding precisely to those of RCC-8. However, if we consider two 2D regions, each with a 2D boundary zone around its edge, 44 possible relations are obtained. These correspond to our 46 egg–yolk relations: there are two less because the 9-intersection model as employed by Clementini and Di Felice cannot distinguish between 30 and 31, or between 37 and 38. In both these cases, the 9-intersection gives the same result for the two configurations. In particular, the boundary zones of each pair of regions considered intersect, and no attempt is made to use the *dimensionality* of the intersection in their paper.

Clementini and Di Felice also eliminate 19, 28, 34 and 42 because they assume that the region of indeterminacy is very small in relation to the entire region. Figure 12.8 illustrates these four configurations: in each case each region's 'yolk' lies within the other's 'white'.

Notice that these four regions form a cluster in our analysis, which seems a nice result: leaving out a set of regions which form such a cluster appears more reasonable than leaving out part of such a cluster. The set of complete crispings associated with this cluster is the entire RCC-5 set: for this set of configurations, nothing can be determined about the possible complete crispings.

Clementini and Di Felice do not use any notion similar to that of one region being a crisper version of another. Indeed, one can imagine the might not want to use such a notion, since it appears that they are principally interested in modelling relations between pairs of regions, each representing as geographic entity with a (relatively narrow) 'transitional zone' around its edge. They group their relations into clusters, but this is done in a relatively *ad hoc* manner, by an informal analysis of the relations and the assignment of intuitively reasonable names to the clusters they identify thereby. Their clustering (our numbering scheme is retained for consistency) is displayed in Figure 12.9. Notice that the links in their diagram are

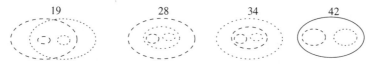

Figure 12.8   The four relations not included by Clementini and Di Felice.

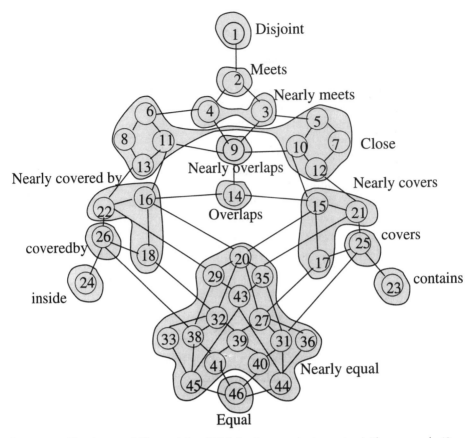

Figure 12.9   The clusters of Clementini and Di Felice (our numbering system). There are only 40 nodes, since 19, 28, 34, 42, 30 and 37 are not considered in their theory.

'closest topological distance' (i.e. minimum difference in the 9-intersections) rather than our 'is crisper than.' It is not certain what intuitive basis there is for the 'closest topological distance' relation. The two approaches produce similar but not identical clusterings.

12.6   Work in progress

This chapter reports work that is still in progress; this final section sketches some of the directions in which we are developing the ideas discussed.

The interpretation of vague regions in terms of pairs of limiting RCC regions, appears to have considerable promise in reconciling our intuitions about spatial vagueness with a consistent formal treatment. However, there is at least one aspect of vagueness which this approach, at first sight, may appear to ignore. Suppose we want to consider the relationship between a hill and the neighbouring valley, or a galaxy and the surrounding intergalactic space. It may not be possible to specify precisely *where* the boundary lies, but surely the two regions in each case are most naturally considered to be EC (Externally Connected, or touching without overlapping), in terms of RCC theory. In such a case, the choice of a complete crisping for one noncrisp region limits our range of choices for another noncrisp region (the two are nonindependent). We need to distinguish between the 'within-model' relation CR – one region being a crisping of another – from the relation between more and less sharply defined models of a real-world spatial configuration. Given a model of the hill-and-valley relationship in which the two are represented as noncrisp regions, we need to be able to specify that in any more sharply defined model in which the hill and valley are represented as crisp regions, these will be EC to each other; that is, they will meet along a common boundary but will not overlap. We believe the axiomatization of the relations between vague regions independent of the egg–yolk model we are developing will provide us with the necessary formal apparatus to express such constraints. This axiomatization, together with other formal aspects of the work, will be detailed in a forthcoming report.

We may also investigate variant notions of crispness where, for example, crisping would not be allowed to change topology (e.g. the number of maximal one-piece subregions) or shape (as measured by concavities; see Cohn *et al.*, 1995). Another aspect to be considered is temporal versus atemporal crispings: the indefiniteness of a hill/valley might perhaps be considered essentially atemporal, whereas a river might be considered essentially temporal (at any time the river is well defined, but it may change, though erosion or flooding, over time).

12.7 Conclusions

The analysis presented here provides an interesting and potentially useful way of representing and reasoning about regions with indefinite boundaries. We have informally specified some properties which we believe such regions have, and then formally axiomatized them. We have also proposed a technique for representing such vague regions as pairs of traditional crisp regions. The advantage of this approach, if it carries through fully, is that reasoning about vague regions can be performed just by translating to RCC theory and then using standard tools (e.g. composition tables) without having to build special-purpose reasoning mechanisms for the calculus of regions with indefinite boundaries.

Notes

[1] We use the word 'theory' in its logical/mathematical sense, meaning a set of formal axioms which specify the properties and relations of a collection of entities, not in the natural scientist's sense of an empirically testable explanation of observed regularities.

[2] In a many-sorted logic, the universe of discourse is divided into subsets called 'sorts', and quantifiers may be restricted to these sorts. Function and predicate symbols are defined only on particular combinations of argument sorts.

[3] Although it is planned to extend the work, using the RCC-8 relations as a basis, in the near future.

[4] There may be applications where this independence assumption would be unjustified. For example, if we wanted to permit vague regions to have a precise, fixed size, with only their position being vague. Here, {EQ ,PO } might be the full set of alternatives for complete crispings of some pairs of vague regions.

[5] We hope also to perform the complete analysis independently of the egg–yolk interpretation, but we have not yet done so.

[6] There are 42 in Lehmann and Cohn (1995), but this is because the database application considered there makes it unnecessary to distinguish different yolk–yolk relations if the eggs are EQ.

[7] Note that although the eggs and yolks are shown as one piece regions, there is of course nothing to stop them being multipiece, which might be useful in modelling some situations; e.g. one might want to consider that the Tower of London, Buckingham Palace and Westminster Palace and Abbey were always part of London whatever else was, and could thus form a trio of disconnected yolks for the London 'egg'.

[8] It is interesting to note that these sets are essentially the minimal hitting sets of Reiter (1987).

REFERENCES

BENNETT, B. (1994) Spatial reasoning with propositional logics, in: Doyle, J., Sandewall, E. and Torasso, P. (Eds) *Proc. 4th Int. Conf. on Principles of Knowledge Representation and Reasoning*, pp. 51–62, Morgan Kaufmann.

CLARKE, B. L. (1981) A calculus of individuals based on connection, *Notre Dame Journal of Formal Logic*, **23**(3), 204–218.

CLARKE, B. L. (1985) Individuals and points, *Notre Dame Journal of Formal Logic*, **26**(1), 61–75.

COHN, A. G. (1987) A more expressive formulation of many sorted logic, *Journal of Automated Reasoning*, **3**(2), 113–200.

COHN, A. G., RANDELL, D. A. and CUI, Z. (1995) Taxonomies of logically defined qualitative spatial relations, in: Guarino, N. and Poli, R. (Eds), *Formal Ontology in Conceptual Analysis and Knowledge Representation*, Dordrecht: Kluwer (special issue of *Int. J. Human–Computer Studies*).

EGENHOFER, M. (1991) Reasoning about binary topological relations, in: Gunther, O. and Schek, H. J. (Eds), *Proc. 2nd Symposium on Large Spatial Databases, SSD'91*, Zurich, Switzerland. *Lecture Notes in Computer Science* **525**, 143–160, Berlin: Springer.

EGENHOFER, M. and FRANZOSA, R. (1991) Point-set topological relations, *Int. J. Geographical Information Systems*, **5**(2), 161–174.

EGENHOFER, M. and HERRING, J. (1991) *Categorizing Binary Topological Relationships between Regions, Lines and Points in Geographic Databases*, Technical report, Department of Surveying Engineering, University of Maine.

EGENHOFER, M. J. and AL-TAHA, K. K. (1992) Reasoning about gradual changes of topological relationships, in: Frank, A. U., Campari, I. and Formentini, U. (Eds), *Theories and Methods of Spatio-temporal Reasoning in Geographic Space, Lecture Notes in Computer Science 639*, pp. 196–219, Berlin: Springer.

FREKSA, C. (1992) Temporal reasoning based on semi-intervals, *Artificial Intelligence*, **54**, 199–227.

GOTTS, N. M. (1994) How far can we 'C': Defining a 'doughnut' using connection alone, in: Doyle, J., Sandewall, E. and Torasso, P. (Eds) *Proc. 4th Int. Conf. on Knowledge Representation and Reasoning*, pp. 246–257, Morgan Kaufmann.

LEHMANN, F. and COHN, A. G. (1995) The EGG/YOLK reliability hierarchy: Semantic data integration using sorts with prototypes, *Proc. 3rd Int. Conf. on Information and Knowledge Management*, pp. 272–279, Gaithersburg, MD: ACM.

RANDELL, D. A. (1991) *Analysing the Familiar: Reasoning about Space and Time in the Everyday World*, PhD thesis, University of Warwick.

RANDELL, D. A. and COHN, A. G. (1989) Modelling topological and metrical properties of physical processes, in: Brachman, R., Levesque, H. and Reiter, R. (Eds), *Proc. 1st Int. Conf. on the Principles of Knowledge Representation and Reasoning*, Los Altos, pp. 55–66, Morgan Kaufmann.

RANDELL, D. A. and COHN, A. G. (1992) Exploiting lattices in a theory of space and time, *Computers and Mathematics with Applications*, **23**(6–9), 459–476 (also appears in *Semantic Networks*, Lehmann, F. (Ed.) pp. 459–476, Oxford: Pergamon, 1992).

RANDELL, D. A., COHN, A. G. and CUI, Z. (1992a) Naive topology: Modelling the force pump, in: Struss, P. and Faltings, B. (Eds), *Advances in Qualitative Physics*, pp. 177–192, Cambridge, MA: MIT Press.

RANDELL, D. A., CUI, Z. and COHN, A. G. (1992b) A spatial logic based on regions and connection, *Proc. 3rd Int. Conf. on Knowledge Representation and Reasoning*, San Mateo, pp. 165–176, Morgan Kaufmann.

REITER, R. (1987) A theory of diagnosis from first principles, *Artificial Intelligence*, **32**, 57–95.

PART FIVE

Data Models for Indeterminate Objects and Fields

P. A. BURROUGH

Introduction

The five chapters in Part 5 have been brought together because, in their different ways, they all address the problems of dealing with indeterminate objects and fields in a computerized information system. They all are located in that extremely important, but often diffuse zone, between the intuitive recognition of a phenomenon, and the choice of a suitable formal data model with which to represent it. In Chapter 13 David *et al.* present the current state of the art view from the Technical Committee 287, which was set up in 1991 by the European Committee on Standards. The aim of TC 287 is to define standard conceptual models for storing and exchanging geographic information, based on current and future practice and needs. The aim is not to be restrictive, but to provide flexibility to deal with known and unknown situations.

The members of TC 287 soon discovered a problem in the lack of common terminology, both within and across nationalities and disciplines. They have developed the ideas of conceptual geometry, and geometric and structure primitives. Unlike many other approaches, these are not arranged hierarchically, but in parallel. The approach is explained in detail in Chapter 13. TC 287 is also very concerned with providing information about the quality of geographic information, and a conceptual model for quality is provided in terms of abstract views of the universe, the views through measuring devices, the semantic information model used, continuous and discrete attributes, linage, accuracy, completeness, consistency, and currency.

How do geographic objects with indeterminate boundaries fit in with the ideas of TC 287? David *et al.* conclude that uncertainty and fuzziness are problems of the application domain and the semantic level of identifying objects and fields. We are led to conclude that if objects and fields are acceptable models, then TC 287 has a standard theoretical data model that can be enlarged to include fuzziness. However, it is up to the application scientist to decide just what are, and what are not, objects and fields. It is also important to note that, as explained by Ferrari and Freksa in

Part 3, that if the conceptual descriptions of similar phenomena do not match, no amount of technical standardization will bring the meanings closer together.

In Chapter 14, Molenaar starts where David *et al.* end, i.e. with the application domain or discipline, the context and time. Molenaar points out that GIS are increasingly being used to follow the dynamics of temporal changes rather than merely being tools for inventory and display. He notes that objects in a GIS are representations of conceptual entities that aid understanding of a given process on the Earth's surface, but for many reasons that understanding can be incomplete.

Molenaar explains the various levels of 'data modelling' from conceptual models down to machine representation in bits and bytes. He points out that these levels intersect application disciplines, geoinformation theory and computer science. In order to satisfy all groups there should be suitable syntactic structures for handling spatial information, and an example of such a structure is presented in Chapter 14. The proposed syntax explores many-to-one relations between sets, object classes and their attributes, and the geometry of 'area objects'. The representation of area objects in vector and raster conventions is explained in terms of graph theory, with the conclusion of syntactic similarity of vector and raster maps, *as long as* the raster elements are cells covering the area they represent and are not just grid-centre coordinates to which attributes are attached.

Having explored the semantic syntax for well-defined objects, Molenaar explains how the semantics of uncertainty (either stochastic or fuzzy) should be incorporated into the data models. A three-way link between syntax–semantics–uncertainty has to be created. Molenaar explores the three situations of fuzziness, imprecision and insufficient evidence, and suggests ways in which uncertainty about static spatial objects should be treated. He concludes that this is only a beginning: there is much to do.

Whereas David *et al.* and Molenaar start from an 'object' viewpoint, and to a greater or lesser extent explore how exact objects can be handled under conditions of uncertainty or fuzziness, in Chapter 15, Laurini and Pariente begin by rejecting object-oriented data models as being insufficiently realistic for modelling natural phenomena. For them, an ability to manage complex continuous fields is essential, and it is sensible to include the principles of dealing with continuous fields in a formal object-oriented approach. Operators are needed to manipulate objects and attributes and also to estimate rates of spatial change of attributes. Laurini and Pariente present a syntax for dealing with morphological discontinuities, continuity and differentiability in surfaces and preserving attribute values at sample points or zones. They present an object-oriented model which aggregates relationships between the class *field* and the classes *sample_point*, *statistical_area* and *discontinuities*, which relate to previous geographic primitives of points, areas and lines, respectively. The class *field* is presented as a new data type (primitive) in its own right, defined by a name, a statement it represents, its dimension, the intervals in each dimension and location-specific validity, sets of sample points, constraints and discontinuities. They then present the first specifications of a field-oriented language and give some simple applications.

Whereas Chapters 14 and 15 suggest theoretical and practical extensions of spatial theory to deal with inexact objects and fields, in Chapter 16 Hadzilacos explores an alternative strategy which is to see how well conventional GIS theory and methods, in particular the idea of the layer or overlay either in vector or raster representation, could serve to model inexact phenomena. His investigations lead him to the conclusion that layers are indeed very sensible ways of separating fuzzy objects. He also concludes that it is probably more sensible to extend current GIS to deal with continuous variation and continuous membership functions than to start again from scratch. He believes that moving from an exact to a fuzzy domain does not invalidate conventional geometry – it is merely an extension. As in fuzzy set theory there is an implicit suggestion that most current GIS and RDBMS deal neatly with the ideal special case – exact geometry, exact attributes, exact topology. Extending this to a general model does not invalidate the theory behind the model, but just makes life more interesting and the representations more realistic.

In Chapter 17 Brändli approaches the problem of determining and extracting terrain features from digital elevation models (DEMs). By definition, a DEM is a continuous hypsometric surface. By implication it contains entities – ridges, peaks, pits, valleys – but these are difficult to extract automatically because they do not have clear boundaries and many definitions of the features contain ambiguities. Brändli presents a top-down, object-oriented approach to solving the problem with three levels: the continuous surface with a low level of detail, contiguous patches with a moderate level of detail, and graphs with a high level of detail. A conceptual model is developed for the representation of semantic and topological properties of specific terrain feature classes. As the demand for more local information increases, so the models move from general continuous surfaces to contain more detailed, structured and ordered information. The more general information creates a framework in which both the more specific and the more uncertain or ambiguous aspects of the definition of terrain features can be handled.

Taken together, these five chapters demonstrate that not only is there a need for extending GIS database theory to take account of indeterminate objects, but that such extensions can be made by extending, rather than by replacing, current theory. In the views expressed by these authors the current models that are restricted to well-defined, crisp entities or to smooth, continuous fields, are merely simplified special cases of a richer spectrum of general spatial 'objects'. These conclusions are encouraging because they suggest that no major conceptual difficulties stand in the way of the development of a truly general GIS which can provide a full flexibility of data types for many kinds of spatial description and analysis. The difficulties lie more perhaps with users (who must change their attitudes to indeterminately structured data) and existing software systems that rigidly enforce a simplified view of the world.

CHAPTER THIRTEEN

Conceptual Models for Geometry and Quality of Geographic Information

B. DAVID

Institut Geographique National, Recherche et Developpement, Saint-Mande, France

M. VAN DEN HERREWEGEN

Institut Geographique National, Belgium

F. SALGÉ

MEGRIN group and IGN, France

13.1 Introduction

The European Committee on standards (CEN, Comité Européen de Normalisation) set up in October 1991 a Technical Committee (CEN/TC 287) whose remit is to create standards in the field of geographic information. In this context geographic information means any information which can be related to a location on the Earth. This location can be expressed either by a set of coordinates such as latitude, longitude, and elevation, or by an indirect reference to a position by means of, for example, postal addresses or identifiers of administrative areas. This standardization effort addresses not only the requirements of topography or map-making, but also those of scientific domains such as environment, land planning, geology, social sciences, and the like, wherever location is central. Having considered the experience of previous standardization efforts in Europe and in the US, specifically those related to exchange standards, TC 287 thought it valuable to adopt a top-down approach, i.e. first to devise concepts, and second to define transfer mechanisms.

In its digital context, geographic information lies somewhere between CAD-CAM (due to its graphical component), information technology (due to its semantics), and telecommunications (due to the requirements of interoperability of systems). TC 287 has therefore analyzed the existing relevant standards and created a reference model which identifies the diverse backgrounds in which standardization will take place, and what standards are lacking and therefore need to be developed.

Inherited from information technology, any geographic information representing the reality of a phenomenon is supported by a set of hypotheses, including a conceptual model of the data, both of which allow modelling of the perceived reality. This modelling work can be compared with the creation of an abstract view of the universe by simplification, discretization and sampling of reality. In its simplest abstraction, geographic information consists of a geometric level, which allows the positions of and the spatial relationships between real-world entities to be identified, and a semantic level, which clarifies their meaning and their functional relationships.

With the goal of being general and able to take account of any modelling approach, TC 287 has defined a general conceptual model for the geometric part of geographic information. It is assumed that the semantic part can be dealt with by using a standard modelling approach as defined within information system technologies. This geometric part is presented in Section 13.2 of this chapter.

Geographic information is a digital representation of the abstract view of the reality, i.e. it provides digital data that fit the specifications fixed by the modelling approach. One outstanding question that remains – does the geographic information represent reality? – can be split into two rather different sub-questions: is the modelling approach relevant to the observed phenomenon? Do the data meet the specifications? Together, the answers to these questions allow us to evaluate the most important user requirement: fitness-for-use. Quality as defined here is one of the most important work items of TC 287, and is addressed in Section 13.3.

Even with the aim of being general, the standards being developed represent the state-of-the-art in the operational use of geographic information. In Section 13.4 we therefore try to identify further studies that would allow better use of geographic information within the next generation of GIS.

13.2   The conceptual model for geometry

13.2.1   Scope and history of TC 287/SWG 2.1

The scope of the Sub-Working Group 2.1 (SWG 2.1) of TC 287, entitled 'Geographic Information – Data description – Geometry', is officially defined as follows:

> To define and represent geometric primitives and their constructs in geographic information using the formal languages defined in the work item Techniques. This will include the identification of the types of geometry relationships, the rules for expressing them and the definition of any constraints. It will include, if deemed necessary, non-geometric elements.

It is obvious that in order to fulfil that task some research, at least at the technological level, has to be performed. On the other hand, the Technical Committees within CEN have not been set up to suggest areas of research, but to standardize such areas for which some general need is felt. In order to overcome this ambiguity, the SWG 2.1 has adopted the following approach:

1. to compile and study a list of available documents dealing with 'geometry' (or conceptual data models), including existing 'standards' such as NTF, EDIGéO, ATKIS, GDF, SDTS, etc., as well as recent publications or studies on the structure of geographic information, and
2. to create a new conceptual schema for 'geometry', trying to avoid as much as possible the major drawbacks of existing formats, and at the same time profiting from the experience acquired during the creation and/or implementation of those exchange formats.

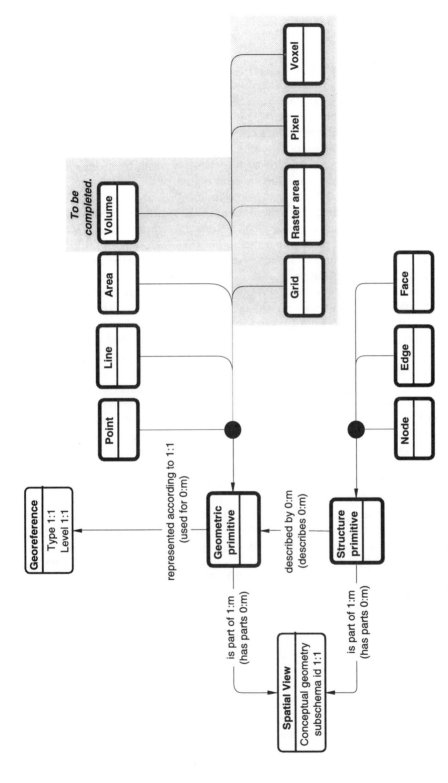

Figure 13.1   Main components of conceptual geometry subschemas.

Needless to say, one of the major problems encountered during the discussions was the lack of a common terminology.

13.2.2   Conceptual geometry subschema

A conceptual geometry subschema can be interpreted as part of an application conceptual schema. Since the real world can be perceived in different ways, several application conceptual schemas and, as a consequence, several conceptual geometry subschemas can be considered. The main components of a conceptual geometry subschema – the geometric primitives, the structure primitives, the georeference system, and the spatial view – are illustrated in Figure 13.1.

In *geometric primitives*, both vector data (point, line, area, etc.) and raster data (grid, pixel, etc.) are considered. Typical *structure primitives* are the node, edge and face. In contrast with most existing conceptual data models, the geometric and structure primitives are treated in parallel rather than hierarchically. This will allow us later to describe a feature by means of zero or many geometric primitives and/or zero or many structure primitives.

The *georeference system* is part of a position information system where not only spatial position, but also temporal changes are taken into account. In the spatial position, a distinction is made between indirect positioning (using non-coordinate reference systems such as addresses) and mathematically defined coordinate reference systems. In this chapter we will only deal with mathematically defined reference systems to define geometric primitives.

As stated earlier, several conceptual geometry subschemas can exist. In the *spatial view* the content (geometric and/or structure primitives) and the rules (relations) of a specific conceptual geometric subschema are defined. Each specific conceptual geometry subschema however has to comply with a *general* conceptual geometry subschema (G0), which should be powerful enough to face all kinds of applications. Apart from this restriction, producers and users of data are free to develop and to document any type of conceptual geometry subschema.

In order to fit some frequent needs, it would be very convenient to dispose of some predefined conceptual geometry subschemas. So far, SWG 2.1 has considered the conceptual geometry subschemas shown in Table 13.1. In the following, only the geometric and structure primitives of the general subschema will be discussed. All the other subschemas are part of the general subschema. For their content and rules, see document N114 of CEN/TC287/WG 2.

Table 13.1 Conceptual   geometry   subschemas considered by SWG 2.1

Code	Name
G0	General subschema
G1	Full planar graph topology
G2	Planar graph line network
G3	Non planar graph line network
G4	Spaghetti
G5	Triangulated irregular network
G6	Raster image

13.2.3   General conceptual geometry subschema

Geometric primitives

By *geometry* we understand the representation of the spatial position of a feature, including its shape and size. The *spatial position* is the position in space based either on coordinates of two or three dimensions in a mathematically defined system or by indirect positioning. In this chapter we will only deal with mathematically defined reference systems. In this case, the definition of a geometric primitive becomes: a *geometric primitive* is an element which directly by means of coordinates and mathematical functions describes the geometry of a feature.

Vector data.   In vector data, a distinction is made between points, lines, areas and volumes. A *point* is a 0-dimensional geometric primitive. The spatial position of a point is described by one set of coordinates referring to one georeference system. A *line* is a bounded continuous one-dimensional geometric primitive. A line can be closed or not. Furthermore, a line can be either a 'line segment', a 'line string' or a 'line arc':

- a *line segment* is a line described by two sets of coordinates and the shortest connection between them;
- a *line string* is a line described by an ordered sequence of sets of coordinates and the shortest connection between them;
- a *line arc* is a line described by an ordered sequence of sets of coordinates and connections between them that are defined by one set of mathematical functions.

An *area* is a bounded continuous two-dimensional geometric primitive, delimited by one outer non-intersecting boundary and zero or more non-nested non-intersecting inner boundaries.

The distinction between the geometric primitives 'line' and 'area' has been made for the following reasons:

1. In the case of the geometric primitive 'line', the line itself (even if it is closed) is the carrier of the most important information. The area inside a closed line is of only secondary importance.

2. In the case of the geometric primitive 'area' the opposite is true. The boundary line in this case is of secondary importance, and the area itself is more important. The boundary is only there to delimit the area.

A *boundary* is a closed one-dimensional non-intersecting element defined by a boundary line. Just as in the case of the geometric primitive 'line', a boundary line can be composed of a boundary segment, a boundary string and a boundary arc.

Grid and raster data.   Up to now, a distinction has only been made between the geometric primitives 'grid' and 'raster area':

- A *grid* is a regular point distribution defining a 2D rectangular pattern on a 2D rectangular part of the base surface. Any of the four corners of the grid can be considered as the origin. Subsets of the geometric primitive 'grid' are the grid elements 'grid column' and the 'grid row'. A *grid element* is the smallest element of a grid. It is identified by a row number and a column number from which its spatial position can be computed.

- A *raster area* is a two-dimensional geometric primitive which is a 2D rectangle regular tessellation of a 2D rectangular part of the base surface. Subsets of the geometric primitive raster area are the 'raster column' and the 'raster row'. A *pixel* is a two-dimensional geometric primitive which is a cell of a raster area. It is defined by a row number and a column number from which its spatial position can be computed.

Structure primitives

A *structure primitive* is an element that describes the relative position of features. Three types of structure primitives have so far been considered: node, edge and face.

Node. A *node* is a 0-dimensional structure primitive. A distinction is made between isolated nodes and connected nodes.

- An *isolated node* is a node that is not related to any edge.
- A *connected node* is a node that is related to one or more edges. A further distinction is made between terminating nodes and intermediate nodes.
- A *terminating node* is a connected node terminating an edge. A terminating node can at the same time be the 'start' node and the 'end' node of an edge.
- A *intermediate node* is a connected node related to an edge without terminating it. It can however be the terminating node of another edge.

The following spatial relations have been defined for the structure primitive 'node':

- an isolated node belongs to 0 or m faces;
- a terminating node starts 0 to m edges;
- a terminating node ends 0 to m edges;
- an intermediate node is coincident with 1 to m edges.

Edge. An *edge* is a one-dimensional structure primitive connecting one start node and one end node. Both terminating nodes can be coincident. The following spatial relations have been defined with respect to an edge:

- an edge has one start node and one end node; these two nodes can coincide;
- an edge may contain 0 to m intermediate nodes;
- an edge has 0 to 1 next right edge;
- an edge has 0 to 1 next left edge;
- an edge has 0 to 1 previous right edge;
- an edge has 0 to 1 previous left edge;
- an edge has 0 to m left faces;
- an edge has 0 to m right faces;
- an edge is a component of 0 to m rings.

Face. A *face* is a minimum two-dimensional structure primitive defined by one outer ring and zero to many inner rings. A *ring* is a closed one-dimensional element

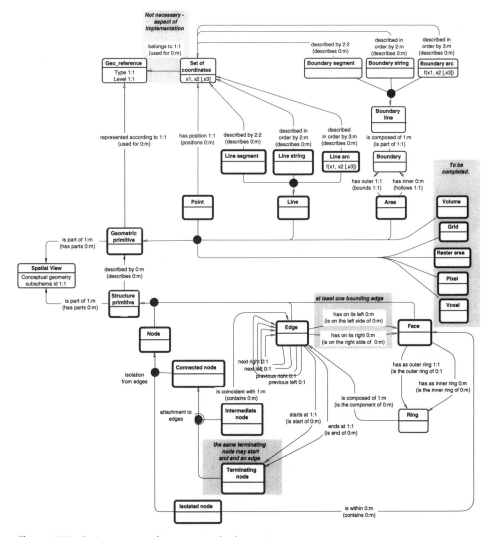

Figure 13.2   Basic conceptual geometry subschema G0.

defined by edges. The following spatial relations have been defined with respect to faces and rings:

- a ring is composed of 1 to *m* edges;
- a ring is the outer ring of 0 to 1 face;
- a ring is the inner ring of 0 to *m* faces;
- a face has one outer ring;
- a face has 0 to *m* inner rings;
- a face can contain 0 to *m* isolated nodes.

The general conceptual geometry subschema G0 is summarized in Figure 13.2. For the graphical expression of this conceptual schema, the STANLI notation has been used.

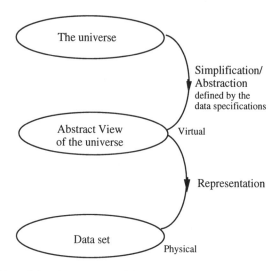

Figure 13.3   Definition of the abstract view of the universe.

13.3   The conceptual model for quality

13.3.1   Status of the work

This section summarizes the work of SWG 2.2 of TC 287 that is currently in progress. A first draft was issued in August 1994 and the work was continued by a project team between October 1994 and June 1995. The team has produced a draft of the standard, which has been circulated to the members of TC 287 for comments and approval.

For the definition of the quality of a data set, it is necessary to define with what the data set should be compared. The data set can not be compared to the real world because not all information from the real world is represented in the data set. The data set has to be compared with the view of the world seen through the data specifications. This concept is developed in Section 13.3.2 and is the most original part of the work presented here. Section 13.3.3 presents the quality aspects defined in the conceptual model.

13.3.2   Abstract view of the universe and data specifications

Abstract view of the universe

In order to be able to define the quality parameters of a data set it is first necessary to define precisely the process that allows us to define the data set from the real world (also called the universe). This process is decomposed into two steps: the first step is called abstraction and is defined by the data specifications, and the second step is called representation and produces the data set. To be able to distinguish between these two steps, the concept of abstract view of the universe is defined. Because reality can not be described in all its complexity, the data specifications define a view of the universe which is a theoretical data set and is called the abstract view of the universe. In other words, the abstract view of the universe is a view of reality defined by the data specifications. Figure 13.3 illustrates these definitions.

The quality of a data set defines both the ability to abstract the universe into its abstract view, and the fidelity of the data set compared with the abstract view of the universe.

Remarks

(1) The universe is always viewed through a capture device, such as an aerial photograph, an existing map that has to be digitized, an existing database, or something else in the universe.

(2) A sequence of several processes may be necessary to produce a data set. In this case the two steps (abstraction and representation) are iterated. In map digitizing, for example, we look at two processes: first, the making of a map, and, second, its digitizing. In the map-making process, the universe is abstracted following the map specifications that define the 'abstract view of the universe for the map'. Then the data are represented on the map.

The second process is map digitizing. The real world is viewed through the existing map. Digitizing specifications define the 'abstract view of the universe for the data set'. The data set is obtained as the result of the representation of this second process. This iteration is shown in Figure 13.4. In this example, the quality of the final data set may be defined either from the map or from the universe. The resulting quality will be very different in the two cases. Therefore, the specifications must clearly define the process to which the quality information refers.

(3) The abstract view of the universe may be abstracted from several data sources such as old maps, aerial photographs, or field surveys. However, the abstract view is

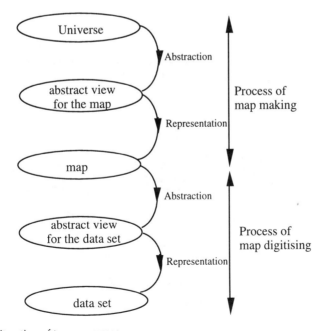

Figure 13.4 Iteration of two processes.

always unique for a given data set because it is defined by the data specifications of
the data set, and not by the capture specifications on the different sources. This case
is shown in Figure 13.5.

The semantic information model

Geographic information is described by features, attributes, and relationships.
Similar features (or relationships) are described by a feature class (or relationship
class). Attributes are associated with a feature class and with a relationship class.
There are four types of attributes:

- Quantitative (also called continuous) attributes, which correspond to quantities
  that can be measured in a given unit. Such values can be expressed as real
  numbers but may also be discretized by defining a resolution. Examples include
  width, temperature, height, etc. The definition of a quantitative attribute must
  include the reference system and the unit.

- Qualitative (also called discrete) attributes, which do not correspond to quantities but usually to a finite set of values that can be enumerated. Examples
  include classes, codes, etc.

- Geometry, which was detailed in Section 13.2. The definition of a geometric
  attribute must include the reference system and the unit.

- Description: text, graphs, photographs (less structured in this model, or with an
  irrelevant structure for this model).

By conversion, any features has at least two attributes: the name of its class and an
identifier; and any relationship has at least four attributes: the name of the relationship, an identifier, and at least two feature identifiers which are connected by the
relationship.

Data specifications

The abstract view of the universe is defined by the data specifications of the data set.
They should contain at least the following information:

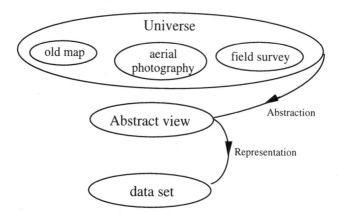

Figure 13.5 A unique abstract view of the universe even when there are several data sources.

- a precise definition of the feature class;
- the conceptual schema for the feature class;
- reference system and units for quantitative attributes and geometry;
- selection criteria;
- accessibility of the information (for example, whether it is classified military information, or is technically difficult to obtain);
- geometric representation of the entities;
- constraint rules;
- aim for quality parameters;
- resolution;
- time of reference;
- update policy;
- description of the lineage information.

13.3.3 Quality aspects

The definition of the quality of a data set is expressed using an informal hierarchy of quality aspects shown in the pyramid in Figure 13.6. The first level of the pyramid, called minimum lineage information, shows the minimum quality information that should always be attached to a data set. This includes the date and method of data capture, the source document, the organization that captured the data, and the name and function of the responsible person within that organization, its address,

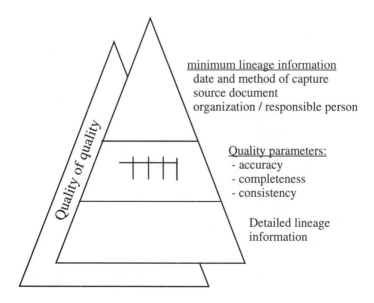

Figure 13.6   The pyramid of the quality aspects (at a given time *t*).

etc. This first level gives only minimum qualitative information on the quality of the data set.

The second level of the pyramid, called quality parameters, gives quantitative information on the quality of the data set. The three quality parameters (accuracy, completeness and consistency) are defined and explained later in this section.

The third level of the pyramid, called detailed lineage information, gives more detailed information such as the definition of the date of capture or source document per feature in the data set. This detailed lineage information depends on the objective of the data set; it can not be generalized, but must be described precisely in the specifications of the data set.

Any quality aspect must be reported. It can be either qualitative, such as how the quality aspect was determined; or quantitative, such as the confidence interval on the quality. A quality parameter can be decomposed into several components. For example, an interesting component of accuracy is the *ability for abstraction*, which is a measure of how well a feature of the universe can be defined in the abstract view of the universe. Ability for abstraction is defined by a distribution of possible abstractions on the abstract view of the universe, so that the parameters are the same as those for accuracy. For example, the edge of a wood can be fuzzy. Several lines may be abstracted on the abstract view of the universe and one is no better than the others. Ability for abstraction is the accuracy of the definition of this line.

Finally, all these aspects are given at a given time t. For example, the completeness of a data set varies through time and a given value of completeness is only valid at the time it was measured.

Lineage

Lineage describes the history of the data, including a description of the source material (dates and methods of data capture), transformations applied (type, functions, parameters), and responsible organizations (name and address of the producer or editor). The data come from different sources and are transformed by different transformations, as illustrated in Figure 13.7. Information can be given on each transformation and each source.

Accuracy

Accuracy describes the stochastic errors of observations and is a measure that indicates the probability that an assigned value is accepted. Accuracy can apply to various kinds of data: feature, relationship, and attribute (quantitative, qualitative, geometry and description). The accuracy of feature and relationship data measures an error on the name of the feature class or on the name of the relationship. Therefore its description is similar to that of a qualitative attribute.

The accuracy of a qualitative attribute is the probability that a correct value has been assigned to the attribute, including the probability of correctly assigning alternative values. In general, it is a vector. $\{i \text{ or } \omega_i, P(X = \omega_i)\}$, where the ω_i are the possible values, and X is the assigned value. There are several ways to reduce this description.

The accuracy of a quantitative attribute is a measure that indicates how far the assigned value is from the best estimation of the true value (the mean value), and is described by a probability distribution. There are several ways to reduce this

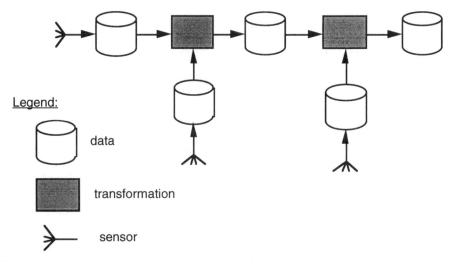

Figure 13.7   Example of a process of production of a data set.

description (standard deviation, histogram of deviations, confidence interval and level, etc.). Accuracy values are given in the same unit and reference system as the measured value.

For geometry, the description of the accuracy of points is similar to that of quantitative values. On lines and areas, a distance between lines should be used (such as the maximum lateral distance or the Hausdorff distance). The description of the accuracy of grid points and pixels is similar to that of a qualitative attribute.

For descriptions, there are two types of error: wrong logic or wrong syntax (e.g. wrong spelling), and wrong semantics (e.g. the spelling may be correct, but the meaning is wrong). The measure of the accuracy of a description is not relevant to this standard.

Completeness

Completeness is the symmetric difference between the abstract view of the universe and the data set at a given moment. It is given by two ratios: *missing data* (also called omission) and *over-completeness* (also called commission). The missing data is expressed by the ratio of the number of occurrences existing in the abstract view of the universe that are not present in the data set, divided by the number of occurrences in the abstract view of the universe. The over-completeness is expressed by the ratio of the number of occurrences in the data set that does not exist in the abstract view of the universe, divided by the number of occurrences in the data set.

Completeness can only be evaluated on a *set* of features, or relation instances. This set can be either the full data set or a subset of it.

Consistency

Consistency includes static consistency, which is the result of the validation of semantic constraints of data and relations; and dynamic consistency, which is the result of the validation of process. In the context of standardization of data transfer, only static consistency will be considered.

Validation is a check of a constraint defined in the specifications. It includes domain validation (on features, attributes, and geometry, and their sets) and relational validation (on relations), as described in the conceptual schema.

Currency

Currency deals with change through time. It describes correctness of completeness and accuracy of a data set at a certain time t (for example, now). Currency probably means two different things to the producer and the user of a data set. For the producer, currency means an updating policy which should be part of the data specifications, whereas the user wants to know whether the data set is still correct for his/her use at the time he/she wants to use it. This depends on both the update policy and the *change frequency*. The change frequency can be roughly estimated for a given zone, for example by a number of changes per class feature, or aggregated into a unique figure by defining weights for each kind of change.

13.4 Conclusions

The models being developed by TC 287 take into account the real requirements of operational applications. The introduction of fuzziness into such systems is bound to increase. In addition, time and 3D modelling are likely to be increasingly used.

Fuzzy boundaries in reality can nevertheless be modelled. To the geometric primitives, one can attach attributes with the aim of defining transition areas, and quality criteria which provide information on the degree of positional approximation to be linked to the primitive: to some extent thresholds in semantic space can be converted into lines in geometric space with a corresponding geometric accuracy.

Nonetheless, the main limitation of such approximations lies in the ability to measure the relevant parameters imposed by the model. Sooner or later the limitations of the model itself will appear and then the fuzziness problem stops being a problem of computer science and starts again to become a problem in the application domain. The experience gathered in the standardization efforts allows us to predict that once the geometric level is stabilized (and here, are there other options than the classic graphical primitives, vector or raster?) and understood in the area of application, the problem lies fully at the semantic level in identifying objects and fields in reality that need to be modelled, and the way they can be abstracted. Further research is needed; soon the diverse approaches to the modelling of reality from the semantic point of view will need to be compared in order to identify the commonalties between the objects and fields being defined. The dichotomy between objects and fields may then disappear, the former being the result of assigning thresholds to the latter.

The key point of standardization is to create standards to facilitate the reusability of geographic information over space and time. TC 287 has recognized that although central to geographic information, the geometric level is not the bottleneck, and efforts should be made to improve understanding of the semantic level and the way the meanings of phenomena are modelled in their interactions over time.

CHAPTER FOURTEEN

A Syntactic Approach for Handling the Semantics of Fuzzy Spatial Objects

MARTIEN MOLENAAR

Centre for Geo-Information Processing (CGI), Wageningen Agricultural University,
The Netherlands

14.1 Introduction

The objects represented in a geo-information system will always be conceptual entities that play a role in some terrain description. This role is made up by the different relationships they have with other objects and their behaviour in time under external influences or due to intrinsic factors. Consequently one should understand the object definitions in the context of such a role pattern and behaviour pattern. This context has several aspects (Molenaar, 1993).

The first aspect is the discipline or disciplines of the users, i.e. are they working in a cadastral environment, soil mapping, demography, etc. Each discipline has its own definitions of terrain objects, classes and attributes. These definitions will depend on the mapping discipline, as well as on the scale or other aggregation level of the mapping, i.e. whether the mapping is made at a local, regional, national, or even continental level. At each level different sets of elementary objects will be relevant, while the elementary objects at one level can sometimes be aggregates of elementary objects at a lower level; for example, at a municipality level a GIS may contain houses, streets and parks, while at a national level a GIS may contain towns and urban areas.

Another aspect of the user's context is the aim of the mapping, and for what tasks the information will be used, e.g. for monitoring the terrain situation, for the identification and analysis of processes, for planning purposes, etc. Each field of activity generally has its own requirements for the type of terrain description and thus for the data to be handled by a GIS, although there are often overlaps between these requirements. A further information aspect of the user's context is the time of the terrain description. In many disciplines and user environments the relevance of data changes with time. In agriculture, for example, the requirements for soil information have changed in recent years. Originally the main interest was to analyze the suitability of soils for sustaining different crops, but the interest has changed to include the capacity of the soil to bind chemical elements that could harm the environment. In the cadastral world we see that, whereas the original tasks were to protect land titles and to raise taxes, they now often play an increasing role in the analysis of economic processes, such as the dynamics of prices of land and buildings, and of the number of sales and mortgages.

These considerations imply that the objects represented in a geo-information system are not simply entities that are 'out there'; rather, they are representations of conceptual entities that play a role in some descriptive model of the Earth's surface. This model defines the semantics of the objects and their role patterns. Each user context will have its own semantics. Many applications have some kind of indeterminacy in their descriptive models in the sense that it is not always clear how the conceptual entities in the model should be linked unambiguously to real-world phenomena. This might be due to the fact that the measuring procedures for data acquisition have stochastic components, but it might also be that the data categories of the model have fuzzy definitions, or that inferences are based on insufficient data. This means that uncertainty will always be an intrinsic aspect of information so that if data models are developed for geo-information then they will not be complete if they are not able to deal with data uncertainty.

Burrough and Frank (1995) describe many different ways of perceiving the world, and indeed in the application fields of GIS there are many descriptive models for phenomena and processes at the Earth's surface with different levels of complexity and at different levels of aggregation. These models express the way the world is perceived and thought about. If these models are somehow to be represented and operationalized in (geo-)information systems then they have to be mapped onto data and processing models that can be handled by computers. Geo-information theory should play an essential role in the development of such geo-information models. The role of this discipline, among the other disciplines involved in spatial modelling and spatial data modelling, can be best illustrated with reference to the different levels of data modelling. When this role is clear it will be possible to explain at which modelling level spatial syntaxes should be formulated, and how they can help to bridge the gap between the spatial models formulated for terrain descriptions and the logical data models for DBMS.

14.1.1 Levels of data modelling

The representation of objects and their relationships in a geo-information system will be in the form of data that should be structured according to some data base concept. The mapping of the original terrain description onto a data model that can be handled by a computer will be done in several steps, each one at a different level according to their conceptual distance from the actual machine architecture. At the lowest level the data are structured in bits, organized in bytes, blocks and pages; these are structures that a machine can handle. The mapping of the data onto these structures is called the physical data model or implementation model. This step is represented by the innermost circle of Figure 14.1. The data modelling at this level is conceptually rather remote from the way most users of information systems think. That is why some data models at a higher level than the physical model have been

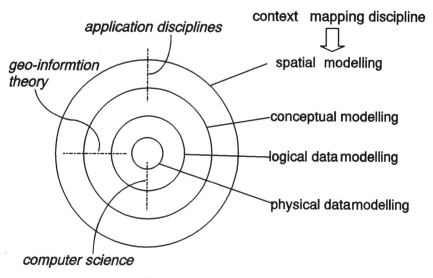

Figure 14.1   Levels for data modelling.

developed in recent years, such as the relational model or object-oriented models. These database models or logical data models are easier for database users to understand. The logical data models are represented by the second inner circle of Figure 14.1.

Before an application can be mapped onto a logical data model the relevant data and their mutual relationships should first be identified. Hence one should decide which entities are important for a particular application, and how they should be represented with their mutual relationships. In a GIS application this means that the terrain features should be represented with their thematic and geometric descriptions. It must also be decided whether their geometry will be expressed in a vector or raster structure. This modelling step is conceptual data modelling, which is represented by the second outer circle of Figure 14.1.

A conceptual model for geo-information should satisfy the requirements of some GIS application; this could be a mapping activity such as soil or vegetation mapping, or some activity such as spatial management or spatial analysis as for demographic studies, physical planning activities, etc. Spatial aspects will play an important role in the terrain descriptions for such applications. The formulation of such descriptive models will be called spatial modelling, this is represented by the outermost circle of Figure 14.1.

14.1.2   The role of geo-information theory

Several disciplines have developed data models at these different levels. Their involvement often overlaps at different levels, but it is possible to make a distinction between these disciplines by the levels they emphasize.

- *The application disciplines* are mainly involved in the formulation of spatial models. Starting from this level they search for adequate conceptual models to develop GIS applications. The application disciplines are also involved in

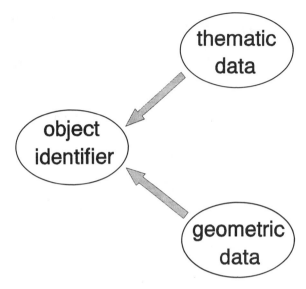

Figure 14.2   The basic structure for representing terrain objects in GIS.

mapping their problem onto some logical model, but they seldom go beyond that level.

- *Geo-information theory* analyzes the relationship between conceptual and spatial models for a large variety of applications. This is to study the common characteristics of conceptual models for classes of applications and to understand how these can best be mapped onto some logical models. This should lead to the formulation of design criteria and strategies that can be used by experts to support and advise users on the development of their applications.

- *Computer science specialists* are primarily interested in the development of powerful concepts for database models and their implementation. These should be useful for a large variety of applications, also beyond the GIS environment. These specialists are mainly involved at the modelling levels represented by the two inner circles of Figure 14.1. This does not mean that they should not pay attention to the conceptual model level and even beyond – of course they should do that to check whether their models are relevant and adequate.

That means that geo-information theory should define syntactic structures for handling spatial information. This chapter introduces a syntax for handling information about terrain objects, with their object classes and their attributes, the geometric data and the different types of relationships among these objects. This syntax should be able to handle different semantics defined within different user contexts; furthermore, it should be possible to implement this syntax in information systems through the definition of the data types and link types among these data types. Section 14.2 introduces a notation for such a syntax.

A terrain description contained in a GIS will be called a map. If the geometry of a map *M* has vector structure then *M* will be called a *vector map*; similarly, if the geometry has a raster structure then *M* will be called a *raster map*. Many other types of tessellation are possible for a two-dimensional space, but we restrict our discussion to these two. Many of the statements made later in a comparison of the raster and vector geometry can also be extended to those other tessellations.

For our further developments we will assume that representations of terrain objects in a geo-information system have the basic structure shown in Figure 14.2. This representation consists of an object identifier to which the thematic data and the geometric data are linked. With the notation developed in Section 14.2 it will be possible to explain that vector maps and raster maps have basically the same syntax. This notation will also help us to understand how uncertainty can be expressed when semantics are introduced into the model.

A two-dimensional terrain description could contain point, line and area objects. The following elaborations are restricted to the representation of area objects. A further restriction with respect to the concept of objects as found in modern computer science is that object dynamics or object behaviour will not be discussed here, and no attention will be paid to composite or aggregated spatial objects. The main aim of this chapter is to formulate a syntax that can handle the semantics of state descriptions of fuzzy spatial objects. No explicit attention will be paid to the time or scale dependency of the object states; rather, we take it that any uncertainty due to these aspects can somehow be expressed as the uncertainty of the object state as such. The basic concepts of the syntactic approach can be sufficiently explained under these restrictions.

14.2 A syntax for object representations

14.2.1 Many-to-one relations between sets

A conceptual data model identifies a number of data types and relations between them. A syntax for such a model will be formulated in a mathematical sense, and the data types will be interpreted as discrete sets between which relations will be defined (Gersting, 1993). This model is graphically represented in Figure 14.3. Each ellipse represents a set identified by a label; an ellipse labelled S represents a set S with the elements s_i: $S = \{\ldots, s_i, \ldots\}$ and an ellipse labelled M represents a set $M = \{\ldots, m_j, \ldots\}$. In Figure 14.3(a) the elements of subsets of M are related to elements of S through some binary relation $\Re \subset M \otimes S$. Each subset will be labelled with a

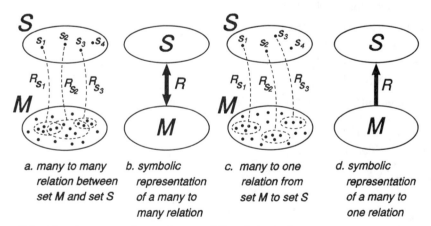

a. many to many relation between set M and set S

b. symbolic representation of a many to many relation

c. many to one relation from set M to set S

d. symbolic representation of a many to one relation

Figure 14.3   Many-to-many and many-to-one relations between two sets.

symbol R identifying the type of relation that generates the subset, and an index s_i identifying the generating element or owner in set S. For some subsets R_{s_k} and R_{s_l} we might find that $R_{s_k} \cap R_{s_l} \neq 0$, in which case $\Re$ is a *many-to-many relation* because the subsets of M are overlapping; see Figure 14.3(a). This situation is represented symbolically in Figure 14.3(b), where the two-headed arrow means that the relation is many-to-many.

Let $R = \{\dots, R_{s_i}, \dots\}$ be the collection of all subsets of M generated by $\Re$, if all these subsets are disjoint, i.e. if for all combinations of subsets $R_{s_k} \cap R_{s_l} = 0$, then the relation $\Re$ is *many-to-one* from set M to set S. This situation is illustrated in Figure 14.3(c) and represented symbolically in Figure 14.3(d); the arrow represents the fact that the relation is many-to-one from set M to S. This means that in Figure 14.3(d) $\Re$ relates 0 or more elements of set M to each element of S (see Ullman, 1982; Molenaar, 1991a,b).

If an element $s_i \in S$ generates a subset $R_{s_i} = \{m_p, \dots, m_q\} \subset M$, then s_i is the owner of R_{s_i}, and therefore the set S is called the owner set of R. The elements of M are members of the subsets generated by R; therefore M will be called the member set of R. For the subset R_{s_i} a membership function will be defined so that:

$$R[m_t, s_i] = 1 \Leftrightarrow m_t \in R_{s_i},$$

$$R[m_t, s_i] = 0 \Leftrightarrow m_t \notin R_{s_i}.$$

The fact that for $s_i \neq s_j \rightarrow R_{s_i} \cap R_{s_j} = 0$ implies that

$$R[m_t, s_i] = 1 \rightarrow R[m_t, s_j] = 0.$$

The choice of membership relationships of this form was made for two reasons; they can easily be used to compute cardinality functions of sets, and they can easily be adapted for the situations where we have to deal with fuzzy sets (Klir and Folger, 1988). These relationships could alternatively be expressed as predicates, or as (labelled) binary relationships. In the latter form they can be compared to the regular entity relations of Chen (1976).

The membership functions can be used to formulate some elementary queries for an information system in which these relationships have been implemented (see Molenaar, 1991a,b):

1> To which subset R_s does m_t belong?
   > give s for which $R[m_t, s] = 1$.
2> Which elements m belong to R_{s_i}?
   > give m for which $R[m, s_i] = 1$.
3> Does m_t belong to R_{s_i}?
   > give $x = R[m_t, s_t]$.

The complements of the first two queries will be found by using $R[m, s] = 0$.

If each element m_j of M occurs in exactly one subset $R_{s_i} \in R$, then $\Re$ is a *many-to-one mapping* from M to S. In that case the collection of subsets $R = \{\dots, R_{s_i}, \dots\}$ defines a *partition* of M.

14.2.2 Object classes

Let $\mathbb{U}_M$ be the set of all objects occurring in a map M; then $\mathbb{U}_M$ is called the *universe of discourse* of M, or, in short, *the universe* of M. Objects in this universe can be distinguished because they have different characteristics; for most GIS applications these differences will be primarily thematic. A thematic description of the objects is then required to express the differences, and this can be done by means of attributes that take values per object. The complete description of an object in a GIS then consists of the thematic component expressed by the attribute values and a geometric component. Two objects can only be distinguished if their descriptions are not equal.

Class intension and extension

Carnap (1956) explains how classes are defined by the properties that are characteristic of their members, i.e. the objects of a class have properties that distinguish them from objects that do not belong to the class. This implies that for each object of a universe $\mathbb{U}_M$ a decision must be made to assign it to the class or not. Criteria should therefore be formulated for each class to specify which objects should be considered as its class members. These criteria express (following Carnap's terminology) the *intension* of the class, and they can be operationalized by a decision function that can be applied to the objects to test whether they fulfil the criteria or not. If an object O_i passes the test for a class C_j, then it will be a member of that class and will be expressed by the membership function

$$M[O_i, C_j] = 1 \quad \text{if } O_i \text{ is a member of } C_j,$$
$$= 0 \quad \text{otherwise.}$$

For each object O_* we define the function by specifying the classes to which the object belongs:

$$CLASS(O_*) = \{C_{**} | M[O_*, C_{**}] = 1\}$$

($_*$ and $_{**}$ represent unspecified index values). If the classes are defined so that they are mutually exclusive, then for each object the extension of this function contains only one class. For O_i we find in that case

$$CLASS(O_i) = \{C_j\}.$$

The *extension* of a class is the set of all the objects that belong to it, hence

$$EXT(C_j) = \{O_i | M[O_i, C_j] = 1\}.$$

It is possible that a class C has been defined to which no objects belong, then the extension is empty, i.e.

$$EXT(C) = \varnothing.$$

Suppose that for a demographic study a database is to be generated with descriptions of the populated areas of a country. The following classes are defined:

- *villages*,   intension: places with $100 <$ no. of inhabitants < 1000,
- *towns*,   intension: places with $1000 <$ no. of inhabitants $< 50\,000$,
- *cities*,   intension: places with $50\,000 <$ no. of inhabitants $< 15\,000\,000$,
- *mega-cities*,   intension: places with $15\,000\,000 <$ no. of inhabitants.

There are some countries that have mega-cities, but for most countries this class will have no members so that the extension is empty.

Objects, classes and attributes

The general situation in a terrain description is that there are sets of objects that are so different that these differences can not be expressed by the attribute values alone. Each set should then have its own description structure, i.e. its own list of attributes; these sets will be called *object classes* or just *classes*. The classes are typified by the fact that the objects that belong to the same class share the same descriptive structure.

A class C_j then has a list of attributes:

$$LIST(C_j) = \{A_1, ..., A_r, ..., A_n\}.$$

Each attribute will have a name, and for each attribute a domain will be specified defining the complete set of attribute values. The scale type of the domain indicates whether these values are from a nominal scale, an ordinal scale, an interval scale or a ratio scale. An attribute can then be specified by a three-tuple:

$$A_i = \{NAME(A_i), SCALETYPE(A_i), DOMAIN(A_i)\}.$$

The attribute structure of an object is determined by the class to which it belongs, so that each object has a list containing one value for every attribute of its class. An object inherits the class attribute structure, i.e.

$M[O_i, C_j] = 1$ implies that $LIST(O_i) = \{a_1, ..., a_r, ..., a_n\},$

                     with $a_r = A_r[O_i]$ is value of A_r for object O_i,

                     and $A_r \in LIST(C_j),$

                     and $a_r \in DOMAIN(A_r).$

Figure 14.4 illustrates the relationship between objects and classes and the attribute structures. The thematic description of an object can now be specified by the class of the object and the list of attribute values. The class specifies the names of the attributes, hence:

$$THEM(O_i) = \{CLASS(O_i), LIST(O_i)\}.$$

Classes are semantically distinct if they have different attribute structures:

    $i \neq j$ implies that $LIST(C_i) \neq LIST(C_j).$

We will assume that this is the case, which means that objects belonging to different classes have different descriptions. In the relational database model this would mean that a table can be defined for each object class.

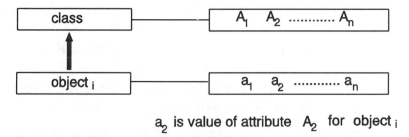

Figure 14.4 is value of attribute A_2 for object i

Figure 14.4   Diagram representing the relation between objects, classes and attributes.

14.2.3   The geometry of area objects

The representation of area objects in a vector map

The geometric structure of a vector map can be described by means of cell complexes. For a two-dimensional map these consist of 0-cells, 1-cells and 2-cells (Frank and Kuhn, 1986). The 0-cells and 1-cells play similar roles as nodes and edges, respectively when the geometry of the map is interpreted as a planar graph; the 2-cells can then be compared to the faces related to the planar graph through Euler's formula (Gersting, 1993; Wilson, 1972). The terminology of the planar graph interpretation will be used here.

The geometry of an elementary area object is represented by one or more adjacent faces. If a face F_f is part of an area object O_a this will be represented by:

$$PartAO[F_f, O_a] = 1.$$

If there are overlapping area objects then each face might be part of several objects, but also each object will consist of one or more faces. Therefore this is a many-to-many relationship. Because all area objects are related to faces this is a mapping from the set of area objects to the set of faces. Overlapping objects can be found through the common faces.

The relationships between edges and area objects can be found through their relationships with the faces, in combination with the relationship between the faces and the area objects. Each edge will always have one face on its left-hand side and one on its right-hand side. These relationships will be expressed by the following functions:

- Edge e_i has area object F_a on its left-hand side $\rightarrow Le[e_i, F_a] = 1$.
  For any $F_b \neq F_a$ we then get
  $Le[e_i, F_b] = 0$.
- Edge e_j has area object F_a on its right-hand side $\rightarrow Ri[e_j, F_a] = 1$,
  and again for $F_b \neq F_a$ we then get $Ri[e_j, F_b] = 0$.

If an edge e_i is part of the border of F_a then only one of the functions Ri and Le is equal to 1, but not both. So if we define the function

$$B[e_i, F_a] = Le[e_i, F_a] + Ri[e_i, F_a],$$

then when e_i is part of the boundary of F_a we find $B[e_i, F_a] = 1$.

The relationships between edges and area objects can then be found as follows:

$$Le[e_i, O_a] = MAX_{F_f}(LE[e_i, F_f] \times PartAO[F_f, O_a]).$$

There is only one face for which $Le[e_i, F_f] = 1$. If for that face $PartAO[F_f, O_a] = 1$ then the product is 1, in all other cases the product will be 0. Consequently, if edge e_i has area object O_a on its left-hand side $\rightarrow Le[e_i, O_a] = 1$. Similarly, we can write

$$Ri[e_i, O_a] = MAX_{F_f}(Ri[e_j, F_f] \times PartAO[F_f, O_a])$$

if edge e_j has area object O_a at its right-hand side $\rightarrow Ri[e_j, O_a] = 1$.

The transitions from the indirect relationships between edges and area objects to the direct relationships are illustrated in Figure 14.5, where the many-to-many relations are represented by the double-headed arrows. The relations between faces and

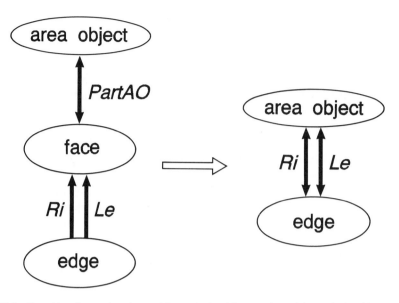

Figure 14.5   Transition from edge–face–object relationships to edge–object relationships.

area objects are many-to-many, so that will also be true for the derived relations between edges and faces. The combination of these two functions gives for edge e_i:

$$B[e_i, AO_a] = Le[e_i, O_a] + Ri[e_i, O_a].$$

If an edge e_i is part of the border of O_a, then only one of the functions Ri and Le is equal to 1, but not both. So for such an edge we find $B[e_i, O_a] = 1$. If e_i has O_a on both its left- and right-hand sides, $B[e_i, O_a] = 2$, in which case it is running through O_a. If $B[e_i, O_a] = 0$ there is no direct relationship between e_i and O_a (see Molenaar, 1991a,b).

The graph structure of rasters

Objects might also be represented in a raster geometry. To understand the similarity between such a representation and that of the previous section the cells of a raster are interpreted as faces. A graph can be linked to a raster according to the generic structure of Figure 14.6; $edge\{n_{i,j}, n_{i+1,j}\}$ is the border between $cell_{i,j-1}$ and $cell_{i,j}$ and $edge\{n_{i,j}, n_{i,j+1}\}$ is the border between $cell_{i-1,j}$ and $cell_{i,j}$. In fact, we have

$$Le[edge\{n_{i,j}, n_{i+1,j}\}, cell_{i,j-1}] = 1, \qquad Ri[edge\{n_{i,j}, n_{i+1,j}\}, cell_{ij}] = 1;$$

and

$$Le[edge\{n_{i,j}, n_{i,j+1}\}, cell_{i,j}] = 1, \qquad Ri[edge\{n_{i,j}, n_{i,j+1}\}, cell_{i-1,j}] = 1.$$

The only difference between this interpretation of a cell raster and the planar graph with its faces of the previous section is that the cells are rectangular faces so that the relationship between the number of nodes, edges and cells is fixed. Due to this structure, each edge has a unique pair of adjacent faces, so that instead of specifying the edges by their beginning and ending nodes, they can be specified by the cells at

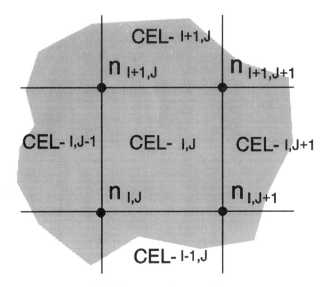

Figure 14.6   A generic structure for linking a graph to a cell raster.

their right- and left-hand sides, so that

$$Edge\{n_{i,j}, n_{i+1,j}\} \equiv Edge\{cell_{i,j}, cell_{i,j-1}\}$$

and

$$Edge\{n_{i,j}, n_{i,j+1}\} \equiv Edge\{cell_{i-1,j}, cell_{i,j}\}.$$

With this interpretation cell rasters appear to be special cases of a planar graph. This conclusion implies that terrain objects can be represented in rasters in the same way as in vector maps with a planar graph structure.

Area objects represented in a raster

The geometry of an area object is represented in a raster by one or more adjacent cells. If a $cell_{i,j}$ is part of an area object O_a this will be represented by

$$PartAO[cell_{i,j}, O_a] = 1.$$

The relationships between edges and area features can be found via their relationship with the cells, for the left-hand side relationship:

$$Le[Edge\{n_{i,j}, n_{i,j+1}\}, O_a \equiv Le[Edge\{cell_{i-1,j}, cell_{i,j}\}, O_a] = PartAO[cell_{i,j}, O_a]$$

or

$$Le[Edge\{n_{i,j}, n_{i+1,j}\}, O_a] \equiv Le[Edge\{cell_{i,j}, cell_{i,j-1}\}, O_a]$$
$$= PartAO[cell_{i,j-1}, O_a],$$

and for the right-hand side relationship:

$$Ri[Edge\{n_{i,j}, n_{i,j+1}\}, O_a] \equiv Ri[Edge\{cell_{i-1,j}, cell_{i,j}\}, O_a]$$
$$= PartAO[cell_{i-1,j}, O_a]$$

or

$$Ri[Edge\{n_{i,j}, n_{i+1,j}\}, O_a] \equiv Ri[Edge\{cell_{i,j}, cell_{i-1,j}\}, O_a]$$
$$= PartAO[cell_{i,j}, O_a].$$

The combination of these two functions again gives

$$B[e_*, O_a] = Le[e_*, O_a] + Ri[e_*, O_a].$$

The boundary of O_a consists of the set of edges with $E_a = \{e_* \mid B[e_*, O_a] = 1\}$.

The syntactic similarity of vector and raster maps

It should be stressed that for the discussion about the similarity of the syntactic structure of raster and vector representations, a raster is regarded as a grid of cells. Then the difference between raster maps and vector maps is due only to the fact that a raster can be described by a planar graph where the faces have a fixed geometry, i.e. the faces are rectangular so that there is also a fixed relationship between the faces and the edges and nodes. This implies that the position information for the faces or cells can be inferred directly from the cell number if a proper coding has been applied. The cell numbers can then be directly related to the numbers of the nodes of the cell boundary and thus a short-cut can be made in the computation of the cell location. In a similar way, short-cuts can be made in establishing the topological (adjacency) relationships between cells. In vector maps the faces do not have a fixed geometry so that the relation between faces and edges and nodes has more degrees of freedom. The identification of the location of the faces requires that the related boundary nodes are found through the edges. Similarly, the topological relationships between faces should be established through their boundaries. The difference is then only a difference in computational efficiency for the location and the topological relationships of the geometric elements, because rasters are special cases of planar graphs in this respect.

The expressive power of both geometries is the same, however, because the syntactic structure for relating geometric elements to (area) objects is the same. This relation for both geometries is based on the function $PartAO[F_f, O_a]$. This function expresses the relationship between a face f and an area object a; in a raster map a cell will be substituted for face f. Through the faces the relationships between the edges and the area objects can be established, i.e. it is possible to identify the edges to describe the boundaries of the objects. For an edge i and an object a these relationships are represented by the functions $Le[e_i, O_a]$ and $Ri[e_i, O_a]$. The discussions concerning the relationships between objects and geometry in Section 14.3 are therefore restricted to these functions. These statements about the syntactic similarity of vector and raster geometries can also be extended to other types of tessellation.

If the interpretation of this chapter of a raster as a collection of cells is not strictly maintained, the comparison of vector and raster structures might come out slightly different. The discussion in Section 1–10 of Tomlin (1990) is an example of this. There the raster elements are considered to be grid points that carry the information about a rectangular zone from which they are the centre points, whereas here the raster elements are cells that cover exactly the area they represent. These interpretations are semantically different in the sense that the relation between the elements of the raster data model and the real world are different, resulting in a different interpretation of the geometric relationship between vector and raster representations.

14.3 Semantics and uncertainty

14.3.1 Uncertainty aspects of spatial objects

A syntax for geographic information systems should have a formalistic (discrete) mathematical structure that can be handled by finite state machines. The syntax developed in Section 14.2 to handle the semantics of spatial models explains how information about spatial objects can be formulated. In its basic form the syntax links object identifiers to information about object classes, thematic attributes and geometric data. Through this syntax statements can be formulated about objects and their mutual relationships, but such an elaboration is beyond the scope of this chapter.

The topic of discussion here is how the semantics of fuzzy spatial objects can be handled in the presented syntax. This is important because statements about real-world phenomena will always have a certain level of indeterminacy. This is certainly true if the syntax of the statements and the information categories that occur within it have (discrete) mathematical definitions, as in our case, because real-world phenomena can generally not be described as discrete categories. So we are forced to a discretization that can not be done with absolute precision or certainty, since there will always be some doubt as to whether the mapping of the real world onto the discrete categories has been done correctly or adequately. Furthermore, as explained in Section 14.1 data handled in a GIS environment refer to objects that are in fact conceptual entities within some semantic framework defined by the data user's context. More often than not, the users are not very specific about the definition of their objects and object classes, etc.

Thus statements about the real world are to be formulated and understood within a certain context (Molenaar, 1993). For such statements not only is it important to know how the semantics can be handled by the syntax, but the uncertainty related to the statements should also be considered. A linguistic model should therefore be able to handle the interrelations of the three-tuple: syntax, semantics and uncertainty, as illustrated in Figure 14.7. The central part of this figure represents a linguistic model that takes care of these interactions; data models and data processing models can then be formulated as implementations of this linguistic model in an information systems environment.

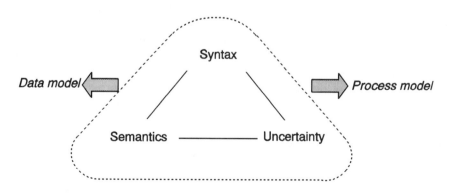

Figure 14.7 A linguistic model should interrelate syntax, semantics and uncertainty.

Uncertainty is related to statements such as $x \in S$, i.e. element x belongs to (sub)set S (see Molenaar, 1993). Statements of this type can be grouped into three categories, related to the information categories defined in previous sections:

- S is a subset of the universe $\mathbb{U}_M$, i.e. S is an object class in the sense of Section 14.2.2 and $x \in S$ means that an object x will be assigned to class S,

- S is a subset or a value of some attribute domain; in this case the formula means that a value will be assigned to some attribute of some object,

- S is a region covered by some object, S is then a subset of the point set representing the geometric space of the map; in this case the formula means that some face is part of that region.

The uncertainty of such statements implies the risk that they might lead to wrong decisions, with the consequence that inadequate actions will follow. This of course should be avoided, or at least the risk should be reduced to an acceptable level. Therefore we should try to understand the causes of the uncertainty, and also what different kinds of uncertainty there are. The formula $x \in S$ has three components, and uncertainty can be related to each one of them (Klir and Folger, 1988).

First, the definition of a subset S may be fuzzy. No sharp criteria can be formulated to determine whether an element x belongs to S or not. The theory of fuzzy subsets gives mathematical rules for handling this type of uncertainty. Fuzzy subsets are distinct from classical 'crisp' subsets in the sense that for crisp subsets the membership function $M[x, S] = 1 \rightarrow x \in S$ (i.e. x belongs to S) or $M[x, S] = 0 \rightarrow x \notin S$ (i.e. x does not belong to S), whereas for fuzzy subsets $0 \leq M[x, S] \leq 1$; hence x may belong a little bit to S. The rules formulated in this theory are mainly qualitative in nature because it is difficult to evaluate $M[x, S]$ for fuzzy subsets (Klir and Folger, 1988).

Second, the description of x may be not precise. In GIS, x stands for a terrain object for which the geometry and the attribute values should be evaluated. In many cases this will be done through measuring procedures or through the processing of measured data. Measuring operations generally introduce stochastic components in the observed data. These stochastic components propagate through the processing steps applied to the data. The uncertainty introduced here can then be dealt with mathematically by means of stochastic models, which means that the uncertainty can often be expressed in terms of variances and probabilities.

Third, there may be insufficient evidence to assign an element x to a subset S. The problem is now that it is not clear whether a particular element x fulfils the criterion or not. In such situations the subset S might be crisp and the description of the element x might be precise, but still we are not sure whether we should decide whether $x \in S$ or $x \notin S$. The mathematical rules for handling this type of uncertainty are given by the theory of evidence or the theory of fuzzy measures (Klir and Folger, 1988). Sometimes it is possible to evaluate these fuzzy measures, such as when they can be expressed as likelihood functions or as Bayes probabilities. But more often this is not the case, so that the rules are again of a qualitative nature.

14.3.2   Handling the uncertainty of spatial objects

Before the object data are entered into the database a decision should be made as to their description structure. For their geometric description a choice should be made between the raster and the vector structure. We saw that this is mainly a choice between the computational efficiency of the geometry of the raster structure versus the geometric flexibility of the vector approach. There are no syntactic differences between these two approaches. For the thematic description of objects the relevant classes with their attribute structure and the attribute domains should be specified.

When the descriptive structure (i.e. the syntax) of the spatial object representation has been defined, the object data must be expressed in this structure. We saw that the syntax of this chapter identified three categories of assignment statements that might involve uncertainty: the assignment of object classes, the assignment of attribute values to objects, and the assignment of a spatial description to objects. These three categories will be briefly reviewed here, with the emphasis on the geometry.

The assignment of objects to object classes

The intension of the object classes can be expressed by a decision function that tests the objects and then assigns them to the classes if they pass the test. If an object O is assigned to a class C then $M[O, C] = 1$; otherwise $M[O, C] = 0$. If the test does not give a definite result then the assignment of object O to class C will be uncertain, and this will be expressed by a fuzzy class membership function that does not take a definite value 0 or 1. In that case we find $0 \leq M[O, C] \leq 1$. Section 14.3.1 explained that there could be several types of reasons for this uncertainty. The function specifying the classes of O defined in Section 14.2.2 should now be modified so that:

$$CLASS(O) = \{C \mid M[O, C] > 0\}.$$

The extension of this function might contain several classes even if they are considered to be mutually exclusive. In that case the object can in principle belong to only one class, but the uncertain decision function does not give a definite conclusion as to which one. The object inherits the attribute structures of the classes to which it belongs, so the attributes should be evaluated. It might very well be that the data required for the evaluation of the attributes contain sufficient information for a definite class assignment in the case of mutually exclusive classes.

The assignment of attribute values to objects

The domains of the attributes may contain values or value classes that are not clearly defined. Another possibility is that the values of the domain are clearly defined, but there is no sufficiently reliable measuring procedure to obtain these values with high accuracy. The evaluation of an attribute A for an object O should then be expressed by a two-tuple $A[O] = (a, m_u)$, where a is the estimated attribute value and m_u is a measure of the uncertainty that a is the correct value. The uncertainty measure m_u should be of a type related to the type of uncertainty in the sense of Section 14.3.1.

The assignment of spatial descriptions to objects

The geometric description consists of the topology of the objects, their shape and their position. The *topology* can be expressed through the relationships between the objects, the faces and the edges. For both raster and vector geometries the basic link between (area) objects and geometry is made through the function $PartAO[F, O]$; in vector geometry F represents a face, and in raster geometry it represents a raster cell. If the objects O are fuzzy in the sense that their spatial extension is uncertain, then this can be expressed by the fact that the function does not only take the value 0 or 1, but that it can take any value $0 \leq PartAO[F, O] \leq 1$. This means for raster geometry that the resolution, i.e. the cell size, should be chosen so that the value of this function is homogeneous per cell. In vector geometry the faces should be defined so that the function is homogeneous per face. The uncertainty of the topo-logical relationships between edges and objects can be established through the cells or faces, the uncertainty of the function $PartAO[F, O]$ will propagate to the functions $Le[e, O]$ and $Ri[e, O]$. This can be done in two steps:

- For each combination of an edge e_i, a face F_f and an object O_a we find

$$Le[e_i, F_f] \times PartAO[F_f, O_a] = MIN(Le[e_i, F_f], PartAO[F_f, O_a])$$

and

$$Le[e_i, F_f] \times PartAO[F_f, O_a] = MIN(Le[e_i, F_f], PartAO[F_f, O_a]).$$

The functions $Le[\,]$ and $Ri[\,]$ still have the values 0 or 1 due to the geometric structure of the object description.

- The second step is again the evaluation of the functions

$$Le[e_i, O_a] = MAX_{F_f}(Le[e_i, F_f] \times PartAO[F_f, O_a])$$

and

$$Ri[e_i, O_a] = MAX_{F_f}(Ri[e_i, F_f] \times PartAO[F_f, O_a]).$$

If an edge has a face at the left- or right-hand side that has an uncertain relationship to an area object O_a, then this is propagated to the functions $Le[\,]$ or $Ri[\,]$, respectively.

These expressions can be simplified for raster geometry because of the fixed relation-ships between cells and edges. These fuzzy relationships between edges, faces and objects can be used to derive fuzzy topological relationships between area objects, such as fuzzy adjacency, fuzzy overlap and fuzzy containment.

The *shape and position* information is mainly contained in the coordinates of the nodes of the object boundaries. Because of the uncertainty of the functions $Le[\,]$ and $Ri[\,]$ the boundary of objects can not be determined in the simple way of Section 14.2.3. For fuzzy objects, boundaries can only be established for specified certainty levels. If this is done for an object O, then the faces can be selected for which $PartAO[F, O] > C(=$ certainty level), and the object boundary at this level can then be found through the edges of these faces. The precision of the position and shape of these boundaries is then a function of the accuracy of the coordinates of the nodes. This accuracy depends mainly on the measuring procedure and on the idealization accuracy of the face boundary, i.e. the accuracy of the identification of the face boundary. In rasters the node positions have been defined by the cell geometry, so that the position and shape accuracy of the boundaries at specified certainty levels follows directly from the accuracy of the raster definition.

14.4   Conclusions

A clear distinction should be made between the models that describe how scientists from different disciplines perceive the world, and the representation of these models in an information systems environment. The many types of models identified by Burrough and Frank (1995) should be mapped onto discrete data sets with their mutual relations and related data processing models if they are to be implemented in an information system. Geo-information theory should play an important role in this step and formulate strategies for mapping the spatial models onto data models. A general characteristic of spatial models is the interrelationship between thematic and geometric data, and the uncertainty aspects of these data.

This chapter has analyzed some syntactic structures of data models dealing with this interrelationship. This was done in a static approach, i.e. the emphasis was on the structure of state descriptions of objects. For dynamic objects these state descriptions should be labelled by a time attribute to indicate the period for which a particular state description is valid. For such objects the actual description may change between periods and this dynamic behaviour may be one of the causes of uncertainty in the object data. It is not very likely, however, that the object dynamics will have any effect on this syntactic structure of the object state description. That implies that when the syntax explained in this chapter is applied to the state description of objects at different periods, then the uncertainty due to object dynamics should be expressed within this syntactic structure, i.e. it should be expressed through the uncertainty of the state descriptions as proposed in Section 14.3. A similar reasoning could be applied when terrain descriptions at different scale levels are considered. It might be that objects at one scale can be considered as composites of objects at some larger scale. The uncertainty of the object descriptions at the larger scale will then affect the uncertainty of the description of the composite objects. If the same syntactic structure is used for the descriptions at different scale levels, then again the tools for expressing the uncertainty will be similar for each level.

There are several possibilities for the geometric description of spatial objects, which are given by the different types of tessellations that are available. The most common tessellations for GIS applications are given by the vector structure (or the general form of cell complexes) and rasters. It seems, however, that these solutions have basically the same syntactic structure for the representation of spatial objects, and this conclusion can in principle also be extended to the other tessellations. The choice between the different tessellations is in fact a choice between the flexibility of geometric description versus computational efficiency for position and topological relationships. The expressive power of these geometric models is the same. This also implies that there is syntactically no difference between these geometries when fuzzy objects are to be represented. The links between these objects and the geometric elements can be expressed as fuzzy relations, and through these fuzzy topological relationships and fuzzy position data can be derived. These conclusions raise the question of whether the existing tools for handling geo-information are indeed incapable of handling fuzzy spatial object data, or whether this deficiency is due to the fact that our concepts have not yet been developed far enough for us to understand how the more complex spatial models could be mapped onto data models provided by computer scientists.

REFERENCES

BURROUGH, P. A. and FRANK, A. U. (1995). Concepts and paradigms in spatial information: Are current geographical information systems truly generic? *Int. J. Geographical Information Systems*, **9**(2), 101–116.

CARNAP, R. (1956). *Meaning and Necessity: A Study in Semantics and Modal Logic*, 2nd edn, Chicago: University of Chicago Press (Midway Reprint Edition 1988).

CHEN, P. P. S. (1976). The entity–relationship model: Toward a unified view of data, *ACM Transactions on Database Systems*, **1**(1), pp. 9–36.

FRANK, A. U. and KUHN, W. (1986). Cell graphs: A provable correct method for the storage of geometry, *Proc. 2nd Int. Symposium on Spatial Data Handling*, Seattle.

GERSTING, J. L. (1993). *Mathematical Structures for Computer Science*, 3rd edn, New York: Computer Science Press.

KLIR, G. J. and FOLGER, T. A. (1988). *Fuzzy Sets, Uncertainty, and Information*, Englewood Cliffs, NJ: Prentice Hall.

MOLENAAR, M. (1991a). Terrain objects, data structures and query spaces, in: Schilcher M. (Ed.), *Geo-Informatik*, pp. 53–70, München: Siemens-Nixdorf Informationssysteme AG.

MOLENAAR, M. (1991b). Formal data structures and query spaces, in: Günther O. *et al.* (Eds), *Konzeption und Einsatz von Umweltinformationssytemen*, pp. 340–363, Berlin: Springer.

MOLENAAR, M. (1993). Object hierarchies and uncertainty in GIS, or why is standardisation so difficult, *Geo-informations-Systeme*, **6**(4), 22–28.

TOMLIN, C. D. (1990). *Geographic Information Systems and Cartographic Modeling*, Englewood Cliffs, NJ: Prentice Hall.

ULLMAN, J. D. (1982). *Principles of Database Systems*, Rockville, MD: Computer Science Press.

WILSON, R. J. (1972). *Introduction to Graph Theory*, Harlow: Longman.

CHAPTER FIFTEEN

Towards a Field-oriented Language: First Specifications

ROBERT LAURINI and DILLON PARIENTE

Laboratoire d'Ingénierie des Systèmes d'Information, INSA de Lyon, Villeurbanne Cedex, France

15.1 Introduction

In the spatial database domain, the data models currently used (relational, r-extended, object-oriented, etc.), are often soon outmoded by the ever-increasing needs of users. For instance, the very discrete aspect of object-oriented models in which spatio-temporal attributes are only specified for selected sample points, does not provide a sufficiently realistic idea of nature. Values are provided only for those sample points, which can lead to problems such as discontinuities when the attribute relates to a continuous field (temperature, hygrometry, elevation, rainfall, wind speed, etc.).

A continuous field management ability is needed (known as field-orientation; see Couclelis, 1992). A database language should be able to declare, update and modify a field as an object belonging to a new kind of abstract data type (Kemp, 1993). Our aim is to include these principles in an object-oriented approach, and to specify some information and precise constraints within the field declaration, to improve the estimation of continuous fields. Operators need not only to manipulate objects and attributes, but they must also estimate the attribute values and gradients of any point, integral and density of any region.

As the advantages of object-orientation (O_2) are considerable (Boursier and Mullon, 1993), existing O_2 definitions (see Delobel *et al.*, 1991; O_2, 1992) can be profitably used in the definition of a spatial modelling language able to deal with continuous fields. There are two aspects to the problem. The first (described in this section) is that of interpolating attribute values from discrete sampled points to a continuous field. The second is that of handling the continuous field in the O_2 database language.

One of the main goals of a continuous field approach is to spread the information that has been measured at discrete sampling locations over the entire space of interest, whenever possible ignoring the boundaries of objects, if they do not affect the structure of the field. This means predicting the value of the field at all unsampled points, for which there are many methods (see Laurini and Thompson, 1992). Pariente and Laurini (1993) have developed a new method of interpolation based on a Hopfield neural net. This method takes all sampled points of a given space into account as well as information on locally operating constraints, as explained below. The interpolation is carried out for points on a regular grid. The finite difference solution and a neural mechanism are embodied in this grid, thereby avoiding most of the common shortcomings of classical interpolation methods and reducing estimation errors by 40–80%.

In this chapter we first present the main constraints on the estimation process. Then the language specifications for defining and manipulating continuous fields are described and illustrated with examples.

15.2   Constraints

A GIS should model reality as closely as possible. When attribute values have only been measured at a subset of all grid points, the values at the remaining grid points need to be estimated as accurately as possible, at the same time preserving salient features of the observed reality. The interpolation model must satisfy the user's needs, but must also respect morphological and mathematical constraints as well as information arising from a statistical study of the environment: all these constraints must be taken into account. The following section describes four main constraints, not all of which are taken into account by existing interpolation methods.

15.2.1   Morphological discontinuities

The system must be able to take into account the possible discontinuities in the studied phenomenon (modelled by means of a continuous field). A number of geographic objects can be considered as discontinuities, such as cliffs, roads, rivers, walls, geological fractures and political frontiers. These objects depend on the category of the phenomenon, and describe the relationships between geographic objects. If the boundary of a geographic entity corresponds to a discontinuity, it will be taken into account in the continuous field estimation process.

15.2.2   Continuity and differentiability

The general trend of attribute values should be continuous, regular and differentiable (in order to avoid a crisp fracture in the values, unless it corresponds to a discontinuity discussed above). Values at sites that are close together in space are more likely to be similar than others, further apart, and to depend on each other from a statistical point of view (spatial autocorrelation).

15.2.3   Preserving sample point attribute values

The estimation method must return the exact value of a sample point, and not an estimated one. Thus exact interpolation methods (returning exact values) are preferred to approximate ones (yielding estimates). Some mathematical methods do not meet this constraint, and preference is given to differentiability rather than precision. It is important to preserve the original values of sample points because of possible problems of database integrity.

15.2.4 Attribute values for points or zones

Attribute values can be either the mean value of a zone (or the standard deviation), or can refer to the attribute value at one particular point. In the first case (attribute of a zone) it is useful to include statistical constraints in the estimation process, allowing constraints to be specified using the standard deviation, or regression parameters, or the mean value of a region. In this way, measurement errors can be reduced.

This (by no means exhaustive) list of constraints for a satisfactory interpolation method has been presented in order to meet precise requirements. In general, classical interpolation methods meet only a small subset of these constraints, which is why new methods of estimation are needed, together with a language including specifications to improve the quality of estimates.

15.3 Language background: overview of the conceptual model

Interpolating the field means that not only must the constraints be taken into account, but also that recognition is given to the class of objects that is associated with a particular constraint (e.g. sample points, statistical areas and discontinuities). Figure 15.1 presents an object-oriented model that aggregates the relationships between the class *Field* and the classes *Sample-Point*, *Discontinuity* and *Statistical_Area*. These latter classes are specialized forms of the classical geographic classes: *Point*, *Geographical_Line* and *Geographical_Area*.

Other classes could be defined, such as the class of spatial models used (grid cell, polygons, triangular irregular networks, a regular grid of points, irregularly spaced

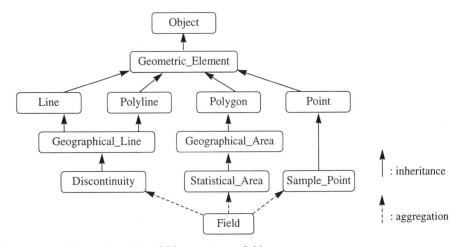

Figure 15.1 Object-oriented model for continuous fields.

points or contour lines) or the class of interpolation method used (see Kemp, 1993) and these classes would be aggregated to the class *Field*. In this chapter, however, we only use the new interpolator derived by Pariente and Laurini (1993), so classes like spatial data models and interpolation methods are not included in the specification of the object-oriented model.

Using O_2 mechanisms, the different definitions are as follows (to simplify them, methods are not specified).

The class *Field*: modelling continuous field of data:

```
class Field
      type tuple
         (      name_of_field : string,
                nature_of_field : string,
                dimension : integer,
                intervals_for_each_dimension :
                                        list(Interval),
                interval_of_validity : Interval,
                samples : set(Sample_Point),
                statistical_constraints :
                                set(Statistical_Area),
                morphological_constraints :
                                set(Discontinuity)
         ) end;
```

Interval is an abstract data type corresponding to intervals of real numbers (such as [2.3, 78.23]; note that it is possible to specify an infinite interval by indicating an interval value R (set of real numbers) corresponding to $]-\infty, +\infty[$).

For each object of the class *Field*, we define:

- a name;
- the nature of the field (temperature, altitude, hygrometry, etc.);
- the dimensions of the field where, e.g. a 3D (or scalar) field corresponds to two spatial coordinates $\langle x, y \rangle$ and one spatial variable (attribute value) $\langle z \rangle$. If the field is a vector field, the dimension is greater, and this field can be described as $\langle x, y, z_1, z_2, \ldots, z_n \rangle$, where $\langle z_1, z_2, \ldots, z_n \rangle$ is the attribute vector of point $\langle x, y \rangle$;
- the list of intervals of definition for each spatial dimension; for example, for a scalar field, the dimensions can be defined as $[-3.4975, +128\,423.1]$ for x, and $[5479.12, 10\,000]$ for y;
- an interval of dates, corresponding to the validity of the different components of the field. Only components with a validity interval corresponding to that of the field will be taken into account (or with an infinite interval of validity);
- a set of sample points;
- a set of statistical constraints;
- a set of discontinuities.

The class *Field* is considered as a new abstract data type, with which the objects can be handled like any other classic object.

The class *Sample_Point*:

```
class Sample_Point inherit Point
     type tuple
     (      #id_point : Point,
            attribute_vector : list(real),
            validness_sample : Interval
     ) end ;
```

Each object of class *Sample_Point* possesses attributes, such as:

- the point coordinates;
- the vector of attribute values (temperature, altitude, etc.); if it is a scalar field, this vector will contain only one element. In the case of a vector field, this vector will contain several attribute values according to the dimensions of the field;
- the period of validity of sampling.

The class *Statistical_Area*: modelling statistical constraints (here, only on the mean value of a zone, but it can be extended to standard deviation):

```
class Statistical_Area inherit Geographical_Area
     type tuple
     (      #id_geographical_area : Geographical_Area,
            mean_of_the_geographical_area : real,
            validness_stat : Interval
     ) end ;
```

The attributes correspond to:

- the name of the geographic area concerned with the statistical constraint;
- the value of the mean to be reached;
- the period of validity of the statistical constraint.

The class *Discontinuity*: modelling morphological constraints:

```
class Discontinuity inherit Line,Polyline
     type tuple
     (      #id_geographical_line : Geographical_Line,
            nature_of_discontinuity : string,
            permeability_factor : real,
            validness_discontinuity : Interval
     ) end ;
```

The attributes correspond to:

- the geographical line;
- the nature of the discontinuity (a wall, river, road, frontier, etc.); and
- the permeability factor (a discontinuity means totally impermeable to a phenomenon, but gradually varying degrees of permeability can also be accepted).

15.4 First specifications of the language

The first specifications of the language follow. Some functions and operators will allow us to handle the objects needed to represent the continuous field, as described

above. The main operations can be classified as basic management (adding, removing or modifying objects, fields or attributes) and queries.

15.4.1 Definition and manipulation

Let *F* be a continuous field of data. *Field* is regarded as a new abstract data type,
providing the definitions of continuous scalar or vector fields.

Declaration of scalar and vector fields with their components (sample points,
statistical areas and discontinuities):

```
Field F ( name_of_field,
          nature_of_field,
          dimension,                  /* of the field */
          (interval1, interval2),   /* intervals of definition
                                         for each dimension *
          [begin_date,end_date],    /* interval of validness *
          (#id_sample_1,...),       /* list of sample points *
          (#id_stat_area_1,...),    /* stt. constraints */
          (#id_discontinuity_1,...) /* discontinuities */
        )
```

Here, the parameters are those defined in the class *Field*. Since *Field* is an abstract
data type, it can be referred to in the definition of other complex geographic objects.
For instance, this is the definition of an object of class *Town*, where some attributes
can derive from fields:

```
Town Baden
(    town_area : Geographical_Area,
     town_hall_location : Point,
     hospital_location : Point,
     town_hall_temperature :
                        TEMPERATURE(town_hall_location),
     hospital_temperature :
                        TEMPERATURE(hospital_location),
     town_population : POPULATION(town_area)
)
```

The temperatures at different locations (town hall and hospital) are derived from the
Field called *Temperature*, as described above. Similarly for the field *Population* one
derives the attribute value *town population*. To improve the estimates of these temperatures and population, a list of sample values and constraints (morphological
and statistical) should be specified in the outline field definitions.

Creating identifiers for new sample points, statistical constraints and discontinuities is done as follows:

```
sample ( #id_point, attribute_vector, validness_sample,
                                     precision_sample )
stat_mean ( #id_geographical_area,
            mean_of_the_geographical_area, validness_stat )
```

```
discontinuity
( #id_geographical_line, nature_of_discontinuity,
             permeability_factor, validness discontinuity )
```

The parameters of each operator are defined by the attributes of the respective classes.

Adding new components (sample points, statistical constraints on the mean and discontinuities) to the field *F*, dynamically:

```
add_sample ( F, #id_sample )
add_mean ( F, #id_stat_mean )
add_discontinuity ( F, #id_discontinuity )
```

Removing components of a field: sample point, statistical constraint, discontinuity:

```
rem_sample ( F, #id_sample )
rem_mean ( F, #id_stat_mean )
rem_discontinuity ( F, #id_discontinuity )
```

Modifying an object (fields or components of a field: sample point, statistical constraint, discontinuity):

```
modify ( object, attribute_to_modify, new_value )
```

Estimating: The process of estimation can take some time (depending on the spatial data model and the estimation method), so the user is allowed to state when he or she wants to perform a new execution, using the function

```
estimate( F )
```

This function will process the estimation, taking into account all the changes applied to the field. When using the neural net interpolation method of Pariente and Laurini (1993), all the different components of the field (sample points, statistical constraints and discontinuities) are used to improve the quality of estimates.

15.4.2 Query operators

A query can relate to the estimated attribute value or the gradient of a point, or the integral or the density of a zone. For the attribute value of a point,

```
F( #id_point ) ?
```

returns a real value, the attribute value of the specified point, if the field is scalar. If it is vector field, it returns a vector. For the gradient of a point,

```
F( #id_point ) ?g
```

returns a vector of three elements, which is the gradient of the specified point. For the integral of a zone,

```
F( #geographical_area ) ?i
```

returns the integral of the domain specified by the area. For the density of a zone,

```
F( #geographical_area ) ?d
```

returns the density of the domain specified by the area.

This list of query operators is by no means exhaustive; other operators could define an algebra for continuous fields, or could operate a zoom on a specified part of the space to improve the resolution.

15.5 Applications

This section provides two applications of the language. One of them deals with a continuous field of temperature, and the other deals with the problem of estimating the populations of neighbouring cities.

15.5.1 A field of temperature

Let *Temperature* be a field. Figure 15.2 maps this field and all its components. A literal description of this field can be made using the language specified previously:

```
Field   TEMPERATURE ("temperature","temperature", 3,
        ([0.0,450.0], [-51.5, 378.384] ), [9.10,11.12],
        /* four sample points */
            ( #sample1, #sample2, #sample3, #sample4 ),
        /* two statistical constraints */
            ( #id_stat_mean_1, #id_stat_std_deviation_2 ))
```

To estimate the temperature at any point in the continuous field *Temperature*:

```
estimate ( TEMPERATURE )
```

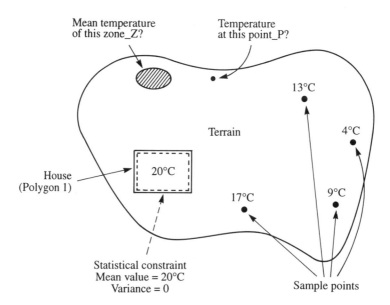

Figure 15.2   A field of temperature.

Now we can query the estimated attribute value of a particular point, or the estimated mean value of the temperature of a zone:

```
TEMPERATURE (point_P) ?
```

returns '14.732°C', the estimated attribute value of point P.

```
TEMPERATURE (zone_Z) ?d
```

returns '17.125°C', the estimated mean temperature of zone Z.

Note that constant attribute values can be easily defined by means of statistical constraints. For instance, the temperature in the house in the above example was previously defined as constant, but is actually considered as a field: two statistical constraints are then defined. The first is a mean value of the house at 20°C, and the second is a standard deviation equal to 0, such as the estimated temperature in the house remains constant.

15.5.2 Population estimate

A file contains the total populations of five neighbouring towns, and the edges of each of these towns. The contents of this file are shown in Table 15.1. Suppose an estimate is needed of the population living in a region determined by a zone which covers more than one unique town. We assume that the population is continuously distributed over the region. Figure 15.3 illustrates the problem. Any cross-section will provide a discretized, staircase form, as shown by the dotted line in the lower part of Figure 15.3.

In many cases population is continuously distributed over the region; it is unlikely that the population density increases or decreases crisply (except at the borders of special zones such as gardens or industrial parks). Let us specify this problem (population estimate) in terms of a continuous field. It is necessary to convert integral values into scalar values, so that the data can be specified in terms of statistical constraints:

```
Field POPULATION (
          "towns_field","population",  3,
          ([0.0, 500.0], [0.0, 500.0] ), [8.11,9.10],
      /* sample points list: no sample point */( ),
      /* the five statistical constraints */
          (
          #id_stat_mean_1,
             /*stat_mean(...Po1A,825/surface(Po1A)...)*/
          #id_stat_mean_2,
             /*stat_mean(...Po1B,137/surface(Po1B)...)*/
          #id_stat_mean_3,
             /*stat_mean(...Po1C,203/surface(Po1C)...)*/
          #id_stat_mean_4,
             /*stat_mean(...Po1D,126/surface(Po1D)...)*/
          #id_stat_mean_5,
             /*stat_mean(...Po1E,475/surface(Po1E)...)*/
          ),
      /* the discontinuity: no discontinuity */ ( ) )
```

Table 15.1 Populations

Town	Population	Polygon
Town A	825	PolA
Town B	137	PolB
Town C	203	PolC
Town D	126	PolD
Town E	475	PolE

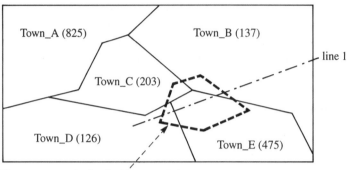

How many people live in this zone (Z)?

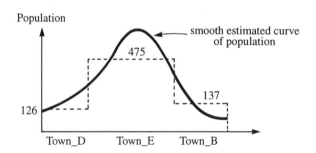

Cross-section following line 1

Figure 15.3 Population estimation.

Here, 'surface' is associated with a *Geographical_Area* object, and returns the surface extents of the specified polygon. To process the field:

```
estimate( POPULATION )
```

Our query is to know how many people live in the area represented by the zone Z:

```
POPULATION(zone_Z) ?i
```

returns '155', the estimated integral (number of people, in thousands) in zone Z. A new cross-section will provide a regular curve between the polygons representing the towns (see Figure 15.3)

15.6 Conclusions

The classic object-oriented approach leads to a crisp discretization of space because attribute values are only provided for selected sample points or zones (which are delimited by boundaries). Many real geographical phenomena, however, are continuous fields that are unbounded or unaffected by discrete objects. In these cases the continuous fields can be thought of as existing independently of these boundaries (which are thereby 'forgotten') over the entire space of interest, and they may vary homogeneously or non-homogeneously.

This chapter has presented the first specifications of a language that is designed to model and manage continuous fields in an object-orientation context, and therefore provides realistic ways (through scalar and vector fields) of dealing with many natural phenomena.

The operation of interpolating from point observations to regular grids can be carried out by many different methods but in order to achieve the best possible results, the interpolation method must take account of a wide variety of constraints from the imposition of statistical areas to morphological discontinuities. The operators in the language permit the definition of fields in terms of a new abstract data type and the manipulation of these fields, thus providing easy accommodation of the methods in existing object-oriented systems. The language has applications in many situations, such as the manipulation of continuous fields with discontinuities and the inclusion of other information about the area of interest.

REFERENCES

BOURSIER, P. and MULLON, C. (1993). GIS evolution: Object-orientation, multimedia, and other trends stemming from research in computer science, *Proc. Urban Data Management Systems '93*, ADV, Vienna, Austria, pp. 11–18.

COUCLELIS, H. (1992). People manipulate objects (but cultivate fields): Beyond the raster–vector debate in GIS, in: Frank, A. U., Campari, I. and Formentini, U. (Eds), *Theories and Methods of Spatio-temporal Reasoning in Geographic Space*, Lecture Notes in Computer Science 639, pp. 65–77, Berlin: Springer.

DELOBEL, C., LÉCLUSE, C. and RICHARD, P. (1991). *Bases de Données: des Systèmes Relationnels aux Systèmes à Objets*, Paris: InterEditions.

KEMP, K. (1993). *Environmental Modelling with GIS: A Strategy for Dealing with Spatial Continuity*, Technical Report No. 93-3, National Center for Geographical Information and Analysis, University of California, Santa Barbara.

LAURINI, R. and THOMPSON, D. (1992). *Fundamentals of Spatial Information Systems*, APIC Series No. 37, San Diego: Academic Press.

O_2 *User Manual*, (1992), Version 3.3, O_2 Technology, France.

PARIENTE, D. and LAURINI, R. (1993). *Interpolation and Extrapolation of Statistical Spatio-Temporal Data Based on Neural Networks*, presented at the Workshop on New Tools for Spatial Analysis, DOSES/EUROSTAT, Lisbon, Portugal.

CHAPTER SIXTEEN

On Layer-based Systems for Undetermined Boundaries

THANASIS HADZILACOS

Computer Technology Institute, University of Patras, Patras, Greece

16.1 Introduction

Coming from the 'rational world of exact objects', there are two extremes on the spectrum of reactions to 'entities with undetermined boundaries'. One is that in order to handle undetermined boundaries we need a totally different philosophical and cognitive approach and a correspondingly different mathematical and computational model. The other extreme reaction is to view them as a natural extension and try to accommodate them in the same models that are used for entities with exact boundaries. In this chapter we take the second view. We try to determine what extensions are needed in order for our models to deal with undetermined boundaries. We would like the situation to be analogous with going from the integers to the rational numbers: some theorems are no longer true; others retain their validity assuming correct extensions of definitions; and some operations (such as division) extend the realm of their applicability. Of course 'natural' extensions to mathematical definitions do not always translate to straightforward extensions of computational algorithms: whereas a floating point addition is computationally comparable to integer addition, division calls for a totally different algorithm.

In taking this approach we build on existing theory and therefore maximize possible reutilization of existing systems. On the negative side, incorporating fuzziness and uncertainty as an add-on, we risk ending up with a system inferior to what we would have if we were to design it from scratch.

In a GIS, spatial entities, phenomena and operations are represented by data structures and algorithmic modules. These data structures and algorithms are using a geometry which is a mathematical model, logically rendering our psychological understanding of space. Philosophical beliefs may support this cognitive grasp of ours. These four levels – philosophical, cognitive or psychological, logico-mathematical and computational – are clearly interrelated. However, in this chapter we focus on each level separately in order not to compromise one due to the limitations of another. For example, the fact that our present computational technology does not allow efficient processing of probabilistic spatial data should not be allowed to restrict our experience of entity boundaries to exact ones only. Of these four levels, the logico-mathematical one stands out with two qualities: It is unambiguous (which the philosophical and psychological levels are not) and is not subject to continual revisions (like the computational one which heavily depends on currently available technology).

This chapter draws on three areas: database theory (both relational and object-oriented), some classical work in geometry, physics and psychology and previous work of the Spatial Database Group at the Computer Technology Institute (CTI) of the University of Patras. The ideas on the fuzziness of entity boundaries and how to deal with it, are influenced by the database theory for incomplete information; Section 16.5.3 contains a short review of relevant literature. Modelling layers as relations and defining a suitable layer algebra are based on standard relational theory. The paper by Burrough and Frank (1995) – with some of whose philosophical points I do not quite agree – provided a good incentive to go back and look at some of the foundations: Euclid, Newton, Piaget and Einstein. Finally, the more specific geographic views have been developed with colleagues at CTI.

The main conclusion of this chapter is that in order to design and build GIS to handle undetermined boundaries we must (a) decide exactly which types of undetermined boundaries the system is going to be able to handle, (b) develop the corresponding mathematical model (i.e. geometry), and (c) design the necessary efficient data structures and algorithms for a computational system. Concrete suggestions are made about the first two issues.

The other important points discussed in the paper are the four levels for modelling space: philosophic, cognitive or psychological, mathematical or logical, and computational; the presentation of layers and objects as orthogonal views of spatial attributes; and the analysis of the concept of entity boundaries. The four levels of modelling are analogous to the levels introduced by Burrough and Frank (1995): 'philosophical and experimental fundaments ... and their subsequent abstraction and coding', although those authors do not always seem to differentiate these levels. Also, Nevalinna (1968) talks about 'the relationships between the empirical, psychological and rational logical ideas whose interplay yields the concepts and theories of space'.

The differentiating characteristic of geographic applications is the requirement to handle instances of spatial attributes, i.e. associating a property to a part of space. These instances can be organized in two orthogonal ways, resulting in layers and objects, respectively. In Section 16.2 we set the general ground for this chapter. The main point is that a spatial model is determined by philosophical beliefs, cognitive views, mathematics and computational technology. Although interrelated, these views should be also considered independently. The limitations of each level should not be carried over to others lightheartedly; Section 16.2.2 presents the motivation for and an informal exposition of our model of space. Section 16.3 discusses layers and objects as orthogonal generalizations of the basic concept of a spatial attribute, which is the association of a property with some part of space. A sketch of the mathematical model is given in Section 16.4, the full exposition of which can be found elsewhere. In Section 16.5 we put forward the view that boundaries may be fuzzy but not undetermined; they may be determined in ways other than 'exact' lines or numbers, such as probability distributions. Furthermore there can be uncertainty in the observation of entities and boundaries; Sections 16.5.3 and 16.5.4

discuss various types of information uncertainty for boundaries of spatial entities. In Section 16.6 we examine the changes that must be made to our model of layers in order to accommodate the fuzziness discussed in Section 16.5.4. Finally, Section 16.7 presents an overall assessment of the results.

16.2 Some comments on spatial models

16.2.1 A naive metatheory of models

A model is a representation, usually a complex one. A representation is a mental symbol (such as a visual image or a word) that signifies an entity, a process or a situation of the objective world – objective meaning outside of the human using the model. In cognitive psychology the semiotic function is defined as the ability to represent: we use (personal) symbols and (conventional) signs such as words or mathematical notation. According to Piaget (Ginsburg and Opper, 1979) this ability is developed during the sixth stage of the 'sensorimotor period' (1.5–2 years old): 'The infant is able to use mental symbols and words to refer to absent objects'.

Why do we model? It seems that 'the capacity to represent mentally an object or action which is not perceptually present ... is required for the child to preserve the original scene in order for it to be evoked at a later time' (Ginsburg and Opper, 1979, quoting Piaget). Thus modelling is a vital intellectual skill. Put another way, modelling is understanding, since we do not understand anything unless we understand it in more ways than one.

There are also significant practical reasons why we model: the Egyptians started modelling fields with rectilinear geometric figures because they could thus measure their areas and re-establish property lines after Nile floods. So, we also model in order to communicate. In modelling, it is important to distinguish four views or levels: *philosophical models* to express our beliefs on space, *cognitive models* to understand space, *geometry* (i.e. mathematical models) to make calculations on space, and finally (and only recently with the invention of automatic computing machines), *computational modelling*. Although its origins date back several centuries when the first algorithms were invented (to manipulate symbols representing numbers for the execution of simple arithmetic operations), concern for computational modelling has been important in the last few decades.

Each level is an 'implementation' of the previous one, usually with restrictions due to the nature of the implementing level. There is also a 'backfiring' effect: restrictions at the implementation level are sometimes carried over as (unnecessary) restrictions to the specification of the level implemented.

For example, consider 'uncertain information'. Does 'certain' information exist, or is all information uncertain to various degrees? Do entities have exact properties that we may or may not be able to measure exactly, but which are nevertheless exact? This is a philosophical discussion on the matter. How do we perceive uncertain information? Are we socially trained to translate inexact boundaries into exact points and lines? This would be the cognitive viewpoint. How should we represent uncertain information? Which information do we consider uncertain? Does this include incomplete information, time-varying data, lack of knowledge and/or interest, irrelevance, etc.? Mathematically uncertain information may be representable by disjunctions of stochastic quantities, which is an exact rendering of uncertainty. Computationally, we can effectively represent and rather inefficiently process only some types of uncertain information.

Each of the four levels serves a different purpose and therefore is suitable in different situations. Mathematics provides a very solid foundation. The physical sciences have convincingly shown that mathematics is an extremely powerful tool for communicating certain precise ideas. However, the expressiveness of the language of mathematics is rather limited, although it has often been hailed as 'the true language of the universe', which it is not. Galileo wrote that 'Philosophy is written in that vast book which stands forever open before our eyes, I mean the universe: but it cannot be read until we have learnt the language and become familiar with the characters in which it is written. It is written in mathematical language, and the letters are triangles, circles and other geometrical figures, without which it is humanly impossible to comprehend a single word' (Newman, quoting Galileo, p. 731).

Philosophical and cognitive viewpoints offer useful insight; they are indispensable for changing scientific paradigms; they are also debatable and quite usefully so. Computational limitations are real engineering facts of life; they must be taken into account when building any spatial system. However, since computational limitations keep moving away, and at a rather fast pace, they should not be hard-wired into our system and certainly not into our conceptual model. In other words, they should not determine the fundamental design choices.

The technological limitations of the early computational systems have led to misunderstandings such as the one by Ciric (quoted in Burrough and Frank, 1995): 'The logic of the information system cannot tolerate any vague or undefined notions'. This is not correct. What is true is that there is a price to pay for vagueness, in the complexity of the system, and in accepting that the 'answers' of the system are not going to always be precise or correct. In other words, an information system which does not contain vague or undefined notions is easier to build. Therefore, when computers were less powerful and computational technology less advanced, such information systems were the only ones built. However, there is nothing intrinsically excluding fuzziness from automated information systems.

16.2.2 The 'accommodate-fuzziness' model

Did we ever say that boundaries are exact or well defined? What does this mean? We know that fully axiomatizable theories based on a set of sound, complete and independent axioms are rather limited in their expressive power; for instance, arithmetic is not subject to such an axiomatization (Gödel, 1931). Such theories are however very elegant; Euclidean geometry is one of them. One of their beauties is that changing just one axiom may result in a world that is at the same time very similar and quite deeply different. The classic example is the geometries that result when the fourth proposition of Euclid's *Elements*, namely that 'given a straight line and a point outside it, there is exactly one line through that point parallel to the line', is replaced by another proposition stating either that no such lines exist (Riemannian geometry) or that more than one such lines exist (Lobatzevsky's geometry). Riemannian geometry is very similar to Euclidean since it can be modelled as Euclidean geometry on the surface of a sphere, and deeply different since it is the geometry of a space with positive curvature, while Euclidean geometry corresponds to a space of constant zero curvature. Today's GIS embody the Cartesian model of space based on Euclidean geometry, which leads to objects with exact boundaries. However, it is clear that several applications would be better served with a less restrictive concept of boundary. Is this a revolutionary upset which requires both rewriting our software from scratch and constructing new mathematical theories, or is it a change that can be accommodated with suitable extensions to the software and the axioms of the underlying geometries? Euclid defines a point as a figure that has no length or width (see definition I-1 in Heath, 1956). Traditionally this definition has been understood as referring to an 'exact' point. Do we have to change it in order to accommodate fuzzy boundaries?

Two observations are in order. First, since length and width have not been defined, the definition of a point carries no mathematical weight; it is just for motivation and intuition. Any concept of a point which is consistent with the axioms would do. In fact, in plane projective geometry, the point and line are dual elements; thus if we interchange points and lines the dual theorems hold true, although often with very different contents. Second, there is nothing in the definition about 'exactness'. Euclidean geometry does not depend on geometric figures having 'exact' boundaries. It is only our usual mental image and the fact that Cartesian geometry represents a point with a pair of numbers that associates Euclidean points to exact locations. So this is what we shall attempt to change. We propose to represent points with pairs of probability distributions, discrete or continuous (of which a constant number is a particular case). That different human views on space can lead to different geographic data systems sounds plausible. It is interesting, however, to explore the reverse direction: how much do today's GIS restrict our view of the world? In particular, would it be possible to handle objects with undetermined boundaries, which is a different cognitive understanding and possibly a different philosophical theory? Our spatial mode has objects, spatial properties and layers. Objects have positions. Positions are geometric figures, representable by areas, lines and points. Points are represented by pairs of numeric quantities which are not necessarily real numbers. They can, for example, be probability distributions.

16.3 Layers and objects

An elementary requirement of a spatial system, computerized or otherwise, is that it should have the capability to represent associations of properties to parts of space; in other words, to express spatial predicates. For example, distance is a property of two points, land_use may be the property of a region, and road_width may be the property of a line (presumably representing the position of a road). Taking the association of a property to a part of space as our fundamental unit, there are two orthogonal ways to extend it. One is to associate all of space with this property; for example, associating land_use with all of space we have a function that specifies the land use at each point (where relevant). The second way to extend our fundamental spatial association is to look at 'all' properties at a given part of space. The first extension leads to layers; the second to objects.

16.3.1 Layers

The use of layers (also referred to as overlays; Tomlin, 1990), in computerized geographic information systems is a carry-over from manual cartography (Burrough,
1986). This is not necessarily a good recommendation: the reasons why a data
organization concept is effective differ widely between manual and automatic
systems. Such a carry-over is often a sign of intellectual laziness or of the immaturity
of a field. However, the ubiquity of this data organization concept warrants some
attention, a first step being its formalization. We are taking a view similar to that of
Delis *et al.* (1994).

   Although definitions of layers abound and practitioners seldom have doubts as
to what constitutes a layer, we are not aware of a standard definition. Burrough
(1986, p. 180) defines a layer as 'a logical separation of map information according
to theme'. De Hoop *et al.* (1993) say that 'a map layer denotes [an abstraction of] a
geographic data set and describes a certain aspect of the modeled real world'. Frank
(1987) defines a (thematic) layer as a 'geometric partition together with the values of
a property'. Dorenbeck and Egenhofer (1991) define a layer as 'a set of equally
shaped cells forming a complete partition'. Delis *et al.* (1994) define layer as a function from a set of geometric figures to any domain.

   There also seems to be some confusion over two concepts of layers: layer at the
conceptual level and layer at the physical level. In Arc/Info terminology the corresponding terms are layer (conceptual) and coverage (physical). Some authors call the
conceptual layers 'thematic layers'. De Hoop *et al.* (1993) make the distinction
between structure and thematic layers; although they give no strict definitions, it
seems that the authors use 'thematic' for conceptual and 'structure' for physical
layers. In all cases, in order to describe a layer we need two elements: a set of
geometric figures and an associated property. The layer tells us the values of this
property on these geometric figures. The geometric figures are also called a geographic data set, or cartographic features, or simply a set of points, lines and
regions. Sometimes more than one property is associated with the same set of
figures. Notice that a liberal concept of 'geometric figures' is needed in order to
accommodate layers representing continuous properties, such as intensity of the
Earth's magnetic field or soil resistivity.

   Thus, it is natural to define a layer mathematically as a function from a set of
geometric figures to a domain of values (or the Cartesian product of several
domains when more than one attributes correspond to one layer), as defined in
Delis *et al.* (1994). It is mathematically equivalent, but we believe conceptually
clearer to define a layer as a relation (in the database sense), the domain of whose
key attribute is a geometric figure. A geometric feature is a shape with absolute
dimensions in absolute positions (all of these 'absolutes' being, of course, relative to
some fixed system of coordinates).

   The set of useful operations on layers is seemingly open-ended (see Burrough,
1986; Tomlin, 1990; and the reference manual of your favourite commercial GIS
package). On closer scrutiny, however, and given the above definition of a layer as a
relation, we can break then up and group the operations as follows:

(i) Operations on a single layer, which add, change or delete non-key attributes,
    but leave the same keys in the relation, i.e. the same geometric figures. We could
    call such operations 'derive computable attribute'.

(ii) Operations which, based on the geometric figures of a single layer, and through geometric constructions of some type (e.g. Euclidean: compass and ruler), create a layer relation with a new set of geometric figures as keys. For example, given a layer with lines and regions (representing rivers and lakes), construct a layer which consists of a buffer zone around the rivers and lakes. We could call such operations 'compute spatial'.

(iii) Reclassification, which operates on a single layer and combines tuples with adjacent keys if the non-key attributes are equal.

(iv) Overlaying, which operates on two layer relations and produces a new layer relation with the geometric overlay as key and the combination of the other attributes from the two layers. For example, given region layers with land use and ownership, construct a layer where each region has a specific use and owner. The meaning of overlaying non-region and especially mixed layers is not always clear or even considered.

Combinations of these four categories in the mathematical sense of function composition, allow the expression of any operation on layers.

If this classification of layer operations proves to be mathematically complete, it will be a very interesting and practically useful computational result, because it points to a modular library of layer operations in which:

- All but one (namely Overlaying) operate on only one layer.

- All but one (namely Reclassification) operate on either geometric or non-geometric attributes, but not both.

Geometric operations are an independent library module, and so are operations on non-geometric attributes. In other words, in order to add layer handling to a relational database system we would need: (a) new data types (for geometric figures) and operations for them; (b) a 'positional join' operation between layer relations for the overlay function; and (c) the transitive closure of neighbourhood in order to implement reclassification. Notice that for (b) and (c) relational algebra or calculus are not sufficient.

16.3.2 Objects

The word 'object' is used in the GIS literature to refer to at least three disparate notions: (1) to refer to real-world things – for these we use the word 'entity' in this chapter; (2) to refer to conceptual and mathematical abstractions (a triangle is an object in Euclidean geometry; a mountain is an object in a geographic application); and (3) the term 'object' is used in a narrower, more exact (albeit at times misunderstood) sense, in the 'object-oriented' approach in computer science: this last sense is outside the scope of this paper. We will try here to understand the conceptual entity 'spatial object' in the second sense of the word.

Spatial objects means boundaries

That A is an entity, implies that we are able to identify it and distinguish it from the rest of the world. That A is a spatial entity means that we are able to distinguish it from the rest of space. Expressing this in a more exact way, we could say that 'A is a spatial object' means, interchangeably, one of the following:

(a) That space is divided in three parts: one part 'inside' A, another 'on the border of A', and the rest 'outside' A. This is a partition of space (in the mathematical sense: mutually exclusive, and covering the whole).

(b) That some properties of interest (or of importance, or convenience) have a sharp change of values when we move from inside A to the outside. It we express these properties as functions of position, they often have a singularity point at the boundary of the entity. To phrase it slightly differently, most of the properties we are interested in have 'similar' values inside A and different values outside A. Equivalently, we could say that there is a sharp boundary for some/most properties we are interested in (and this boundary coincides with the spatial boundary of A).

(c) We are interested in measuring some properties within A, e.g. population. More formally, if $F(p)$ is a spatial predicate we are interested in quantities like $F(p)$, where p Inside(A), or Avrg($F(p)$), and so on. In general, we are interested in processing the set of values $\{F(p)\,|\,p\ \text{Inside}(A)\}$.

Thus the notion of an object is intrinsically related to the notion of a boundary, be it sharp or fuzzy, well defined or not, well determined or not so well determined. Put another way, an object is the set of all spatial properties we are interested in, defined in the part of space called 'Inside(A)' (and possibly 'OntheBorderof(A)'); see Couclelis (1992).

On spatial object complexity

The object-oriented approach, i.e. the use of computational classes of objects, is supposed to be good for dealing with 'complex objects' (notice again the double meaning of the word 'object': complex objects represent real-world entities for whose modelling and computational manipulation we intend to use classes of objects). In which way are these real-world representations 'complex'? The double meaning of the word 'object' is not the sole source of confusion. Real-world entities, like mountains, are always complex. If, however, all we are interested in is a mountain's name and height, then it can be represented by a simple object. If, on the other hand, we also need its erosion history, plant life and fuzzy boundaries, then it is a 'complex object'.

Looking at the various aspects of the complexity of an object we can distinguish:

(a) Non-atomic, i.e. recursively consisting of parts that are objects themselves. There are various notions of 'consists' in the literature such as aggregation (a car consists of a body and an engine) and grouping (a class consists of several students). In the spatial domain we should distinguish more cases (the spatial hierarchy is one), some of which can be extremely difficult to handle computationally, such as the case of 'a line consisting of points', unless we change our computational model of space accordingly (e.g. raster, in this case).

(b) Having interactive parts, i.e. parts such that a change to one of them may provoke changes in others. Again in spatial applications things get an order of magnitude more complicated. For example, consider a water utility network where valves are allowed to open and close. Answering questions related to the consequences of a failure in the network, is next to impossible.

(c) Having complex properties/attributes. For reasons of computational effectiveness, early computing systems assumed that domains are simple, such as numbers, strings or dates. Recently, more complicated domains including free text, sound and video have been added. However, objects may have attributes that are arbitrarily complex: for example, a 'hotel' may have not a room price, but a pricing policy. Similarly, an object 'mountain' may have as 'vegetation' attribute, not a string but a complex time- and space-dependent function. Here we use the term 'spatial attribute' to refer to an attribute whose domain is a function of space.

(d) Finally, objects can be 'active', i.e. changing with no external probe.

16.3.3 Combining layers, objects and relations

We give two simple examples of the combination of layers and objects in spatial applications. We purposefully use the same name for layer, object class and relation, in order to underline the difference in content among the three. A layer is used to model a property with spatial distribution; an object class is used to model a set of similar entities characterized by their positions and other properties, some of which may have spatial distributions, usually taken from several layers. The spatial distribution of objects modelled by points or lines often coincides with the spatial distribution of some property (layer). A relation holds the attributes that bear no spatial dependences.

Example 1: Layer 'tree' is a point layer with attributes tree_type, age, and disease_history. Object class 'tree' also has attributes soil_type and ownership. Notice that layer 'tree' and Object class 'tree' have the same spatial distribution. Relation 'tree' has tree_type as a key, and attributes avrg_production, and harvest_month.

Example 2: Layer 'forest' is a region layer with attributes tree_density, and main_tree_type. Object class 'forest' also has attributes soil_type (which presumably comes from another layer). Note that layer 'forest' and object 'forest' do not have the same spatial distribution, since the layer is also defined as non-forest areas (with value 'agricultural', or 'civic'). Relation 'forest' has forest_type as a key, and attributes degree_of_protection and fire_danger.

A more detailed description can be found in Hadzilacos and Tryfona (1994).

16.4 Mathematical models for geographic databases

This section sketches the models that are fully described in Hadzilacos and Tryfona (1922), Delis *et al.* (1994), Giannakopoulou and Hadzilacos (1994) and Hadzilacos and Tryfona (1994). They constitute different approaches to mathematical modelling of geographic databases, and although not inconsistent, they are far from integrated. They are included here in order to substantiate the point that our approach to fuzziness is consistent with several models for geographic databases. In fact we are not aware of any one that is not.

16.4.1   A layer relation model

Definitions

A domain is a set of values. Examples of domains are the integers, the dates between 15 October 1953 and today, the names of Austrian citizens and sets of geometric figures. There is no restriction that domains be 'simple' in any sense: the probability distribution of life expectancy is just as good a 'value' as the average age.

A relation is any subset of the Cartesian product of one or more domains. The members of a relation are called tuples. When we view a relation as a table, the columns are called attributes. The list of attribute names for a relation is called the relation scheme, often written as $R(A, B, C)$. A collection of relation schemes used to represent information is called a relational database scheme and the current values of the corresponding relations is called the relational database (most of the previous definitions are abstracted from Ullman, 1988). A geographic database is a set of relations, at least one of which is a layer relation.

A layer relation, or simply a layer, is a relation with a primary key whose domain is a set of geometric figures. Relational algebra is based on five set-theoretic operators on relations: union, set difference, Cartesian product, selection (of tuples, or rows) and projection (of attributes, or columns). All other operations of relational algebra, such as intersection, quotient, theta-join, and natural join can be expressed as combinations of the five basic relational operators.

Now let us consider whether relational algebra is sufficient for the operations needed on layers. From the four types of operations described in Section 16.3.1, computing a new attribute is similar to operations done in usual relations; spatial computations (e.g. buffering) are again of the same type, except the domain 'geometric figures' has different fundamental operations than, say, the domain 'dates'. Where the analogy breaks down is with the operations of reclassification and overlaying.

Theorem 1:   Layer reclassification cannot be expressed as a combination of the basic relational algebra operations.

The proof is based on Aho and Ullman (1979), where it is shown that relational algebra cannot express the transitive closure of relations (e.g. cannot resolve the query 'find all ancestors of X'). Reclassification, on the other hand, requires the transitive closure of neighbourhood, since any geometric figure of a layer may be connected to any other if a suitable chain of neighbours exists.

Theorem 2:   Layer overlay can not be expressed as a combination of the basic relational algebra operations.

The proof is based on the fact that in the relational model, if (a, b, c) is a tuple in the result of some relational algebra query then the values a, b and c all appeared in some tuple of the queried database instance. In constrast, the layer overlay results in new values, namely, the geometric figures that are created by the overlaying.

16.4.2   An object class model

Definitions

A domain is an (unrestricted) set of values. An attribute is an association of a name with a domain. An object class is a pair consisting of a set of attributes and a set of methods. A method (mathematically defined as a set of method executions, which in turn are defined to be partial orders of local and message steps with return values), represents a program on the object class. A database is a set of objects which represent part of the real world. Each object belongs to an object class and is characterized by a set of properties which are the attributes of its class. So each object in a database is represented by a set of values each belonging to the domain of the corresponding attribute of the object class. A database schema is a set of object classes.

In spatial databases, objects have one special attribute, the dimensionality of the object represented. Its domain is {point, line, region, none}, or equivalently {0, 1, 2, none}. Objects with dimensionality (0, 1 or 2 are called spatial. They have an additional special attribute, position, whose domain is a finite subset of sets of geometric figures of dimension not exceeding the dimensionality of the object. For example, a spatial object with dimensionality 1 (i.e. linear) can have as 'position' a set of arcs and points.

16.4.3 Boundary indeterminacy

In both preceding models the concept of a geometric figure encapsulates all geographic properties in the database. Thus it suffices to incorporate fuzziness into geometric figures in order to build systems which will be able to handle the indeterminacy of boundaries.

16.4.4 A word of caution on the geometry

The reader will notice that no model of space (in the sense of physics) has been explicitly specified. What has been done is a small extension of the language used to manage the data in order to include spatial objects, methods and relationships. The spatial model is hidden in the geometry. This is mathematically correct, so long as we do not use concepts specific to a single model of space. The models presented hold good irrespective of the assumptions made on properties of the physical space.

In any case, for the present realm of GIS applications we can safely assume that space is two-dimensional, homogeneous, isotropic, infinite, continuous, and infinitesimal (as tacitly assumed by Euclid but never explicitly expressed, see Heath, 1956); or can we?

16.5 'Undetermined' boundaries: A contradiction in terms?

16.5.1 The 'Point-in-Polygon' problem

The issue of 'undetermined' boundaries can be discussed from various points of view. Philosophically, the issue is whether objects exist and, if so, whether they have boundaries. We shall skip this discussion (though it is far from irrelevant) and content ourselves with the observation that – from a cognitive point of view – we perceive objects and boundaries (for reasons of survival, physical necessity, social habit or otherwise) as being sometimes sharp, sometimes fuzzy. Mathematically, the Euclidean abstraction of points, lines and regions provides one model that does not depend on the exactness of object boundaries. Computationally, currently available systems are not able to handle boundary fuzziness efficiently.

Computationally, the 'Point_in_Polygon' (PiP) problem is very closely related to the concept of boundaries. In particular, the class of answers allowed to PiP directly corresponds to the classes of boundaries possible in a system. The 'standard' approach, like the 'classic' view on objects, assumes that the PiP problem has a discrete and certain Yes/No/OnBorder answer. Mathematically, this is a statement about the function PiP (pn, pl) from pairs of points and polygons, namely, that it is total and its range is $\{Y, N, B\}$.

If we only allow PiP(pn, pl) to range in $\{Y, N, P\}$ then:

- area(pl) has an exact answer, i.e. Area(y) $\in R$;
- non-overlapping objects have non-overlapping representations as polygons;
- overlapping objects have overlapping representations as polygons;
- we cannot model non-crisp or uncertain boundaries.

If, on the other hand PiP(pn, pl) is allowed to range continuously in $[0, 1]$ then:

- quantities such as Area(y), Perimeter(y), etc., cease to be real numbers and become probability distributions themselves;
- non-overlapping objects may have overlapping representations as geometric figures;
- fuzzy boundaries can be modelled.

16.5.2   What is a boundary?

A boundary is something that separates two things. When the 'things' are geographic entities or phenomena in two-dimensional space, then a boundary is a 'line' on either side of which some property of interest (or convenience) has different values. Such 'definitions', for all their vagueness, are in agreement with common understanding of boundaries, as corroborated by the use of (figurative?) expressions like 'the boundaries of politeness'. A boundary is something that divides physical, mathematical, social, or other 'space' in two parts, on either of which some property of interest has different values, while on each side it has similar values.

One could argue that object boundaries are interesting because many properties change values there. One could even use this point to argue that therefore objects exist. From the cognitive point of view, we have been trained to recognize objects as constellations of properties. (Of course, at the mathematical level the differentiation between one and several properties makes little sense: the Cartesian product of any number of properties is 'one' property, and given the encoding of any one property we can form several properties out of its (say, bit) projections.) Administrative boundaries, set for reasons of social necessity and often the subject of geographic information systems, tend to be crisply defined, well-known and stable. Natural boundaries, on the other hand, tend to be fuzzy, uncertain and often move with time. In all cases, object boundaries cannot be undetermined as they are inseparable from the objects themselves. It is only the limitation of mathematical or computational systems that restrict themselves to crisp and known boundaries that must leave out other types of boundaries as 'undetermined'.

16.5.3   Uncertain or incomplete information in databases

The need to handle uncertain or incomplete information is present in all areas of data management. In the earlier work on the subject null values or special marks were proposed to deal with missing data, distinguishing between applicable but missing and inapplicable (see Codd, 1979, 1986; Wong, 1982; Vassiliou, 1979, 1980). Several authors (e.g. Winslett 1986), have tried to deal with the problems of updating and generally manipulating incomplete information stored in the form of null values (for example, the DBMS is expected not to equate null values when joining relations). Theoretical models have been built, mainly using logic to handle incomplete information (Imielinski and Lipski, 1984; Reiter, 1986). More recent work (e.g. Liu and Sunderraman, 1990; Ola and Ozsoyoglu, 1993) includes other types of incomplete information such as partial values, indefinite/disjunctive information and 'maybe' information. In summary, the types of incomplete information dealt with by database theory include:

- null values (attribute missing for various reasons or inapplicable in a particular tuple);
- partial values (partially missing values);
- indefinite or disjunction information (i.e. tuple t or tuple t should be in the database);
- 'maybe' information (the semantics of a maybe tuple t is that neither t nor its negation can be asserted from the information present in the database).

16.5.4   The various meanings of fuzziness and uncertainty for boundaries

'Undetermined' boundaries are due to fuzziness or uncertainty. Fuzziness is usually a property of the geographic entity or phenomenon itself, while uncertainty is usually caused by limitations of the observation. Properties of the data representation method and requirements of the computer application are also, although to a lesser degree, reasons for 'undetermined' boundaries. Here is a (partial) list of situations which one could model using spatial objects with fuzzy or uncertain boundaries. Different mathematical models may apply to each case, each of which in turn is best served by different computational models.

1. *Don't know.* This is a case of uncertainty of observation. Sometimes the object boundaries are not known (at the time information about the object is inserted in the database and subsequently manipulated). Uncharted space, such as geological strata, the Zambezi River at the time of Livingston and Stanley (*c.* 1850), or the exact position of Peary's route to the North Pole in 1909 (Davis, 1990), are examples of this situation.

2. *Thick boundaries.* This is another case of uncertainty of observation. Disputed property lines or national frontiers are an example of undetermined boundaries, with a bounded indeterminacy. We can distinguish two subcases; (a) exactly bounded, and (b) stochastically bounded.

3. *Alternative.* Different measurements or observations may lead to situations where disjunctive information on an object boundary must be recorded in the database. It is interesting to note that this is exclusive disjunction, in contrast with the usual case of disjunctive information in databases (see Liu and Sunderraman, 1990). The exclusiveness arises because position (and therefore, boundaries) are, in fact, part of the primary key for the object, even if we use computational tricks (such as object id) to mask this fact. Again this is a case of uncertainty in the observation.

4. *Blend-in.* Situations like the boundary between desert and prairie, or between some types of water pollution are not properly modelled by an arbitrary exact boundary dividing the world in black and white, but are best served by several (possibly non-discrete) grades of grey. This is a case of fuzziness as a property of the entity or phenomenon being modelled.

5. *Time-varying.* Moving boundaries are the natural state of the world, and this is one reason why it is so important to incorporate the time dimension into GIS. However, there are cases in which boundaries are constantly changing back and forth, such as the boundary between sea and land, or the part of the Antarctic which is covered with ice, but the purposes of the application do not warrant the recording of the exact boundaries as a function of time. This is a case of fuzzy boundaries mainly due to application requirements.

6. *Don't care.* In certain applications it is sometimes sufficient to be able to specify some points inside or outside an object with no need to know the boundaries.

Clearly, various combinations of the above cases are possible. For example, the don't know and don't care cases could be complemented with (exact or stochastic) bounded information. The time-varying and the alternative cases could be combined with any of the rest, etc. A somewhat different but more disciplined classification is presented by Helen Couclelis in Chapter 3 of this volume.

16.6 Layers and fuzzy boundaries

We have seen in Section 16.2.2 that Euclidean geometry does not depend on the nature of points. Cartesian geometry represents a point with a pair of real numbers. We propose to associate points with pairs of probability distributions, discrete or continuous (of which a constant number is a particular case.) This allows us to represent most types of boundary fuzziness mentioned in the previous section:

- Blend-in situations, alternatives (finite or infinite), and thick boundaries (both exactly and stochastically bounded) can easily be modelled.

- Time-varying boundaries can be modelled, but with deficiencies (for example, if we have two equal figures revolving around a point, the probability distributions describing their positions would be identical, although the two objects never share the same space).

- 'Don't care' situations could be modelled with partial functions.
- The 'don't know' case can be captured with null values (Vassiliou, 1979).

The question of whether two entities which have the same values for all properties can be distinguished or not, can be posed in the philosophical, the cognitive, the mathematical and the computational contexts. In the last one, the two answers are known as the value approach (according to which entities are completely determined by their values, such as the relational model, which cannot distinguish between two tuples with the same values) and the object approach (which uses an 'external' object identity to differentiate among entities). In the value-oriented approach, a point can only be distinguished by its position. Two points with the same position coincide (not only in the original meaning of the word, i.e. $co + incidere$, but also in the sense of being identical). We would therefore need a computational model capable of fuzziness at the key attribute, which is something all relational models of incomplete information try to avoid.

Relational database models for incomplete information exclude the possibility of a primary key attribute being unspecified or somehow incompletely specified. Codd makes a very strong statement in this respect: 'An important rule for relational databases is that, for integrity reasons, information about an unidentified (or inadequately identified) object is never recorded in these databases (a sharp contrast with non-relational databases). Thus the primary key attribute of each base relation is not permitted to include [null values]' (Codd, 1986).

This is a point of importance for us, since we have modelled layers as relations whose key is a geometric figure; therefore undetermined boundaries directly translate into relations with uncertainty in the key. In any case, when modelling spatial objects there is no way around the fact that some relation or other will represent the geometric figures. If the boundaries are undetermined, then we must deal with non-exactly specified keys. One way to deal with the problem is to invent artificial keys and let geometric figures be a non-key attribute. In a very relevant article on the management of probabilistic data, Barbara *et al.* (1992) also make it a central premise of their model that keys are deterministic. They contend (but do not substantiate) that 'this is not the only choice, but we feel that deterministic keys are very natural and lead to simple relational operators. Furthermore, it is still possible to represent stochastic entities'.

In an object-oriented approach, a point can have a 'system' identity, while 'position' is just an attribute. This way we can represent a known real entity (nothing undetermined here) whose position is not known exactly. In this way we could model situations such as the following: particles x and y rotate around each other in a way that their position is specified by a probabilistic distribution $P(t)$, the same for both of them, while the points never occupy the same position at the same time and are not to be seen as identical although their positions (and possibly their other attributes too) are, in the fuzzy sense, identical.

Notice that fuzzy boundaries do not necessarily imply fuzzy geometry. For instance, a fuzzy triangle has fuzzy angles summing up to exactly two right angles; a region fuzzily divided into two produces two regions whose fuzzy areas sum up to exactly the original area, which could have been fuzzy itself. Again, the time-varying situation is different: motion can very well change the geometry, although this is more so in the extremely small and the extremely large scales (subatomic and cosmic, respectively) which are way outside the realm of present geographic information systems.

Operations on the non-geometric attributes (type (i)) are not affected at all by the change in the nature of points. Pure geometric operations (type (ii)) are enriched in content and are made computationally much more difficult (consider, for instance, a fuzzy buffer operation around a fuzzy line).

Reclassification needs reviewing, because the positions of two figures are no longer sufficient to tell whether they are neighbours or not, and therefore whether they are candidates for unification. Consider, for instance, three rectangles, A adjacent to B and B adjacent to C. The (fuzzy) positions of A and C indicate that they might be neighbours, but the existence of B in between excludes this possibility. Overlay is computationally much more complex, but conceptually clear. Given two polygons, the problem of determining their overlay is reduced to one of finding the intersections of a set of pairs of line segments and answering several instances of the PiP problem. Both problems can be solved with the new model of points.

16.7   So what?

In this chapter we have tried to cut across the philosophical and cognitive questions of the nature of object boundaries and the reasons for the different perceptions of such boundaries. We have looked into situations which would be better served by dealing with fuzzy objects no matter what the answer to the philosophical, cognitive and cultural questions, taking into account input from database theory on indefinite information. Finding several of these (code-named the 'don't know', 'thick boundaries', 'time-varying', 'blend-in', 'alternative' and 'don't care' situations) whose existence is not disputed by either side in the philosophical argument, we then examined the mathematical and computational means of handling them.

From the two possible approaches, namely to build 'fuzzy geographic information systems' from scratch or to extend existing ones, we explored the second avenue. For this, we went back to some of the conceptual foundations of GIS and tried to define some concepts as precisely, broadly and as implementation-independently as possible.

Looking into the concept of boundary, it seems that by extending the possible answers to the 'Point_in_Polygon' problem (from Yes/No to a continuous interval [0, 1]) we can model fuzziness rather well. No change is necessary in the (Euclidean) geometry. A change in the Cartesian representation of points from a pair of reals to a pair of probability distribution functions, covers most, but not all, situations where fuzziness is needed to model the real world. A lot remains to be studied here.

Layers and objects are two orthogonal generalizations of the primary capability of any spatial information system: the association of a property with a part of space. The mathematical definition of a layer as a relation having geometric figures as key, has several advantages: not only is it exact and general but it also brings along with it very useful pieces of theory, such as relational algebra, and incomplete information; it also leads to a grouping of layer operations needed, which is both theoretically and practically important. The development of GIS has followed a well-known recurring pattern in the development of computer science, the systems–theory–systems cycle: first systems were built (such as compilers, DBMS, etc.) with not much regard to theory; about a decade later theoretical work in the foundations of computer science (such as formal languages, relational theory) brought new, revolutionary or evolutionary, ideas for such systems; and finally another decade later a new generation of systems is being developed which incorporates the theoretical results. We seem to be at the beginning of the theory part of the cycle for GIS. This is why it is very important to study which features/components of today's geographic information systems are part of the theory. In this sense, one conclusion that can be drawn from this chapter is a direction for the developers of GIS regarding layers, and this is 'extend' and not 'discard'.

Looking at the limitations of our approach, it seems that handling moving objects is not its strong point. I believe that a suitable extension to the concept of a 'point', along with an extension to the possible answers to the Point_in_Polygon problem, would accommodate this case as well. An interesting observation is that although position is sufficient to tell points apart in the world of exact boundaries, this is not the case in the fuzzy world: an object id is needed (which is a phenomenon with well known physical counterparts for particles.) Several open problems remain: the statement and proof of the completeness of the operations for layer relations; the proof that Cartesian geometry can be used with points as probability distributions rather than real numbers; research into topological relationships of objects described in our model (which may well be related to reasoning about gradual changes in topological relationships (see Egenhofer and Al-Taha, 1992)); and, last but not least, computational algorithms and data structures for the implementation of all this.

Crisp and certain boundaries are a special case of fuzzy and uncertain boundaries, which is the norm. We have been using the special case because our mathematics and computational technology has been insufficient for the general case, but this has had negative effects on the way we perceive the world, most prominently in administration: much human suffering has been caused by beliefs congruent with the non-existence of anything but crisp and certain boundaries. Now that we can produce spatial information systems that can handle fuzzy and uncertain borders effectively (and soon, hopefully, efficiently), we can maybe use such systems to open our minds in situations where we used to think that violations of crisp and known boundaries was a crime.

REFERENCES

AHO, A. V. and ULLMAN, J. (1979). Optimal partial match retrieval when fields are independently specified, *ACM Trans. on Database Systems*, **4**(2), 168–179.

BARBARA, D., GARCIA-MOLINA, H. and PORTER, D. (1992). The management of probabilistic data, *IEEE Trans. on Knowledge and Data Engineering*, **4**(5), 487–502.

BURROUGH, P. A. (1986). *Principles of Geographical Information Systems for Land Resources Assessment*, Oxford: Oxford University Press.

BURROUGH, P. A. and FRANK, A. U. (1995). Concepts and paradigms in spatial information: Are current geographical information systems truly generic? *Int. J. Geographical Information Systems*, **9**(2): 101–116.

CODD, E. F. (1979). Extending the database relational model to capture more meaning, *ACM Trans. on Database Systems*, **4**(4), 397–434.

CODD, E. F. (1986). Missing information (applicable and inapplicable) in relational databases, *ACM SIGMOD Record*, **15**(4), 53–78.

COUCLELIS, H. (1992). People manipulate objects (but cultivate fields): Beyond the raster–vector debate in GIS, in: Frank, A. U., Campari, I. and Formentini, U. (Eds), *Theories and Methods of Spatio-temporal Reasoning in Geographic Space*, Lecture Notes in Computer Science 639, pp. 65–77, Berlin: Springer.

DAVIS, T. (1990). New evidence places Peary at the pole, *National Geographic*, **177**(1), 44–61.

DELIS, V., HADZILACOS, T. and TRYFONA, N. (1994). An introduction to layer algebra, *Proc. 6th Int. Conf. on Spatial Data Handling* (SDH, 94), Edinburgh, UK (also available from atlas.cti.gr with anonymous FTP).

DORENBECK, C. and EGENHOFER, M. (1991). Algebraic optimization of combined overlay operations, *Proc. 10th Auto-Carto Conference*, Baltimore, MD, pp. 296–312.

EGENHOFER, M. and AL-TAHA. (1992). Reasoning about gradual changes of topological relationshps, in: Frank, A. U., Campari, I. and Formentini, U. (Eds), *Theories and Methods of Spatio-temporal Reasoning in Geographic Space*, Lecture Notes in Computer Science 639, pp. 196–219, Berlin: Springer.

FRANK, A. (1987). Overlay processing in spatial information systems, *Proc. 8th Auto-Carto Conference*, Baltimore, MD, pp. 16–31.

GIANNAKOPOULOU, D. and HADZILACOS, TH. (1994). *Modeling Geographic Applications with Objects: Promises and Limitations*, CTI Technical Report-94.03.18, Computer Technology Institute, Patras (available from atlas.cti.gr with anonymous FTP).

GINSBURG, H. and OPPER, S. (1979). *Piaget's Theory of Intellectual Development*, Englewood Cliffs, NJ: Prentice Hall.

GÖDEL, K. (1931). On formally undecidable propositions of *Principia Mathematica* and related systems I, in: van Heijenoort, J. (Ed., 1970) *Frege and Gödel*, Cambridge, MA: Harvard University Press.

HADZILACOS, TH. and TRYFONA, N. (1992). A model for expressing topological integrity constraints in geographic databases, in: Frank, A. U., Campari, I. and Formentini, U. (Eds), *Theories and Methods of Spatio-temporal Reasoning in Geographic Space*, Lecture Notes in Computer Science 639, pp. 252–268, Berlin: Springer.

HADZILACOS, TH. and TRYFONA, N. (1995). Logical data modelling for geographic applications, *Int. J. Geographical Information Systems*, **9**(6), (to appear).

HEATH, T. L. (1956). *Euclid's Elements* (3 volumes), New York: Dover.

DE HOOP, S., OOSTEROM, P. and MOLENAAR, M. (1993). Topological querying of multiple map layers, *Proc. COSIT '93*, Elba, Italy, Lecture Notes in Computer Science 716, pp. 139–157, Berlin: Springer.

IMIELINSKI, T. and LIPSKI, W. (1984). Incomplete information in relational databases, *J. of the ACM*, **31**(4), 761–791.

LIU, K. and SUNDERRAMAN, R. (1990). Indefinite and maybe information in relational databases, *ACM Trans. on Database Systems*, **15**(1), 1–39.

NEVALINNA, R. (1968). *Space, Time and Relativity*, London: Addison Wesley.

NEWMAN, J. R. (1956). *The World of mathematics*, Vol. 3, New York: Simon and Schuster.

OLA, A. and OZSOYOGLU, G. (1993). Incomplete relational database models based on intervals, *IEEE Trans. on Knowledge and Data Engineering*, **5**(2), 293–308.

REITER, R. (1986). A sound and sometimes complete query evaluation algorithm for relational databases with null values, *J. ACM*, **33**, 2.

TOMLIN, C. D. (1990). *Geographic Information Systems and Cartographic Modeling*, Englewood Cliffs, NJ: Prentice Hall.

ULLMAN, J. D. (1988). *Principles of Database and Knowledge-Based Systems*, Vol. 1, Rockville, MD: Computer Science Press (previous edition, 1980, p. 67).

VASSILIOU, Y. (1979). Null values in database management: A denotational semantics approach, *ACM SIGMOD Record*, 162–169.

VASSILIOU, Y. (1980). Functional dependencies and incomplete information, *Proc. 6th Int. Conf. on Very Large Databases*, Montreal, Canada.

WINSLETT, H. (1986). Updating of logical databases containing null values, *Proc. Int. Conference on Database Theory* (ICDT '86), Rome, Italy, Lecture Notes in Computer Science 243, pp. 421–435, Berlin: Springer.

WONG, E. (1982). A statistical approach to incomplete information in database systems, *ACM Trans. on Database Systems*, 7(3), 470–488.

CHAPTER SEVENTEEN

Hierarchical Models for the Definition and Extraction of Terrain Features

MARTIN BRÄNDLI

Department of Geography, University of Zürich, Switzerland

17.1 Introduction

Many spatial processes and phenomena of the physical environment depend on terrain relief either directly or indirectly. In the digital domain, the Earth's surface is commonly modelled by way of digital terrain models (DTMs). In the past few years, digital terrain modelling has gradually evolved into a major constituent of geographical information systems technology. In conjunction with the two-dimensional functions of a GIS, its methods provide a powerful and flexible basis for modelling, analysis and display of phenomena related to topography or similar surfaces.

Despite this high potential for spatial modelling and the increasing demand for terrain data by applications such as environmental management, the inclusion of this sort of data for analysis problems is still a cumbersome and difficult task. There are two main reasons for this unsatisfactory situation. First, there is a problem of integrating DTM data into a GIS. Terrain-related problems can often only be solved by special-purpose DTM systems which require costly data transfer to and from a GIS for analysis. Although substantial research is currently being undertaken on the functional integration of DTMs or integration on the database level (Weibel, 1993), comprehensive solutions will not be available in the near future.

Second, the limited functionality with regard to terrain analysis of GIS or DTM systems restricts GIS users and the range of applications considerably. While it is simple to calculate slope or aspect at specific locations, no solution has yet been found to the problem of extracting specific terrain objects or landscape elements like hills or valleys, which is essential for computer-assisted generalization of terrain, for instance.

These two problems – the integration of terrain data into GIS, and the availability of a comprehensive set of functions for terrain analysis – are interrelated and need to be addressed from a more general position. We believe that the reason for both problems can be attributed to the lack of comprehensive conceptual data models for terrain surfaces, and we therefore propose an approach that concentrates on the development of such models in order to provide a rich and flexible framework for terrain feature extraction and analysis. The conceptual understanding of terrain surfaces should then help in the development of mechanisms for the integration of terrain data into GIS. The objective of this chapter is therefore to outline adequate conceptual data models with respect to the different tasks that exist in digital terrain modelling.

These models must support different levels of complexities. Burrough and Frank (1995) present and examine four idealized views of the world with different representations and complexities of the associated spatial data models. They propose a sequence of models representing the increasing power of the capabilities that are available for modelling the real world. Within the individual views, the authors distinguish between an 'entity view', which is a representation that stresses individual objects, and a representation that models a continuous variation by 'continuous fields'. In order to satisfy the needs of digital terrain modelling and to enable terrain-related applications not just one of the proposed views is needed but the full range and combinations of the complexities of the presented views. In addition, both representations – the entity view and the continuous field – must be provided by a DTM system. The modelling of the terrain surface as a continuum requires the representation of a continuous field, whereas deriving and structuring geomorphological features makes use of the entity view.

Figure 17.1 sketches the scheme which we will use in the development of conceptual models for terrain surfaces. It is outlined in a top-down manner, starting on the top level with a model of the continuous surface. The models of the following levels are successively enriched by increasingly structured features. Essential properties of the approach are the increasing degree of details included in the models, and the change in the surface representation from the continuous field view to the entity view.

Before we address this modelling approach, we first explain why we believe that a modelling approach based on the object-oriented technique is adequate for this purpose.

17.2 Object-oriented modelling

The main objective is to find models that can represent the real world on a conceptual level (i.e. conceptual models). Various domains have contributed to the field of conceptual modelling: artificial intelligence, programming languages, and database research. Recently linguists and cognitive psychologists have also made important contributions (Loucopoulos, 1992). The evolution of data modelling in the field of database design was substantially influenced by the relational model of Codd (1970). However, the concept of relations could hardly express the complexity of the real world. There was a strong need to integrate more semantics into the models, which led to the development of the so called 'semantic' data models (Peckham and Maryanski, 1988). The most prominent representative of this class of models is the entity-relationship model (ERM) of Chen (1976). According to Peckham and Maryanski (1988) the basic concepts of semantic models are objects (or entities), classes, associations, aggregations and generalizations. Although these concepts provide a rich set

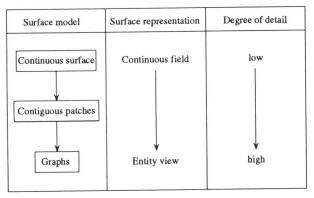

Surface model	Surface representation	Degree of detail
Continuous surface ↓ Contiguous patches ↓ Graphs	Continuous field Entity view	low high

Figure 17.1   Scheme for a top-down modelling approach for terrain surfaces.

of tools, the modelling of processes, controlled by objects, remains restricted. The object-oriented model (OOM) which is an extension of the ERM, builds the basis for the inclusion of behaviour into the model and thus enables process modelling (Herring, 1991). Because of the richness of the available concepts in the object-oriented paradigm, we rate the object-oriented modelling approach as an excellent way to conceptualize terrain surfaces.

It is not the aim of this chapter to repeat the concepts of object-oriented modelling in detail. Comprehensive introductions are provided by Coad and Yourdan (1991a, b) and Rumbaugh *et al.* (1991), for example. We would like to point out that a distinction must be made between object-oriented modelling and object-oriented programming. Following Korson and McGregor (1990) five general concepts exist in object-oriented programming: objects, classes, inheritance, polymorphism and dynamic binding. For high-level design and analysis, which strongly relates to the modelling task, only the first three concepts are needed. The last two concepts are provided for the task of low-level implementation (Korson and McGregor, 1990), which is not discussed here.

As mentioned above, there is a need both for models which can describe continuous fields, as well as for entity-based models. Because the OOM is based on the ERM, it is a model which follows an entity-based view on the one hand, and on the other, through the possibility of integrating behavioural aspects, allows to model the continuous view.

17.3   Modelling a terrain surface as a continuum

The modelling process starts at the top level, with the surface being modelled as a continuum. Mark (1979) identified different views that different groups of specialists (e.g. geomorphologists, surveyors, etc.) have of the structure of the terrain. His aim was not to find conceptual models but to explore the data structures that can best serve the needs of the different specialists. Nevertheless, Mark provided a classification of data structures for digital terrain models which could form a starting point for creating an object-oriented model hierarchy (Figure 17.2). The most widely used structures or representations for DTMs are the regular grid, the triangulated irregular network (TIN), and contours. Although the structure of the regular grid does not correspond to any of the views of the observed specialists, it remains a very

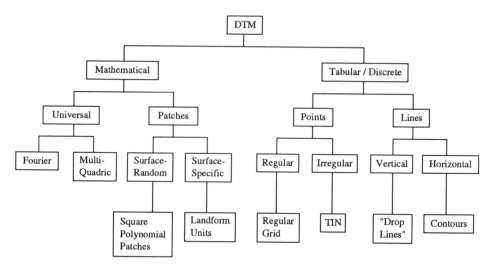

Figure 17.2   A classification for digital terrain models (after Mark, 1979, p.33).

popular structure in DTM systems. TINs, on the other hand, are much more flexible because they can adapt to local variations of the terrain. As a side remark, contours should not be referred to as a DTM data structure, although they are useful as a data source for generating DTMs and can be used for surface visualization.

In Mark's classification the regular grid and the TIN are located in the tabular/ discrete branch of the classification tree. This branch is based on the following underlying concept of space: 'Space is thought of as a collection of an infinite number of dimensionless points which form a continuum' (Frank, 1992, p.413). Both structures, grids and TINs are thus discretizations of this concept.

However, for a conceptual object-oriented model, we want to choose a better realization of this concept of space that does not primarily use data structures as a basis. Therefore we define the surface of a DTM as a field with unique z-values over x and y (Heller, 1990) or, in other words, as a function of x and y. In comparison with Mark's scheme, with our definition the surface is always a mathematical surface. In the words of Frank *et al.* (1986), such mathematical surfaces need 'parameters which describe the surface (usually heights for specific points) and rules for the use of these parameters (usually interpolation methods to determine the values for points other than the given ones)' (p.588). An object-oriented model thus builds the optimal way to include both sample points as static properties and the interpolation as the behavioural property of a class.

The question is then how to specialize a general object 'DTM surface' into sub-classes. The approach we have chosen is to distinguish between different inter-polation procedures and to characterize them according to the range of influence of the involved data points. There is an extensive literature on methods for inter-polating terrain models (see Watson, 1992, for instance); these are usually sub-divided into global and local methods (McCullagh, 1988). Global methods use all data points to estimate parameters for the interpolation function, whereas local methods work in a piecewise manner and take into account only nearby points, leading to better results. Thus, as the main specialization the general object can be subdivided into 'global surface' and 'patch surface' (Figure 17.3). A further special-

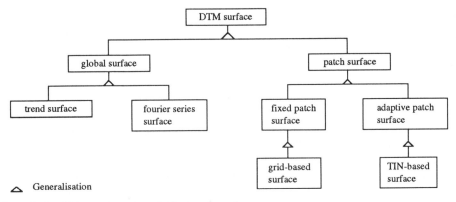

Figure 17.3 Object-oriented model for terrain surfaces.

ization of the hierarchy tree on the global side can, for instance, be into a trend surface class or a class where the interpolation is based on Fourier series. Along the branch of the patch surface classes we can further distinguish between surfaces which are interpolated from nearby points within a fixed scheme, and classes which can be termed adaptive because they select data points according to the nearby terrain variations. An example of the former class is the regular grid, where interpolation takes place within one cell or within a defined neighbourhood of cells. Examples of the latter class are TINs, which can adapt locally to the density of the data points.

Given as set of data points, the class representation with this basic set, and the chosen interpolation procedure as a class method, every point on the surface can be reproduced (or estimated). For a specific user, the underlying data structure will be hidden, and the DTM surface object itself dynamically chooses the appropriate structures and methods to solve a given problem.

This conceptual object-oriented model for DTM surfaces serves now as a basis for the following steps towards the modelling of more specific terrain features. The view then changes from one of a continuum, to one in which the entities move to the foreground.

17.4 Modelling terrain-specific objects

17.4.1 Terrain-specific objects

The literature on general geomorphology (e.g. Chorley *et al.*, 1984) describes a considerable number of different terrain-specific features, depending on climatic and geological conditions. For simplicity, we concentrate on fluvially eroded terrain hoping that the models that will be developed can be generalized to other conditions. An additional problem is that for many of these objects commonsense definitions exist that make them easily locatable in nature, but there is a lack of clear and consistent definitions that could be used for computer-based extraction. Furthermore, there is not only a need for clear definitions of individual terrain features but also a need for conceptual models which express the relationships among the features and which could act as drivers and constraints for automatic feature extraction from DTMs. We therefore try to develop a comprehensive model for structural features which includes the semantic and topological relationships of surface-specific features.

A classification of terrain features with respect to topological type leads to point, line and area features, as listed in Table 17.1. The table shows only the most prominent features without claim of completeness. In the following sections we discuss different conceptual models and approaches for the extraction of selected items from the table. The discussion focuses on what sort of conceptual model (if ever) is used, how realistic it is, and whether it is useful for the task as a driver for the extraction process. We can distinguish between two different approaches. The first approach, which we call the classification approach, is based on the subdivision of space into areas with similar properties, whereas the second approach, termed the graph-based approach, starts by extracting point and line features and connecting them to graph structures. This distinction is comparable to the scheme of Evans (1972) for the recognition of structures in DTMs. He distinguishes between a 'general geomorphometry', whose objective is to characterize and classify continuous surfaces, and a 'specific geomorphometry' which uses methods to extract specific terrain features.

17.4.2 Classification models

The goal of the classification of a terrain surface, or more general, a mathematical surface, is to find basic descriptive elements of the surface morphology (Falcidieno and Spagnuolo, 1991). Classification thus helps to recognize geological and geomorphological patterns in DTMs and, for example, to support analysis for the production of geomorphological maps. An analytical processing of a DTM (either grid- or TIN-based) results in a broad range of terrain-related parameters such as slope and curvature (for a detailed description, see Pike, 1988). The classification approach then groups individual DTM elements into successively broader terrain types, forming a hierarchy of ascending levels (Wright, 1972).

The analytical extraction of parameters yields a good characterization of the continuous surface and thus allows statements about the morphology of the surface. However, it neglects topological and semantic aspects. Several attempts have been made to find suitable concepts for a categorization and description of the surface within an embedded topology and semantics. A prominent concept was introduced by Christian (1960), which is based on land units and land systems. Parts of the land surface which can be described similarly in terms of topography, soils, vegetation and climate, are regarded as belonging to the same land unit. The land system then

Table 17.1 A selection of terrain-specific features in a fluvially eroded terrain

Point features	Line features	Areal features
Pit	Channel	Drainage basin
Peak	Ridge	Valley
Pass	Break line	Hill
River junction	Drainage divide	Terrace
Ridge bifurcation		

is defined as 'an assembly of land units which are geographically and genetically related' (p.76). A land unit in such a system is therefore not individually determined but is related to surrounding units (the context), forming a certain pattern as a whole. Another property of the concept is its independence of scale. The concept can be adjusted to a certain working scale. However, although there exist complete models based on land units (for instance, Dalrymple *et al.* 1968) the realization of this model based on digital data remains a difficult task, because most of these concepts do not consider pure topographical parameters but are based on visual interpretations from field surveys or on photo interpretations. A purely digital approach has to rely on the geometric signature only (Pike, 1988).

An approach which is comparable to the concept of the land system but which is restricted to the geometric signature was developed by Dikau (1989), who defines a hierarchy of space by different levels of types of relief according to different scales. According to that model, an individual level of the hierarchy can be described by form facets (relief units with homogeneous gradient, aspect and curvature) and their association to form elements. Individual form elements can successively be synthesized to relief forms. The classification for a clustering of form facets is performed by calculating profile and plan curvature by distinguishing convex, concave and plan shape. Although this classification enables a promising analysis of the relief, it is not based on a conceptual model of the topology of the classified features. Besides this problem, the clustering of form facets to form elements requires the setting of various threshold values which affects the robustness of the method. The boundaries of the form elements thus are not very clear and are subject to individual interpretation. However, a conceptual model of the topology of form facets and form elements would create a framework which would allow the handling of fuzzy boundaries.

A topological model based on characteristic regions was developed by Falcidieno and Spagnuolo (1991). Starting from a TIN data structure regions of concave, convex, planar and saddle shape are extracted. The regions are tied together by a so-called 'characteristic region configuration graph'. The lines and edges that form the boundaries of homogeneous regions can be considered as characteristic lines, which Falcidieno and Spagnuolo classify as ridges, ravines or generic creases. However, the models presented by these two authors have to be tested and verified on large sets of real data. The sample surfaces used in their paper are rather artificial and do not exhibit the problem cases that are abundant in real data.

At this time, we cannot present a mature conceptual model of a topology based on the classification and characterization of topographic surfaces, since our research is still ongoing. However, the approach presented by Falcidieno and Spagnuolo with an initial subdivision of terrain into concave, convex and plan regions seems promising. In an object-oriented model, these general classes of surface types can be specialized successively into more meaningful terrain features like hills or valleys, for instance. A partitioned surface of meaningful patches could then serve as a framework for the modelling of network structures, which will be discussed in the next section.

17.4.3 Graph-based models

In contrast to the structuring of the surface into patches (land units), this approach starts by extracting so called 'critical points' of the surface and connecting them to surface networks or surface graphs. A topological model based on critical points was first introduced more than a century ago (Caley, 1859). The model starts by tracking slope lines which are derived from contour maps. Tracing slope lines up- or downhill ends at critical points where the first derivative of the surface is zero. Three types of critical points appear on a continuous surface: peaks, pits and passes. The connections between these critical points lead to the definition of critical lines such as ridge and course lines. The path between a pass and a peak forms a ridge line, whereas the path between a pass and a pit forms a course line. According to Mark (1988) ridge lines in this topological model are exactly equivalent to drainage divides, and course lines often but not always correspond to stream channels. The early work of Caley (1859) was picked up and extended by Warntz (1975) by the definition of embedded areal features. Warntz included hills as divergent and dales as convergent areal features which overlap and are independent of each other. The sum of all slope lines reaching the same peak forms an individual hill, and, accordingly, slope lines that reach the same pit form an individual dale. This work was continued by Pfaltz (1976), who focused on graph-theoretical issues leading to the development of the so-called 'surface network', and by Wolf (1984) who used the model of the surface network for cartographic generalization.

These conceptual models about the topology of a surface can be considered as elegant from a graph-theoretical point of view but are far from a realistic representation of topography. The most important problem that arises is the basic assumption of the model about the terrain surface as a 'synthetic' surface. An essential deviation of a terrain surface from the behaviour of a mathematically defined surface is the rare occurrence of pits in fluvially eroded terrain. If pits are missing in the topology of the presented graphs it becomes partly useless because all passes have to be connected to an artificial exterior pit. A second serious problem of the presented graphs is the lack of ability to express river junctions or river bifurcations and the according features along ridge lines, although these are very essential features in fluvially eroded terrain. Wolf (1990) tried to integrate these non-critical but essential points into the surface network by embedding this purely topological structure into a geometrical environment and termed the structure 'metric surface networks'. In order to include river junctions and ridge bifurcations Wolf defined them either as pit–pass combinations or as peak–pass combinations. From an implementation point of view this might be a reasonable work-around, in terms of a conceptual model, although these features should have their own existence and should be modelled accordingly. A third critique of such surface networks relates to the emphasis which is put on passes as critical points. There are of course important passes in real terrain, but the overwhelming majority of passes represent only minor features along extended ridges.

These shortcomings of an approach which is based on critical points necessitates a better solution. An alternative model was designed by Werner (1988) which is based primarily on ridge and channel segments and their associated networks. Werner determined ridge and channel lines based on contour maps. In contrast with the above-mentioned models, however, peaks or passes belong to these lines but they are not the points which bound them. According to Werner, both types of segments – ridge and channel segments – build networks equivalent to planar trivalent trees and do not differ from each other concerning graph-theoretical properties. Based on this definition Werner coined the term 'interlocking networks' because every outer segment of one network is surrounded by two outer segments of another (Figure 17.4).

Of course, even this model has its weak points. The specification of trivalent trees is too rigid and is thus not applicable to real terrain. Therefore the formulation of a more general graph model seems more appropriate. Before specifying a more general model we have to address the problem of the definition of ridges which is important for both concepts presented above: the surface networks of Pfaltz and Wolf and the interlocking network of Werner.

17.4.4 The ridge problem

We address the problem of the definition of ridges by first discussing stream channels which form a dual. One of the principal areas in network analysis in geomorphology has been the analysis of stream channels (Mark, 1988). Because of this particular interest, which is shared by hydrologists and geomorphologists, the automatic extraction of streams or stream channels from DTMs has been well studied and conceptually solved. The main approach taken to extract stream channels is the so-called hydrological approach, which simulates the flow of water over the surface (Band, 1986). Stream channels then occur at places where there is a high concentration of water. Locations with a higher water accumulation than a specified threshold are extracted and connected as the channel network of the study area.

In contrast, the extraction of ridges is conceptually far from being solved. Ridges play an important role in computer-assisted generalization of terrain maps, for instance. A similar method compared to the extraction of channels by inverting the height matrix has been used (Weibel, 1989), causing problems because ridge lines are not treated as the dual of stream channels in that approach. The definitions of ridges in the literature provide little or no help in developing a conceptual approach for the extraction as elegant as that for streams. Mark (1988) defined ridges as 'the places in landscape where the least geomorphological activity occurs' (p.73). A similarly fuzzy definition is provided in the glossary in Stamp (1961): 'A long and narrow stretch of elevated ground; a range or chain of hills or mountains'. Another

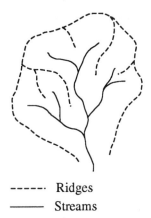

------- Ridges
———— Streams

Figure 17.4 A pair of interlocking ridge and channel networks with alternating outer links (after Werner, 1988, p.256).

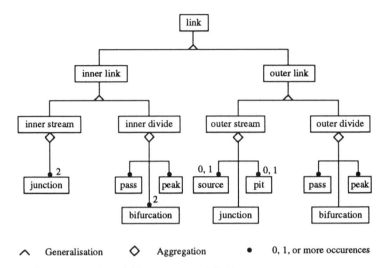

Figure 17.5 Object-oriented model for streams and divides.

more operational definition which considers ridges as line features is given by Mark (1988): 'Also, ridges often correspond with drainage divides, forming the boundaries of drainage basins' (p.87). Beside the lack of a clear and consistent definition there is the problem of identifying ridges (or divides) in nature as objects of clearly linear shape, which is the case for instance if the terrain consists of plateaux. Nevertheless, we model ridge lines as linear features for the time being, because it allows the extraction following the hydrological approach based on divide lines. An extension of the model proposed below, which treats ridges as having both linear and areal shape, will be developed in the near future. The same holds for the treatment of stream lines or valley bottoms which can be poorly identified on wide flood plains, for instance.

On the basis of the hydrological approach for the extraction of stream channels, it can be proven that ridge lines must be a subset of drainage divides. We therefore model ridges always in relation to drainage divides.

17.4.5 A graph-based conceptual model

Based on the above observations about the extraction of stream channels and definitions of ridges and drainage divides, a revised object-oriented conceptual model can now be formulated. It is primarily based on the idea of Werner (1988) of the interlocking network but by relaxing the rigid assumptions with respect to trivalent trees and by incorporating drainage divides.

An approach to building an object representation of extracted stream and divide lines has already been proposed by Lammers and Band (1990), although they concentrate on the representation of the stream net and the related drainage basins and hill slope facets. The incorporation of divide lines is done in an implicit manner. The model proposed here tries to treat stream, divide and ridge lines as objects of equal importance, which allows their use for different fields of applications.

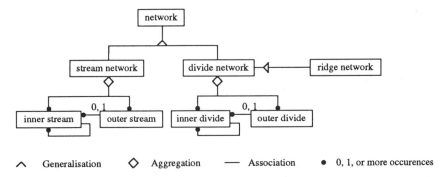

Figure 17.6   Object-oriented model for a stream and divide network.

The general classes of the object-oriented model are nodes and links which are formed by an association of different nodes. One level higher is the network, consisting of links and nodes. The final general class is the interlocking network which associates two complementary networks. Figure 17.5 shows different specializations of links into inner and outer links and subsequently inner and outer stream and divide links with the associated nodes. It is important to note that 'critical points' such as peaks and passes are not the boundary points of divide links in this model. The links are only bounded by junction and bifurcation nodes.

Figure 17.6 represents the composition of the divide and stream networks. Both networks are aggregations of associated inner and outer links of the corresponding type link. The network of ridge links, as a special case, is modelled as a subclass of the divide network. The class is supplied with methods that determine which part of the divide network effectively belongs to ridge lines, by applying filter methods, for instance, depending on a specific application.

Beside the handling of point and line features, the model allows the integration of drainage basins as areal features. The modelling of the drainage areas is performed on the level of the stream links. Every link can be associated with the area which drains to this specific link. The area is bounded by the divide links which can be derived by traversing the interlocking network.

Comparing the features of the conceptual model presented in this section with Table 17.1, most of the listed features could be incorporated. However, break lines which do not conform to stream lines or ridge lines are not yet included, nor are area features like hills and valleys. We are currently working on the completion of this conceptual model and on the automatic extraction of the features included in the model.

17.5   Conclusions

We have outlined an object-oriented model for digital terrain modelling which follows a top-down approach. The different stages of the modelling approach can be considered as levels of increasing degrees of structuring starting from a continuous surface without any distinct entities, to a highly structured model based on graph-like structures which expresses topological and semantic relations of terrain-specific features. Along this hierarchy of structuring there are successive changes in the conceptual view of geographic space from a view of a continuous field to a view of a set of related entities.

At the top level of this top down modelling approach we presented an object-oriented model for a surface as a continuum which is independent of the underlying data structures such as grids or TINs. A general class 'DTM surface' was specialized into subclasses according to interpolation methods and the range of data points the interpolation methods take into account. However, the hiding of the underlying data structure for an application user has its consequences. Although the TIN data structure is widely accepted as the more adaptive structure for digital terrain modelling, analysis algorithms based on TINs are far from mature, in contrast with those that exist for grids. For an efficient analysis a system should therefore provide efficient and adequate conversion functions in order to choose the data structure which can solve a given problem. An object-oriented environment provides an excellent framework for this task.

Derived from this top level are the medium levels of the modelling framework. These can be characterized as subdivisions of the surface into sets of contiguous patches. The partitioning process is controlled by the extraction of topography-related parameters. Although there exist several classification schemes based on such parameters, a conceptual model of the topology and semantics is missing. An approach which seems promising, and on which we are currently working, is an object-oriented model which starts with basic shapes such as concave and convex form elements and then successively finds specializations which lead to more meaningful features of the observed terrain.

At the lowest levels of the top-down modelling we finally find the most detailed models, mainly based on point and line features which are connected to graph-like structures. The presented conceptual model is a synthesis and enhancement of various compact descriptions of the surface based on critical lines and points mainly by relaxing overly rigid assumptions. Research on this level is concentrated on completing the model and on the implementation of the geometric extraction of the desired objects.

We would like to point out two essential properties of this top-down approach based on object-oriented techniques. First, terrain-specific features, independent of the level of abstraction, can be stored either implicitly as methods of less detailed levels, or explicitly as persistent objects. This fact makes the framework flexible to tailor the objects to a given application. Second, the topology and semantics of the objects allow us to handle the individual objects in a specific context. Both properties – the semantic and topological relations between objects and the embedding of increasingly structured features – should enable us to handle a certain fuzziness which is inherent in the definition of terrain-specific features.

The presented approach can contribute to a fundamental problem of the extraction of terrain features from DTMs. All the definitions for the individual terrain features are derived from the view of a continuous geometry. However, the representation of DTMs is restricted to a certain degree of discretization, and therefore the methods for the automatic extraction also have to be based on discrete principles. In many cases, results of the extraction process unavoidably contradict the definitions with respect to continuous geometry, and are susceptible to artifacts and instability. The top-down approach outlined in this chapter provides a possible solution to this problem. There is an increasing degree of discretization in the top-down direction, but every level is constructed based on the preceding level. In the case of problems and artifacts, there is always a possibility to backtrack the extraction by climbing up one level of the model hierarchy.

Acknowledgements

The work reported in this chapter is part of a larger project which has been funded by the Swiss National Foundation under contracts 20-28976.90 and 20-37649.93. Most thanks are due to Dr R. Weibel and M. Heller, who supported this project with many technical and conceptual hints. Dr R. Weibel kindly reviewed the text.

REFERENCES

BAND, L. E. (1986). Analysis and representation of drainage basin structure with digital elevation data, *Proc. 2nd Int. Symposium on Spatial Data Handling*, Seattle, WA, pp. 437–450.

BURROUGH, P. A. and FRANK, A. U. (1995). Concepts and paradigms in spatial information: are current geographical information systems truly generic?, *Int. J. Geographical Information Systems*, 9(2): 101–116.

CALEY, A. (1859). On contour lines and slope lines, *Philosophical Magazine*, 18, 264–268.

CHEN, P. (1976). The entity-relationship model: Toward a unified view of data, *ACM Trans. Database Systems*, 1, 9–36.

CHORLEY, R. J., SCHUMM, S. A. and SUDGEN, D. E. (1984). *Geomorphology*, London: Methuen.

CHRISTIAN, C. S. (1960). The concept of land units and land systems, *Proc. 9th Pacific Science Congress*, 20, 74–81.

COAD, P. and YOURDON, E. (1991a). *Object-Oriented Analysis*, 2nd edn, Englewood Cliffs, NJ: Prentice Hall.

COAD, P. and YOURDON, E. (1991b). *Object-Oriented Design*, Englewood Cliffs, NJ: Prentice Hall.

CODD, E. F. (1970). A relationship model for large shared data banks, *ACM Computing Surveys*, 20, 153–189.

DALRYMPLE, J. B., BLONG, R. J. and CONACHER, A. J. (1968). A hypothetical main unit landsurface model, *Zeitschrift für Geomorphologie*, 17, 60–76.

DIKAU, R. (1989). The application of a digital relief model to landform analysis in geomorphology, in: Raper J. (Ed.), *Three-Dimensional Applications in Geographic Information Systems*, pp. 51–77, London: Taylor & Francis.

EVANS, I. S. (1972). General geomorphometry, derivatives of altitude and descriptive statistics, in: Chorley, R. J. (Ed.), *Spatial Analysis in Geomorphology*, pp. 17–90, London: Methuen.

FALCIDIENO, B. and SPAGNUOLO, M. (1991). A new method for the characterization of topographic surfaces, *Int. J. Geographical Information Systems*, 5, 397–412.

FRANK, A. U. (1992). Spatial concepts, geometric data models and geometric data structures, *Computers and Geosciences*, 18, 409–417.

FRANK, A. U., PALMER, B. and ROBINSON, V. B. (1986). Formal methods for the accurate definition of some fundamental terms in physical geography, *Proc. 2nd Int. Symposium on Spatial Data Handling*, Seattle, Washington, pp. 583–599.

HELLER, M. (1990). Triangulation algorithms for adaptive terrain modeling, *Proc. 4th Int. Symposium on Spatial Data Handling*, 1, Zurich, Switzerland, pp. 163–174.

HERRING, J. R. (1991). The mathematical modeling of spatial and non-spatial information in GIS, in: Mark, D. M. and Frank, A. U. (Eds), *Cognitive and Linguistic Aspects of Geographic Space*, NATO ASI Series, pp. 313–350, Dordrecht: Kluwer.

KORSON, T. and MCGREGOR, J. D. (1990). Understanding object-oriented: A unifying paradigm, *Communications of the ACM*, **33**, 40–60.

LAMMERS, R. B. and BAND, L. E. (1990). Automating object representation of drainage basins, *Computers and Geosciences*, **16**, 787–810.

LOUCOPOULOS, P. (1992). Conceptual modeling, in: Loucopoulos P. and Zicari R. (Eds), *Conceptual Modeling, Databases, and CASE*, pp. 1–26, New York: Wiley.

MCCULLAGH, M. J. (1988). Terrain and surface modeling systems: Theory and practice, *Photogrammetric Record*, **12**, 747–779.

MARK, D. M. (1979). Phenomenon-based data-structuring and digital terrain modeling, *Geo-Processing*, **1**, 27–36.

MARK, D. M. (1988). Network models in geomorphology, in: Anderson, M. G. (Ed.), *Modeling Geomorphological Systems*, pp. 73–97, New York: Wiley.

PECKHAM, J. and MARYANSKI, F. (1988). Semantic data models, *ACM Computing Surveys*, **20**, 153–189.

PFALTZ, J. L. (1976). Surface networks, *Geographical Analysis*, **8**, 77–93.

PIKE, R. J. (1988). The geometric signature: Quantifying landslide–terrain types from digital elevation models, *Mathematical Geology*. **20**, 491–511.

RUMBAUGH, J., BLAHA M., PREMERLANI, W., EDDY, F. and LORENSEN, W. (1991). *Object-Oriented Modeling and Design*, Englewood Cliffs, NJ: Prentice Hall.

STAMP, L. D. (1961). *A Glossary of Geographical Terms*. London. Longman.

WARNTZ, W. (1975). Stream ordering and contour mapping. *Journal of Hydrology*, **25**, 209–227.

WATSON, D. F. (1992). *Contouring: A Guide to the Analysis and Display of Spatial Data*. Computer Methods in the Geosciences, Oxford: Pergamon.

WEIBEL, R. (1989). *Konzepte und Experimente zur Automatisierung der Reliefgeneralisierung*, Dissertation, Geographisches Institut, Universität Zürich.

WEIBEL, R. (1993). On the integration of digital terrain and surface modeling into geographic information systems, *Proc. Auto-Carto II*, Minneapolis, MN, pp. 257–266.

WERNER, C. (1988). Formal analysis of ridge and channel patterns in maturely eroded terrain, *Annals of the Association of American Geographers*, **78**, 253–270.

WOLF, G. W. (1984). A mathematical model of cartographic generalization, *Geo-Processing*, **2**, 271–286.

WOLF, G. W. (1990). Metric surface networks, *Proc. 4th Int. Symposium on Spatial Data Handling*, **2**, Zurich, Switzerland, pp. 844–856.

WRIGHT, R. L. (1972). Principles in a geomorphological approach to land classification, *Zeitschrift für Geomorphologie* (Neue Folge), **16**, 351–373.

PART SIX

Practical Issues of Dealing with Objects with Indeterminate Boundaries

P. A. BURROUGH

Introduction

The five chapters in Part 6 bring us to the real issues of handling real-world geographical phenomena. All are taken from physical science, where the problems of what to do with natural phenomena in soil science, physical geography, geology and hydrology are perhaps more acute than in other areas of GIS application such as cadastral applications or utilities.

In Chapter 18 Philippe Lagacherie *et al.* explain how soil surveyors have traditionally forced their mapping units into the choropleth model, with the attendant simplifications of ignoring within-unit variability and enforced crispness of the boundaries. Digital versions of soil maps have been predominantly vector-based, not because sharply delineated soil polygons are reality, but because this kind of representation has always been used to draw soil maps. They demonstrate with examples from soil variations in the south of France how fuzziness and uncertainty can arise naturally in soil transitions in the landscape (boundaries). Pointing out that it is not sensible to represent transition zones as infinitely thin lines, they show how the 'boundaries' can be replaced by fuzzy transition zones to indicate the possibility of a site being in, or out of a polygon, as was indicated by Burrough in Chapter 1. They demonstrate that fuzzy boundaries can easily be manipulated and displayed in a standard GIS (ARC/INFO), confirming the feelings of both Burrough and Hadzilacos that these kinds of manipulations do not necessarily require the wholesale redevelopment of GIS data structures and analysis tools.

In Chapter 19, from his position in the world of remotely sensed imagery, Poulter starts very firmly in the area of discretized continuous fields but his task is to distil 'objects' from these fields. He starts by pointing out that modern sensors are capable of delivering a high level of detail, with pixel sizes for commercial use now down to below 2.5 m, but that this level of resolution has a price because of the increasing heterogeneity between pixels. In short, the finer the level of spatial resolution, the greater the amount of variation that can be seen. As the level of detail increases, so it becomes easier to delineate man-made structures such as roads, field boundaries and urban areas, but for natural variation this increased resolution only reveals more details in the transition zones between different land cover classes. Automatic edge-finding algorithms break down when variation is continuous or noisy so the automatic extraction of natural boundaries from remotely sensed images is a difficult process. Poulter describes an object hierarchy of land cover classes that is useful for describing the main features of a landscape to a database. Such a representation is flexible enough to permit users to decide on the level of abstraction needed for their application. A fully integrated database with raster and vector capabilities with a complete description of the variation of any feature in the attribute information would be most suitable. Statistical measures and regression analysis can be used to extract information from 'mixed pixels', so that information on uncertainty can be combined with other data to give a better idea of land cover diversity. Like Lagacherie, Poulter thinks that the best way to represent uncertainty around the boundary of any spatial object is in a coded raster format where the pixel value represents a membership 'probability' derived from some previous analysis. The intersection of membership functions from adjacent areas can provide a locus for a vector boundary (as Burrough showed in Chapter 1). By combining raster and vector approaches in one database, Poulter suggests, it would be easy to combine the handling and display of both crisply delineated objects and diffuse objects, such as the combination of administrative boundaries and complex variations in land cover.

In Chapter 20 Sarjakoski takes up the problem of how to count the number of natural objects of a certain class in a given area, using examples of the lakes, islands and rivers in Finland. He points out that the first problem is to define the phenomenon in question, and like Brändli with his problems of definitions of ridges, valleys and passes, he shows that deciding what is, and what is not an island or a lake, is far from easy. Added to the definition problem is a problem of resolution – at what size is an area of land surrounded by water too small or too ephemeral to be called an island? Is a puddle in a ploughed field a lake? In practice people use size or map resolution to cut off the class of objects, but this is completely subjective. Sarjakoski explores the inconsistencies in the definitions of lakes, islands and rivers, and demonstrates how uncertainty in the number of these entities can be derived from definitions, aggregation, identification, interpretation, temporal variation and size. It appears that the distribution of the lake sizes is fractal-like over some four orders of magnitude. Of course, fractals are excellent examples of one class of indeterminate objects (as mentioned by Burrough in Chapter 1), but they are not exclusively so. It is perhaps worth noting that the participants in this meeting preferred not to become involved in discussion about fractals, preferring alternative ways of dealing with indeterminate natural objects.

In Chapter 21 Kavouras describes the work of the geoscientist concerned with predicting the grade of mineral ores in 3D volumes underground. He cannot 'see' the phenomena that he is dealing with, but can only infer what is happening from data taken from drill holes (point observations). Because it is impossible to delineate subsurface objects, such as ore bodies, exactly from a few point observations, geostatisticians treat the observations as a realization of a 'regionalized variable' that behaves as a continuous random field with a known spatial covariance structure. The methods of geostatistics are used to interpolate from observations to make predictions of ore concentrations (or other continuous variables) for blocks of land that can be treated as a whole for mining extraction or mapping. Contiguous blocks of land that have similar values of the predicted ore concentration can then be thought of as an 'ore body', although the boundaries need not be exact. The definition of an ore body is an example of entity definition from continuous fields, using interpolation by geostatistics to create the fields from point observations. Geostatistics also permits an estimate of uncertainty in the predictions and the fields to be made, depending on the number of observations and their spacing, and the spatial covariance structure of the attribute in question. These estimates can be used to build zones of confidence around geographic objects – points, lines, areas or volumes – which can be of practical use for mining or drilling for natural resources.

Chapter 22 by Albrecht has been deliberately kept to the last because he faces the greatest problems in dealing with indeterminacy and uncertainty in geographic phenomena. He is involved in building a monitoring information system for the Wadden Sea (the tidal mudflat coasts from northern Netherlands through to northeast Germany) and faces the problems of dealing with geomorphological and biological phenomena that vary continuously in time and space. The phenomena of the Wadden Sea defy conventional ideas of crispness and stability: sandbanks and shoals change form with time, processes of deposition, erosion and reshaping occur at all conceivable scales, and there is ambiguity and confusion in the definitions of phenomena, which depend not only on the point of view of the persons working in the area but also on the processes that have taken place. Given that data collection proceeds by remote sensing and point sampling, and because of expense and difficulty of access the data are always incomplete, there is little point in setting up a database on conventional 'object–entity' terms. Instead, Albrecht explains that the original data should be stored in as primitive a level as possible, as point observations or as discretized or interpolated fields with appropriate metadata to guide users. All questions of spatial topology and 'objects' are left to the analyzing model and presentation of results so that each group of scientists can work in their own domain free from the restrictions of prior classifications imposed by others.

The general conclusions to be drawn from these five very interesting case studies is that many, if not all, natural phenomena vary indefinitely at multiple scales over time and space. To ignore this fact is to ignore an essential part of reality. Given the hard evidence, why do we still persist in modelling environmental processes in GIS using static, two-dimensional imposed artifacts that are poor models of the real world? Clearly, the time has come for change in the ways we perceive, model and code the natural processes that are ultimately responsible for shaping our lives.

CHAPTER EIGHTEEN

Fuzziness and Uncertainty of Soil Boundaries: From Reality to Coding in GIS

P. LAGACHERIE, P. ANDRIEUX and R. BOUZIGUES

Laboratoire de Science du Sol, INRA, Montpellier, France

18.1 Introduction

Traditionally, soil surveyors have represented variations within the soil cover by choropleth maps. This model of spatial variation implicitly supposes that the soil cover can be stratified in homogeneous subareas, the soil units, separated from each other by infinitely sharp boundaries at which should be concentrated all the soil variations. Naturally, the choropleth map must be considered as an approximation and a simplification of a more complex pattern of variation. Many authors have shown that the within-soil-unit variability is generally far from negligible. On the other hand, although variations at soil boundaries are most often greater than within the soil units, they are rarely, if ever, infinitely abrupt. This reality is partly presented in the soil survey reports given to users with the soil maps. However, as the soil map is much more appealing and (apparently) easier to use than a report, some users fail to take into account the additional information on soil variation which cannot be visualized through the map. Moreover, the reports do not explicitly propose methods for dealing with this variation.

As they deal with choropleth maps, soil surveyors logically consider vector GIS as the right tool for storing and manipulating soil information. However, the information on soil variations included in soil survey reports is again neglected although some soil information systems (i.e. Gaultier *et al.*, 1993) include attributes for describing the within-soil-unit variability. In particular, soil information systems still have no attribute that indicates how sharp or gradual a soil boundary is, implying that it is infinitely sharp. This approximation of reality leads to well known difficulties in GIS processing, i.e. in point-in-polygon and overlay operations (Burrough, 1986). Furthermore, some geostatistical analyses have shown (Stein *et al.*, 1988; Voltz and Webster, 1990) that taking into account the discontinuities in the soil cover strongly improves interpolation of soil properties. In the future, soil surveys could be used to localize these discontinuities. For this purpose too, it is important to identify the soil boundaries which correspond to an actual discontinuity from those which only indicate a gradual change.

In this chapter we examine the nature of the indeterminacy affecting a soil boundary with a view to coding it in a GIS. Then we discuss how information on this indeterminacy can be collected, and which theoretical framework must be selected.

18.2   Soil boundaries: Fuzziness and uncertainty

In our opinion, to analyze the indeterminacy of the soil boundaries a clear distinction needs to be made between two aspects that are currently invoked in the field of artificial intelligence (Badia and Martin-Clouaire, 1989; Haton et al., 1991): uncertainty and fuzziness. Uncertainty corresponds to a lack of knowledge about an object or a fact; for example, 'it is probable that John is 16 years old' is an uncertain statement. Fuzziness occurs when the considered object or fact itself cannot be precisely defined. 'John is about 16 years old' is a fuzzy statement. Naturally, fuzziness and uncertainty can occur simultaneously as in the statement 'it is probable that John is about 16 years old'.

The indeterminacy of soil boundaries, and more generally of any delineations of natural objects, clearly encompass these two aspects. First, when you are close to a soil boundary, you ought to say 'it is probable that I am on the soil unit 1' instead of 'I am on the soil unit 1', since the errors of delineation could have modified the allocation of your site to a given soil unit. Furthermore, even if you could be sure that the position of the soil boundary is absolutely error-free, you would also have to deal with the fuzziness of the two adjacent soil units which often gradually merge at their boundary.

To illustrate this, we present four cases studies selected from published French soil surveys (Figure 18.1). These case studies show sufficiently contrasting situations so that, in the following, we can oppose 'low fuzziness' to 'high fuzziness', or 'low uncertainty' to 'high uncertainty'. However, we must not forget that these statements must be always referred to a given scale of observation; for example, a given boundary can be considered as fuzzy in the detailed survey of an agricultural field; the same boundary will become abrupt for users dealing with the whole region. In the following, 10 m will be the threshold below which a transition between two soil units will be considered as abrupt or accurately positioned. Of course, another scale of observation could have been chosen with another transition level.

18.2.1   Low fuzziness versus high fuzziness

In Figure 18.1, the case studies 1a (Legros and Boyer, 1970) and 1b (SESAER, 1985) show fuzzy soil boundaries. These boundaries are, in fact, midway lines within which the critical soil properties for the soil unit definition (the depth of each soil layer in 1a, or stoniness in 1b) gradually change. In other words, there is a transition area in which the soil does not perfectly fit the definition of one of the two adjacent soil units. In that case, soil surveyors have a choice between two courses of action: either to 'forget' the soil variability occurring at the transition zone, or to modify the

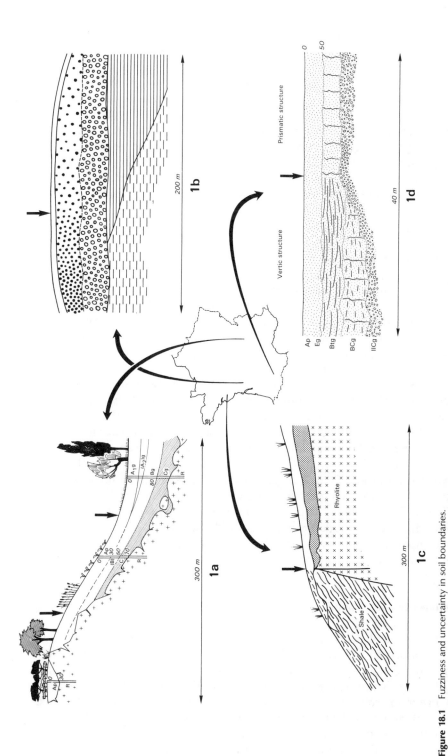

Figure 18.1 Fuzziness and uncertainty in soil boundaries.
1a: *High fuzziness, low uncertainty*: gradual variation of the soil layer thickness delineated by marked changes in vegetation and land use (Legros and Boyer, 1970).
1b: *High fuzziness, high uncertainty*: gradual variation of the stoniness of an intermediate soil layer with no observable surface features (SESAER, 1985).
1c: *Low fuzziness, low uncertainty*: abrupt variation of the whole soil profile (geological change) detectable by a break of slope and a variation in the stoniness of the surface. (Vaquié, 1985).
1d: *Low fuzziness, high uncertainty*: undetectable but abrupt variation in the soil structure in deep soil layers (Favrot et al., 1992).

description of the soil units so that this variability can be taken into account. To take the first course would lead us to accept that a part of the studied area is not correctly described; to taken the second would probably cause an overestimation of the within-soil-unit variability. Soil surveyors generally choose a middle course between these two extreme attitudes. They also try to minimize the problem by selecting an optimal boundary location (Vink, 1963). Some statistical methods have been proposed to do this (Webster and Wong, 1969; Hawkins and Merriam, 1973, 1974, cited in Burrough, 1986).

On the other hand, case studies 1c (Vacquié, 1985) and 1d (Favrot *et al.*, 1992) show that boundaries are not systematically fuzzy. Clear-cut boundaries can occur, e.g. with a geological change (rhyolitic intrusion within shales in 1c) or with a pedological change (differences in soil structure in old alluvium deposit in 1d). Naturally, we could find intermediate cases with fuzziness ranging between those of the four presented in Figure 18.1. Therefore, each boundary must be characterized by a degree of fuzziness, which varies with the nature of the boundary.

18.2.2 Low uncertainty versus high uncertainty

However fuzzy or abrupt the soil boundary is, its delineation in the field induces also a new indeterminacy. Theoretically, the denser the observations, the more accurate the boundary will be positioned. This was recognized early on by soil survey organizations. Many of them produced norms of quality (i.e. USDA, 1951; GEPPA, 1967) which fix, for a given scale of the soil map, the acceptable degree of uncertainty in positioning the boundary.

However, uncertainty varies not only with map scale but also from one boundary to another. Since the number of observations is not unlimited, boundaries are sketched using whatever surface criteria are available. Consequently, the uncertainty affecting a soil boundary will be diminished if this boundary coincides with observable features. For example, these features can be a break in slope and the stoniness of surface layers (case 1c) or the vegetation and land use (case 1a). Conversely, if the soil surveyor has no reliable surface feature to guide the sketching of boundaries, as in case studies 1b and 1d, the uncertainty affecting the position of the boundary will be large. This variation between boundaries probably decreases when the density of observations (and the map scale) increases, since the soil surveyors will have more latitude in compensating for the lack of observable features by locating observation points close to the boundary.

18.2.3 Fuzziness and uncertainty

Although fuzziness and uncertainty are very different concepts, they are not often clearly distinguished by soil surveyors. This is probably because they are positively correlated: a fuzzy limit has a good chance of not having reliable surface features and thus also of being an uncertain boundary (case 1b). Conversely, if there is an abrupt change in the soil cover, a variation of other features will often be observed, or, at least, will induce some visible changes at the soil surface (case 1c). Nevertheless, the correlation between fuzziness and uncertainty is far from perfect. Soil surveyors often have to delineate abrupt changes in the soil cover which are hardly detectable by surface criteria (case 1d). On the other hand, even if a soil boundary can be precisely delineated with clearly defined observable features, it must not be systematically interpreted as a clear-cut boundary (case 1a). Consequently, it would be strongly preferable, in our opinion, for soil surveyors to make a clear distinction between fuzziness and uncertainty when estimating the indeterminacy of a boundary.

18.3   Coding soil boundary indeterminacy in a GIS

The coding of soil boundary indeterminacy encompasses two different but strongly interconnected problems: (i) what information on soil boundary indeterminacy can be reasonably collected?, and (ii) which theoretical framework must be selected to model indeterminacy? A third problem consists in building a suitable GIS architecture so that undetermined data can be efficiently manipulated. In this chapter we focus on the first two of these problems in the case of a soil boundary.

18.3.1   The information on soil boundary indeterminacy

The literature dealing with soil boundaries is poor. Even in soil survey manuals (e.g. USDA, 1953; Jamagne, 1967; Maignien, 1969; Servant, 1972; Dent and Young, 1981) no more than two pages are devoted to the problem of soil boundaries. Consequently, to sketch and to describe a soil boundary remains an empirical activity and there are no guidelines for codifying the boundary description once it is drawn on the map in the same way as there are for the description of soil profiles (Bertrand et al., 1979) or soil series (Gaultier et al., 1993). This explains why neither soil survey reports nor existing soil databases can be directly used to provide information on how to code the soil boundary indeterminacy in a GIS. In the following, we discuss the problem of collecting this information. As in the previous section, we will clearly separate fuzziness and uncertainty.

The fuzziness of a soil boundary can be quantified by estimating the width of the transition zone between the two adjacent soil units. In the case of a large-scale soil survey (i.e. $>1:25\,000$), a soil surveyor can generally provide this value since a substantial number of soil observations are positioned near the boundaries to delineate them. This would be more difficult in the case of small- or medium-scale surveys in which boundary delineation is mostly interpretated from external features of the landscape. Additional transects of soil observations across boundaries could possibly be planned to refine the estimation if necessary. Nevertheless, whatever the number of observations, this estimation will never be perfect since there will always remain an ambiguity about the location at which to end a transition zone.

Table 18.1 shows an example of soil boundary fuzziness coding of part of the $1:5000$ soil map of the Roujan catchment area (Andrieux et al., 1993) located in the Mediterranean coastal plain in the south of France. The boundaries are identified by their two adjacent soil units. Magnitudes of width of the transition zones between these units are estimated with a precision of 5–10 m. They range from 5 m to 50 m. Figure 18.2 shows the soil boundaries of the Roujan catchment area with their transition zones. A substantial part (37%) of the total area is included in these transition zones. The most abrupt soil boundaries correspond to geological changes

Table 18.1 Magnitudes of the width of transition zones between soil units (in metres).

soil units	1	2	3	4	5	6	7	8	9	10	11	12	13	14	15	16	17	18	19
1																			
2	40																		
3	–	50																	
4	50	30	–																
5	10	10	–	15															
6	–	5	–	–	5														
7	30	10	–	–	10	30													
8	–	10	–	–	15	–	20												
9	5	10	–	10	5	20	10	30											
10	–	–	–	–	–	30	30	30	30										
11	–	–	–	–	–	–	–	–	5	5									
12	–	–	–	–	–	–	–	–	5	5	30								
13	–	–	–	–	–	–	–	–	–	–	30	–							
14	–	–	–	–	–	–	5	–	50	5	20	40	30						
15	–	–	–	–	–	–	–	–	–	–	–	50	–	40					
16	–	–	–	–	–	–	–	–	20	–	–	30	–	30	40				
17	–	–	–	–	–	–	–	–	–	–	–	–	–	–	10	–			
18	–	–	–	–	–	–	–	–	–	–	–	–	–	30	40	50	30		
19	–	–	–	–	–	–	–	–	–	–	–	–	–	–	10	30	–	30	

Figure 18.2   Transition zones between the soil units of the Roujan catchment area soil map (Herault, France).

(Pliocene to Miocene) or topographical discontinuities. The most fuzzy ones are located in areas where soils have formed on colluvial and alluvial deposits.

The estimation of the soil boundary uncertainty is not straightforward. Ideally, one would need to know the 'true' soil boundary location and then to repeat the delineations so that an error can be calculated. It is naturally impossible to do this during a classical survey and, to the authors' knowledge, no published paper has dealt with this kind of experiment. Norms of map quality which are provided by the soil survey organization could be used as an approximation of the soil boundary uncertainty. They fix the magnitude of the admitted uncertainty of a soil boundary for a given soil map scale. Once again, this degree is expressed as a width of an uncertain zone centred on the boundary. However, these norms are old, ambiguous and they have never been controlled by experiment. More attention must be paid in the future to their reassessment. Modern soil boundary uncertainty norms should take into account the density of soil observation, the geographical positioning technology (GPS, aerial photographs, etc.) and the absence or presence of observable surface features (see Section 18.2). This also supposes that in the future, soil surveyors would be required systematically to record this information for each boundary they sketch.

18.3.2   The choice of a theoretical framework

Modelling map boundaries within a GIS has received much attention in the recent years. Most authors (e.g. Maffini *et al.*, 1989; Dunn *et al.*, 1990; Bolstad *et al.*, 1990) have focused on the problem of representing the positional errors that occur when digitizing boundaries. This corresponds to the concept of uncertainty that was considered earlier. All these works have demonstrated that the uncertainty of boundaries caused by errors of digitizing can be efficiently modelled through a probabilistic approach. The general principle consists in choosing *a priori* a probability distribution curve (generally a bell-shaped one) and experimentally to estimate its parameters. Regular grids of probability of each map unit present could be easily deduced from this curve to feed a probabilistic error model (Goodchild *et al.*, 1992).

However, applying this probabilistic approach to the case of a soil boundary is far from straightforward. First, the works cited above deal solely with precisely sharp boundaries (e.g. administrative boundaries). We showed in Section 18.2 that not only uncertainty but also fuzziness must be considered for modelling indeterminacy. Because a probabilistic approach will not satisfactorily represent both fuzziness and a combination of fuzziness and uncertainty, it can only be applied to a limited set of situations (e.g. cases 1c and 1d of Figure 18.1). Second, even if soil boundary uncertainty norms are revised, they will only provide an estimate of the magnitude of uncertainty. This vagueness may be incompatible with the use of a probabilistic approach.

To the authors' knowledge, no other theoretical framework has been proposed to model the indeterminacy of 'complex' boundaries such as a soil boundary. However, as indeterminacy includes fuzziness, the use of fuzzy set theory seems to be relevant. Burrough (1989), Burrough *et al.* (1992) and McBratney *et al.* (1992) have shown that this theory can be an efficient tool for manipulating soil data. In the following, we propose a way to model the soil boundary fuzziness within this theoretical framework. We will focus on the problem of coding. Readers are referred to Haton *et al.* (1991) and to the papers cited above for information on techniques based on fuzzy set theory for retrieving and combining fuzzy data.

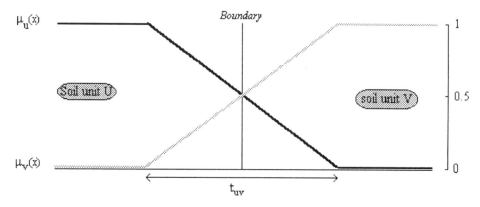

Figure 18.3 Distribution of grades of membership representing the fuzziness of the boundary between soil units *U* and *V*.

In order to model the soil boundary fuzziness defined above, let us consider each soil unit of a soil map as a fuzzy set $U = \{x, \mu_u(x)\}$. In this case, x denotes a point in geographic space which belongs to U, and $\mu_u(x)$ is a number in the range 0–1 which reflects the grade of membership of x in U. The fuzziness of the boundary between the soil units U and V will be indicated by the two distributions of grades of membership, $\mu_u(x)$ and $\mu_v(x)$, shown in Figure 18.3. At points located far enough from the boundary, we will have either $\mu_u(x) = 1$ and $\mu_v(x) = 0$ (if x is included in U), or $\mu_v(x) = 1$ and $\mu_u(x) = 0$ (if x is included in V). For points located close to the boundary, $\mu_u(x)$ and $\mu_v(x)$ will take their values between 0 and 1 and will increase as the distance from the centroid of the concerned soil unit decreases. We propose to admit that these variations could be linear so that we can build the $\mu_u(x)$ and $\mu_v(x)$ functions with a single parameter corresponding to the width of the transition zone. The magnitudes of the width given in Table 18.1 could then be used. For instance, the variation of $\mu_u(x)$ should be expressed as follows:

$$x \in U \quad \text{and} \quad d_{uv}(x) \geq \frac{t_{uv}}{2} \Rightarrow \mu_u(x) = 1,$$

$$x \in V \quad \text{and} \quad d_{uv}(x) \geq \frac{t_{uv}}{2} \Rightarrow \mu_u(x) = 0,$$

$$x \in U \quad \text{and} \quad d_{uv}(x) < \frac{t_{uv}}{2} \Rightarrow \mu_u(x) = 0.5 + \frac{d_{uv}(x)}{t_{uv}},$$

$$x \in V \quad \text{and} \quad d_{uv}(x) < \frac{t_{uv}}{2} \Rightarrow \mu_u(x) = 0.5 - \frac{d_{uv}(x)}{t_{uv}},$$

where $d_{uv}(x)$ is the distance from the boundary separating U and V, and t_{uv} is the width of the transition zone.

In the particular case where a point x of the soil unit U is enclosed in the transition zones of more than two boundaries (i.e. x is close to more than one adjacent soil unit), the above formula will yield several values of $\mu_u(x)$. We propose to keep the value which minimizes $\mu_u(x)$, i.e. to take into account only the influence of the adjacent soil unit which has the highest grade of membership at x.

This method of coding the fuzziness of a soil boundary can be programmed within existing commercial GIS. We used the GIS ARC/INFO_ for building a procedure (AML) which computes the grade of membership of each soil unit on the Roujan catchment area. A total of 19 2 × 2 m grids (one grid for each soil unit) were produced, each cell of a grid containing a grade of membership of a given soil unit. The width of the transition zone is read from an item of the Arc Attribute Table, which describes each soil boundary.

Figure 18.4 shows an example of the 2 × 2 m grid which displays grades of membership of the soil unit 9. It clearly appears that the degree of fuzziness of soil unit 9 does not remain constant along its whole perimeter. In fact, this perimeter must be seen as a set of boundaries, each boundary being defined by a couple of soil units and having its proper magnitude of fuzziness. Therefore fuzziness is not a property of a soil unit, but a property of a soil boundary which is defined as the contact between two soil units.

We note that representing the fuzziness of the soil boundaries through raster format wastes a great amount of computer memory. More attention must be paid to GIS architectures that could offer more efficient solutions.

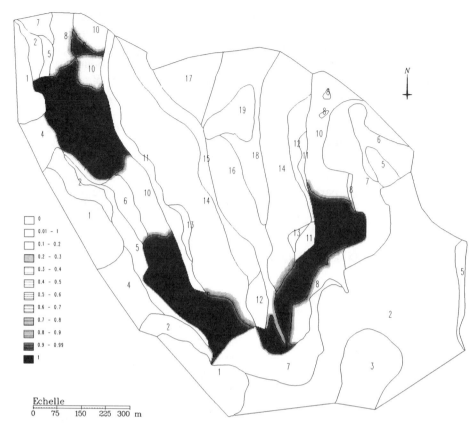

Figure 18.4 Grades of membership of soil unit 9 of the Roujan catchment area soil map (Herault, France).

The use of fuzzy set theory does not totally solve the problem of coding the soil boundary indeterminacy since it deals only with fuzziness, and not with uncertainty. However, fuzzy set theory can be seen as included in a more general theoretical framework, the possibility theory (Dubois and Prades, 1988) in which uncertainty can also be modelled. Various workers in non-geographic fields (e.g. Buisson, 1987, cited in Haton *et al.*, 1991; Badia and Martin-Clouaire, 1989) have demonstrated the utility of possibility theory for combining fuzziness and uncertainty. Therefore this theory appears to be a promising tool which must be applied in the future to the problem of boundary indeterminacy.

18.4 Conclusions

Research on the significance and use of the 'natural' boundaries stored in geographical information systems, particularly soil boundaries, is still in its infancy. A first step could be to model their indeterminacy. In the case of a soil boundary, in this chapter we have proposed ways to achieve this purpose. Our main conclusions can be summarized as follows:

(i) We must make a clear distinction between uncertainty and fuzziness of a soil boundary. Uncertainty corresponds to an ill-known position of the boundary, fuzziness indicates ill-defined soil attributes close to the boundary.

(ii) Soil surveyors must pay more attention in the future to the collection of data on soil boundary indeterminacy. This will probably lead to an evolution of soil survey practices. Furthermore, efficient norms of quality are strongly required.

(iii) Soil boundary indeterminacy cannot be satisfactorily modelled through a probabilistic approach. The use of fuzzy set theory and possibility theory looks promising.

On the other hand, we must be aware that indeterminacy does not only concern soil boundaries. Residual within-soil-unit variability, errors of measurement and vagueness of soil unit morphological characterization induce other indeterminacies which are not necessarily located close to a boundary. All of these aspects must be merged in the future in a single model of indeterminacy.

Acknowledgements

Financial support for this research was provided by the INRA project AIP 'vabrisation de la ressource en eau'. The authors also thank F. Mazzela for contributing the figures.

REFERENCES

ANDRIEUX, P., BOUZIGUES, R., JOSEPH, C., VOLTZ, M., LAGACHERIE, P. and BOURLET M. (1993). *Le Bassin versant de Roujan: Caractéristiques générales du milieu*, INRA UR Science du Sol Montpellier.

BADIA, J. and MARTIN-CLOUAIRE, R. (1989). Choice under imprecision: A simple possibility-theory-based technique illustrated with a problem of weed identification, *Computers and Electronics in Agriculture*, 4, 103–120.

BERTRAND, R., FALIPOU, P. and LEGROS, J. P. (1979). *Notice pour l'enregistrement des descriptions et analyses de sols en banque de données*, Document IRAT–INRA, Montpellier, SES no. 487.

BOLSTAD, P. V., GESSLER, P. and LILLESAND, T. M. (1990). Positional uncertainty in manually digitized map data, *Int. J. Geographical Information Systems*, 4(4), 399–412.

BURROUGH, P. A. (1986). *Principles of geographical information systems for land resources assessment*, Oxford Science Monographs on Soil and Resources Survey no. 12.

BURROUGH, P. A. (1989). Fuzzy mathematical methods for soil survey and land evaluation, *J. Soil Science*, 40, 477–492.

BURROUGH, P. A., MACMILLAN, R. A. and VAN DEURSEN, W. (1992). Fuzzy classification methods for determining land suitability from soil profile observation and topography, *J. Soil Science*, 43, 193–210.

DENT, D. and YOUNG, A. (1981). *Soil Survey and Land Evaluation*, London: Allen & Unwin.

DUBOIS, D. and PRADE, H. (1988). *Possibility Theory: An Approach to Computerized Processing of Uncertainty*, New York: Plenum.

DUNN, R., HARRISON, A. R. and WHITE, J. C. (1990). Positional accuracy and measurement error in digital databases of land use: An empirical study, *Int. J. Geographical Information Systems*, 4(4), 385–398.

FAVROT, J. C., BOUZIGUES, R., TESSIER, D. and VALLES, V. (1992). Contrasting structures in the subsoil of the boulbènes of the Garonne basin, France, *Geoderma*, **53**, 125–137.

GAULTIER, J. P., LEGROS, J. P., BORNAND, M., KING, D., FAVROT, J. C. and HARDY, R. (1993). L'organisation et la gestion des données pédologiques spatialisées: Le projet DONESOL, *Revue de Géomatique*, **3**(3), 235–253.

GEPPA (1967). *Modèle de marche pour une étude de sols*, Ministère de l'Agriculture, Direction des aménagements ruraux, Service de l'hydraulique.

GOODCHILD, M. K., GUOQING, S. and SHIREN, Y. (1992). Development and test of an error model for categorical data, *Int. J. Geographical Information Systems*, **6**(2), 87–104.

HATON, J. P., BOUZID, N., CHARPILLET, F., HATON, M. C., LÂASRI, B., LÂASRI, H., MARQUIS, P., MONDOT, T. and NAPOLI, A. (1991). *Le raisonnement en intelligence artificielle. Modèles, techniques et architectures pour les systèmes à bases de connaissance*, Collection IIA, Paris: Interedition.

HAWKINS, D. M. and MERRIAM, D. F. (1973). Optimal zonation of digitized sequential data, *Mathematical Geology*, **5**, 389–395.

HAWKINS, D. M. and MERRIAM, D. F. (1974). Zonation of multivariate sequences of digitized geologic data, *Mathematical Geology*, **6**, 263–269.

JAMAGNE, M. (1967) Bases et techniques d'une cartographie des sols, *Annales agronomiques*, **18**, 142.

LEGROS, J. P. and BOYER, G. (1970). *Etude pédologique des plateaux du Moyen Vivarais*, INRA Montpellier, SES no. 126.

MAFFINI, G., ARNO, M. and BITTERLICH, W. (1989). Observations and comments on the generation and treatment of error in digital GIS data, *Accuracy of Spatial Databases*, pp. 55–67, London: Taylor & Francis.

MAIGNIEN, R. (1969). *Manuel de prospection pédologique*, ORSTOM initiations, documentations.

MCBRATNEY, A. B., DeGRUIJTER, J. J. and BRUS, D. J. (1992). Spatial prediction and mapping of continuous soil classes, *Geoderma*, **54**, 39–64.

SESAER (1985). *Etude pédologique du secteur de référence du Pays Bas de Cognac et de Matha (Charente–Charente Maritime)*, Opération drainage ONIC, Ministère de l'Agriculture.

SERVANT, J. (1972). *Méthodologie de la carte pédologique*, Sols, paysages, aménagement SES INRA, Montpellier, pp. 5–13.

STEIN, A., HOOGERWERF, M. and BOUMA, J. (1988). Use of soil map delineations to improve (co)-kriging of point data on moisture deficit, *Geoderma*, **43**(2–3), 163–178.

USDA Soil Survey (1951). *Soil Survey Manual*, USDA Agriculture Handbook.

VAQUIÉ, P. F. (1985). *Etude pédologique du secteur de référence du canton de Pouzauges (Vendée)*, Opération drainage ONIC, Ministère de l'Agriculture.

VINK, A. P. A. (1963). *Aspects de pédologie appliquée*, Neuchâtel: La Bacconnière.

VOLTZ, M. and WEBSTER, R. (1990). A comparison of kriging, cubic splines and classification for predicting soil properties from sample information, *J. Soil Science*, **41**, 473–490.

WEBSTER, R. and WONG, I. F. T. (1969). A numerical procedure for testing soil boundaries interpreted from air photographs, *Photogrammetria*, **24**, 59–72.

CHAPTER NINETEEN

On the Integration of Earth Observation Data: Defining Landscape Boundaries to a GIS

MARK A. POULTER

Space Department, CIS Sector, Defence Research Agency, DRA Farnborough, Hants, UK

19.1 Introduction

Civilian sensors in operation today regularly image the Earth's surface with spatial resolutions from 10 m to 7 km. End applications are likely to be revolutionized by the availability of very high resolution data from Russia, and systems proposed by commercial consortiums in the United States and South Africa. With new constellations of satellites offering data at spatial resolutions down to one metre, and revisits to the same location in hours rather than days, many new applications may make use of this information. But what will it mean for users of geographic information? By combining this data with information from other new sensors, such as the multi-frequency synthetic aperture radars (SAR) and imaging spectrometers from the NASA led Earth Observing System (NASA, 1991), land cover maps of unprecedented detail may become possible. However, greater detail has a price. While many hundreds of land use classes may be identified, the resulting databases feature extreme heterogeneity between pixel and classes. The delineation of boundaries between land cover types may not become clearer with this increase in resolution. In fact, recent research has shown that the automated extraction of features from highly detailed imagery presents many new problems to the user (Dowman *et al.*, 1994).

To illustrate this problem, Figure 19.1 demonstrates the high level of detail possible with these new sensors. The image shown is an extract from a Russian 'DD5' image with a ground resolution of approximately 2.5 m. This allows a reasonable precision in the delineation of artificial structures such as fields, roads and the housing estate at top left. However, this same resolution enhances more indeterminate boundaries such as that between the housing estate at bottom right and the neighbouring woodland. Identifying land cover by textural measures or pixel classifications will not automatically define discrete boundaries in all cases.

Figure 19.1   Example of Russian DD5 satellite imagery. Extract supplied by DRA Image Data Facility.

This chapter reports work performed at the Defence Research Agency (DRA), under a research programme initiated by the British National Space Centre (BNSC), to identify suitable techniques and supporting technologies that will enhance the use of remotely sensed data within GIS user applications.

19.2   Earth observation as a source of information

Earth observation offers a cost-effective, timely source of data for the updating of digital topographic maps and the creation of thematic land use datasets. Such information is of value to applications as diverse as cartographic base mapping and forestry resource assessment. To extract this information from the raster imagery, processing techniques are used to partition, or segment, an image into regions and objects which can be subsequently identified or labelled by their spatial, spectral or contextual properties. As a result of this processing, the features of interest may be represented as either discrete vector objects, such as buildings, roads, rivers, etc., or a continuous raster surface with pixels labelled to defined classes.

The generation of discrete vector elements from raster imagery is becoming more sophisticated as the processing power of computers increases and higher-resolution imagery becomes available. However, research by Dowman *et al.* (1994) using 2 m resolution DD5 imagery from Russia to simulate proposed US commercial 1–5 m systems (Corbley, 1994) has shown that while region-based procedures such as contrast segmentation produce good results, current algorithms for the semi-automatic extraction of linear features are not optimized for such high resolutions.

Gooding *et al.* (1993) reached similar conclusions while researching the utility of satellite imagery to update digital maps in an integrated raster/vector environment. The line-following algorithms they developed to follow linear features in single band, grey scale imagery achieved mixed results in operation as the profile of a particular feature changed when different imagery was used. A one pixel wide road on a TM image became two or three pixels wide in a SPOT image, and so wide in high-resolution data like DD5, that the algorithm could find only one edge. Gooding concluded that the representation of particular features in an image greatly affects the performance of the algorithm. Training would be required to 'tune' algorithms to desired features for effective detection at differing image resolutions.

Although these procedures have achieved some success with structured artificial landscapes, natural ecosystems display greater heterogeneity with less dramatic contrast between neighbouring elements. These cannot be easily segmented or represented in a GIS database by this approach. This problem can be illustrated by considering the mapping of vegetation for environmental applications. In many parts of the world, especially where cultivation is poor, there is a gradual gradation from cover such as grassland to scrub to forest where boundaries are not definitive and may vary over time. In these cases, a continuous rather than discrete representation of data has a distinct advantage as it allows properties of uncertainty and variation to be more easily represented through probabilistic classes.

To generate land cover datasets in this format, each image pixel is assigned to a particular thematic class by labelling those that cluster together in locations defined by statistical measures in a multi-dimensional spectral space (e.g. Sabins, 1987). These classes can then be assigned a particular land cover description, such as water, arable crops, deciduous forest, urban, etc.

Figure 19.2 depicts the clustering of pixels into distinct classes. While most applications today treat the assignment of individual pixels as a Boolean case – i.e. a pixel either belongs to a class, or is unassigned – the real world is more complex. At any resolution the pixel is an artifact of the imaging sensor and represents an abstraction whereby differing land cover types within, or adjacent to, the instantaneous field of view may contribute to its value. Some measure of this mixing of land cover classes within a pixel can be determined from the position of a pixel relative to the cluster centre, as defined in Figure 19.2. Conceptually, the assignment of a pixel to a particular class can therefore be viewed as a function of its location against the likelihood of belonging to any class possible in the dataset. A temporal variable can also be included to account for changes in land cover over time, such as periodic floods. This might be written as,

$$(L_1, L_2, \ldots, L_n) = f(x, y, z, t), \tag{19.1}$$

where L is the likelihood, or possibility, of a pixel at a given location belonging to each of n land cover classes.

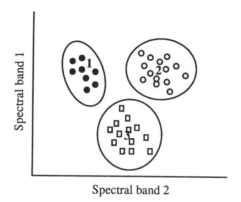

Figure 19.2 Clustering of image pixels in spectral feature space.

The application of artificial neural networks (ANNs) to satellite imagery has been investigated by several researchers (Simpson *et al.*, 1994) as a means of speeding up the extraction of useful information. An example of the application of this technology resulted from a study performed as part of the Earth observation research programme by BNSC, where ANNs were proposed as a method of combining crop classifications and yield predictions into a single operation (Simpson *et al.*, 1994). By reporting a percentage probability for crop type identified on a per-pixel basis, cases of 'mixed pixels' could be processed to increase the accuracy of yield estimates. Potentially this procedure could generate a matrix of probability measures for land cover composition in each image pixel. From a dataset such as this it would then be possible to create a 'probability ' surface. However, many researchers remain sceptical about the use of ANNs as 'black boxes' in this manner, since the exact nature of the manipulation applied to the data is hidden from the user. This raises issues of unquantified errors and reduces the confidence in results.

19.3 Integrated GIS systems

The facilities required by analysts to extract thematic information from satellite imagery are not usually found within the current generation of commercial GIS products. Image processing is typically performed in a separate module or software product with raster and vector information only being brought together for display purposes, such as vector overlays to raster files. Processing of each format has its own interface and communication is by either format conversion and file exchange, or forming complex links between relational database tables.

In the UK, the BNSC recognized that GIS users were finding this decomposed raster/vector approach unsatisfactory, and that the approach was inhibiting increased use to Earth observation data. A possible solution was to closely integrate image processing with GIS analysis in a single environment. As part of the BNSC national space programme, a project managed by DRA was approved to research this approach, including the development of a prototype system under a contract let to a consortium headed by LaserScan Ltd of Cambridge (Poulter, 1993). Requirements for such a system were initially identified by a survey of the remote sensing and GIS user communities (Glidden and Gugan, 1991). A prototype integrated GIS

(IGIS) environment was then developed and evaluated (Poulter, 1994) in incremental stages over two years. Each new stage built on experience gained by operating the previous installation, introducing new functionality and interfaces that enhanced the utility of the system to users from the GIS and remote sensing communities. A series of pilot application projects were performed with the prototype IGIS at DRA to develop and evaluate 'integrated' analysis methodologies that combined data from raster and vector sources in complex analysis to produce faster results. The development of applications is continuing with increasingly complex datasets that include imagery at very high resolutions. The ability to incorporate existing digital maps into the processing of these data has demonstrated many problems surrounding the definition of boundaries.

The procedure of map-guided segmentation (Leason, 1992) allows the use of existing databases to provide *a priori* information to improve the accuracy of classification algorithms (Strahler, 1980). However, the procedure was found to be successful only on images that did not contain complex structures with resolutions related to the scale of the map. With a pixel resolution of 2 m, suitable maps have to be available at scales of 1 : 10 000 or better. Additional problems were encountered with temporal accuracies as the map is likely to be older than the image and any boundaries shown may have since been altered, raising the likelihood of misclassifications. Consideration must also be given to 'mixed' pixels at boundaries where scattering from adjacent pixels and extraneous factors such as footpaths could further reduce the accuracy of classifications.

As a result of these studies it was recognized that there are many problems associated with imposing a discrete boundary on image data that essentially represent a continuous surface. In an attempt to better represent complex areas, a new database structure is now being investigated to improve the integration of both raster and vector structures, and support facilities to handle uncertainty at the boundaries of homogeneous regions. This development is discussed conceptually below.

19.4 Landscape descriptions for geographic databases

With the development of imaging sensors with higher temporal and spectral resolutions, we can expect an increase in the ability of classification techniques to discriminate between surface materials, providing suitable calibration measures have been applied. However, this new-found precision will have a price. At present, when assembling a database, the user expects groups of similarly labelled pixels to cluster together forming identifiable regions with a definitive land use. This will no longer be the case as greater precision of identification increases the variation in the dataset markedly. The user is presented with extreme heterogeneity between pixels, making the clear identification of boundaries an uncertain process.

To reduce this initial complexity, we can aggregate classified pixels into larger entities, object categories and attribute descriptions in a labelled raster file. The advantages of this approach are reduced data volumes and aggregated objects that can be more readily processed by existing tools for geographic analysis. Although suitable for regional applications such as the calculation of 'greenness' indices over continental areas, this process sacrifices precision by reducing diversity and hence does not pass on all the improvements in accuracy to the subsequent analysis.

In generating a GIS database from thematic land cover maps, an appropriate schema must be applied to perform this aggregation of pixels into areas of land use relevant to specific applications (Lo, 1986). The most notable system in use today is that devised by Anderson *et al.* (1976) for the US Geological Survey, which utilizes a layered decision tree of increasingly detailed subsets to describe the objects required by an application. For example, the top layer describes broad entities, such as urban, water and vegetation, while lower levels describe specific subtypes, e.g. vegetation can be divided into 'natural' or cultivated, which in turn can be subdivided further. The rules of classification applied at each branch of the tree can use a number of approaches that maybe functional (i.e. activity oriented), such as agriculture and forestry, or morphological, which emphasizes land cover with descriptions such as sand dunes, grassland, etc. Other rules may apply ecological principles using attribute information for each class of pixel (Stumpf, 1993), or principles of spatial context. The latter considers the relation of individual pixels and structures to each other as the basis for aggregation. For example, if one pixel and those surrounding it are labelled as water, then we are looking at a lake or river. This last example shows where the problems of boundaries and 'mixed' pixels affect database descriptions of land cover most strongly. When dealing with a time series of satellite images as the source of information, many boundaries may have a temporal nature attached to them. Using the example cited, where is the water edge of a river that periodically floods, or a reservoir during a drought?

A layered classification structure to form an object hierarchy of land cover classes is a useful method of describing the main features of a landscape to a database. The representation of land cover is flexible enough for the user to decide what level of abstraction is suitable to their application. This has the advantage of speeding up processing, as only a subset of the full database might be required to satisfy any one query. However, further research is required to determine if such a structure can adequately preserve diversity and represent uncertainty.

A database design to accommodate issues such as these is being considered as part of the BNSC programme undertaken by DRA. Experience with the prototype IGIS system has demonstrated a requirement for close integration of raster and vector structures at the database level. With this capability it should be possible to assemble and fully utilize hierarchical land cover datasets of the type described above. Although the assembly of larger land use blocks from pixelated data is an established technique, it does not always retain the complete diversity of information in the database. The ideal model for a fully integrated database would retain a complete description of any variation in features within an area as part of the attribute information. Map-guided segmentation provides the easiest route to generating the information for such descriptions, but requires an existing object to define the bounding region. By including the use of statistical measures and regression analysis to extract land cover information from mixed class pixels (Foody and Cox, 1991), further improvements in the accuracy of 'within-area' descriptions could be accomplished. For example, a pixel might be labelled forest and grassland at 70% and 30%, respectively. When this measure of uncertainty in class membership is combined with other pixels in the object, a more complete representation of land cover diversity can be established. Use may also be made of statistical measures to describe the distribution of elements, i.e. uniform or random, within a polygon (Baskent, 1993).

Figure 19.3   Conceptual representation of objects with uncertain boundaries.

But what if the boundary itself is undetermined and cannot be readily represent-ed by a discrete vector entity? In this instance the requirement is either to create new objects or to alter existing boundaries to reflect these regions of uncertainty. The conceptual model being considered would represent this uncertainty as a raster 'continuous field' of membership probability straddling a vector boundary. An ideal-ized object showing uniform uncertainty at the boundary is depicted as region A in Figure 19.3. The bold line describes the vector boundary, while the shaded regions represent varying degrees of uncertainty from being definitely inside the object, to being definitely outside. However, the real world is not idealized and the region of uncertainty will vary both internal and external to the vector construction. Region B in Figure 19.3 therefore represents a closer approximation to the real world.

The region of uncertainty surrounding the area object in the database is best represented by a coded raster with membership 'probability' derived from the orig-inal classification and subsequent aggregation processes. As Lowell (1994) points out, the use of probability surfaces within GIS has the potential to provide an ana-lytical structure that will allow certain types of error to be represented and used in subsequent analysis, but notes that current approaches to the production of such surfaces have concentrated on cases involving only two classes. He used a method of discriminant function analysis where a set of functions are applied to a pixel or measure from which a 'discriminant score' is computed for each possible category, or map class, in the database. The pixel or region is then assigned to the category which has the highest discriminant score, which is equivalent to assigning an element to the class to which it has the highest probability of belonging. This approach has similarities to a rule-based aggregation procedure, but with the advantage of producing probabilities directly which could generate a 'certainty' surface.

The advantage of conceptually linking raster, vector and attribute data is the flexibility to preserve small features that are important to an application or analysis, but are below the resolution of a generalized database. Examples of such cases are green areas within an urban extent and small groups of 'urban' pixels representing farm buildings in a predominantly rural area.

19.5   Designing a database for uncertainty

The organization of GIS databases today has been designed to support the carto-graphic display of data using points, lines and polygons as the spatial primitives

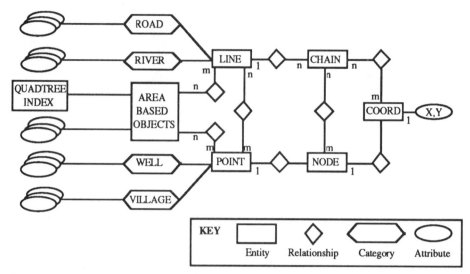

Figure 19.4 An integrated data model of spatial relations. Adapted from Armstrong and Densham (1990).

from which more complex elements are described. Such structuring is widely regarded by some researchers as inadequate for the requirements of physical modelling and analysis. Cartographic representation deals with exact, absolute measures of location and spatial extent, whereas real-world landscapes comprise complex spatial patterns which are influenced by many factors including soil conditions, vegetation structure and type, climate, terrain and both past and present land use practices. The interaction of these factors in producing inexact phenomena, such as changes in in slope and transition zones between plant ecosystems, require more generalized database representations that can vary with the scale of data presentation and the requirements of the end application.

Fortunately, concepts of object orientation in database management (Healey, 1991) provide many of the structures required for a database design that can manipulate differing formats of types of information. To satisfy the requirements of representation described in the previous section, a suitable database would have to store and process vector spatial primitives for the construction of area, line and point geometries, as well as coded raster files for the 'probability' surface associated with each object. Vector constructions would require the definition of one-to-one, many-to-one and many-to-many relationships to form hierarchies that aggregate smaller elements into larger categories and entities, such as combining fields and buildings to define a farm. Each object within a category or hierarchy must also be related to its attendant attributes for within area summaries of land cover. The facility for raster processing must be suitably flexible to allow differing cell resolutions to represent varying measures of uncertainty around objects. For example, the edge of a road is a reasonably well defined phenomenon that can be located to within one pixel, while an intergrade between ill-defined areas, such as bush and forest would extend over many pixels. This form of variability is best processed by a quadtree index, with suitable relations to attribute cells with measures of membership 'probability'.

A conceptual database design that would support these requirements to integrate raster and vector representations is shown in Figure 19.4. This is an adaptation of a model portrayed by Armstrong and Densham (1990) to support spatial decision analysis, that could be prototyped using the advanced object-oriented programming environments now available.

The utility of this design for seamless raster/vector integration in support of GIS applications using Earth observation data sources is being considered within the BNSC research programme. However, this design assumes 'probabilistic maps' of pixel membership already exist and can be suitably structured in the raster representation for subsequent analysis. More research is required into the construction of such maps that combine information from all sources, including Earth observation sensors, existing databases, ground sources and aggregation rules.

19.6 Summary and discussion

The advent of new satellite sensors, imaging the Earth's surface regularly at high spectral and spatial resolutions, provides potential for wider use of satellite data within new and diverse applications. The application of current processing techniques to these data will result in a greater detail of definition when extracting land cover information. However, the price of this detail is datasets that, while being accurate, are often too heterogeneous to form immediately clear boundaries. Higher spatial resolutions may improve the definition of boundaries surrounding well structured, artificial phenomena, such as roads and fields, but will not improve the delineation of more natural features such as grasslands. The rules applied to aggregate pixels and polygons into more structure units specific to each application require further investigation. Ideally, this should be done in close collaboration with the disciplines concerned to ensure the utility of final products to their applications.

This chapter has discussed the generation of thematic land cover maps from satellite imagery, and how those data can be structured into a hierarchy of larger, application-dependent objects through the use of aggregation 'rules'. We have used these requirements to describe a database development that would combine vector representations for objects containing defined boundaries, such as roads and buildings, with thematic data stored in an indexed raster. However, no attempt has yet been made to investigate how this database design would present regions of uncertainty to the user, nor how that information would then be input to physical models and geographic analysis.

Some researchers have put forward concepts of 'fuzzy' classification to overcome some of the problems of a deterministic approach when representing 'real-world' boundaries in an analysis. These classifications seek to determine a 'membership function' which assigns a pixel to a particular land cover class by a measure of possibility (Burrough et al., 1992), while Lowell (1994) considers a conceptual model that uses probability maps rather than thematic maps to conduct geographic analysis within GIS. Using the conceptual database design described in this chapter, we can perhaps see how overlapping regions of uncertainty surrounding neighbouring objects may interact to allow the user to quantify the region of uncertainty. As an example we shall consider the boundary between two isolated regions. Assuming a graded function of membership, as portrayed in Figure 19.3, region A, the interaction between these regions can be represented schematically, as in Figure 19.5(a).

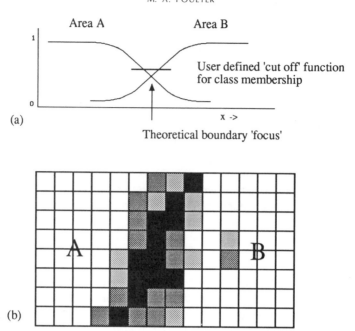

Figure 19.5 (a) Two-dimensional representation of overlapping membership functions; (b) raster representation of the uncertainty in membership for pixel elements between areas A and B. The black areas are undefined pixels below the 'cut-off' function shown in (a).

The intersection of these two 'membership' transfer functions would ideally be the focus at which the vector representation of the boundary in the database would be established. However, allowing for a quality measure of uncertainty, the user may introduce a 'cut-off' value of membership probability. This value may be above the intersection and therefore leave some areas as belonging to neither class. Figure 19.5(b) demonstrates how the raster representation of this boundary might look. The black pixels are those areas below the cut-off level and hence represent neither object, while the grey shades portray varying levels of membership below positive identification. Of course, in the real world these areas may not be just restricted to the boundaries of objects.

As the representation of features and their definition is greatly influenced by the requirements of particular applications, thought must be given to how researchers in diverse disciplines view relationships between objects and entities. This affects how they view the utility of undetermined boundaries to their analysis of geographic datasets. A single style of database representation is therefore unlikely to suit every-one. Once regions of uncertainty at boundaries have been identified they could be stored as an 'overlay' to highlight those areas where a deterministic processing approach would be unsuitable. Alternatively, new 'boundary' objects could be created, inheriting descriptive attributes from the class membership probabilities of each pixel within the region. While the requirements of map displays may be differ-ent for users in disciplines such as geology, geophysics, geomorphology and environ-mental assessment, there is sufficient common ground in processing requirements to justify further research in the following areas to explore the utility of land cover datasets derived from Earth observation source:

- Rules of aggregation combining units of land cover by application-dependent principles to derive object hierarchies and descriptions most suitable to that application.
- The depiction of probabilistic maps involving large numbers of classes.
- The graphical representation of an undetermined boundary once it has been defined.
- Object-orientated databases to support integrated models of land use representation and their subsequent application to analysis within GIS.

Acknowledgements

The work reported in this paper forms part of the UK national space programme sponsored by BNSC. I wish to thank my colleagues in BNSC, DRA, industry and academia for many helpful discussions during the course of this work.

REFERENCES

ANDERSON, J. R., HARDY, E. E., ROACH, J. T. and WITMER, R. E. (1976). *A Land Use and Land Cover Classification System for Use with Remote Sensor Data*, Geological Survey Professional Paper 964, Washington: US Government Printing Office.

ARMSTRONG, M. P. and DENSHAM, P. J. (1990). Database organisation strategies for spatial decision support systems, *Int. J. Geographical Information Systems*, **4**, 3–20.

BASKENT, E. Z. (1993). Quantifying forest landscape structure for integrated resource planning in *Proc. 7th Annual Symposium on Geographic Information Systems in Forestry, Environment and Natural Resources Management*, Vol. 2, pp. 855–862, Vancouver: Polaris Conferences.

BURROUGH, P. A., MACMILLAN, R. A. and DEURSEN, W. (1992). Fuzzy classification methods for determining land suitability from soil profile observations and topography, *J. Soil Science*, **43**, 193–210.

CORBLEY, K. P. (1994). Private satellites now the trend in remote sensing: Two new satellite plans on the table, *Earth Observation Magazine*, January, 22–23.

DOWMAN, I., MCKEOWN, D. and SPRATT, A. (1994). *A Study of High Resolution Satellite Data for Cartographic Feature Extraction*, unpublished report, Defence Research Agency, Farnborough, UK.

FOODY, G. M. and COX D. P. (1991). Estimation of sub-pixel land cover composition from spectral mixture models, in: Dowman I. (Ed.), *Conference Proceedings: Spatial Data 2000*, London: University College London Press.

GLIDDEN, D. J. and GUGAN, D. J. (1991). User requirements for an integrated GIS, in: *Association of Geographic Information Conf. Papers*, pp. 1.15.1, London: AGI.

GOODING R., TAYLOR S. and FISHER J. (1993). *Integrated Map from Image Update System (IMIUS) Automatic Line Following II Prototyping Report*, unpublished document EOS-94/097-2000-RP-002, Earth Observation Sciences Ltd, Farnham, UK.

HEALEY, R. G. (1991). Database management systems, in: Maguire, D. J., Goodchild, M. F. and Rhind, D. W. (Eds), *Geographical Information Systems: Principles and Applications*, Vol. 1, pp. 251–267, New York: Longman.

LEASON, A. N. (1992). *Image Segmentation: Final Report*, unpublished report, LaserScan Ltd, Cambridge, UK.

LO, C. P. (1986). *Applied Remote Sensing*, pp. 228–279, New York: Longman.

LOWELL, K. E. (1994), Probabilistic temporal GIS modelling involving more than two map classes, *Int. J. Geographical Information Systems*, **8**, 73–93.

NASA 1991, *EOS Reference Handbook*, Greenbelt, MD: NASA/GSFC.

POULTER, M. A. (1993). Integrating Earth observation data: A view from the UK, in: *Proc. North Australian Remote Sensing and Geographic Information Systems Forum*, pp. 182–187, Darwin: Australian Government Publishing Service.

POULTER, M. A. (1994). *IGIS Initial System Development Prototype Evaluation*, unpublished report DRA/CIS/(CSC2)/5/26/1/3/REP/2, Defence Research Agency, Farnborough, UK.

SABINS, F. F. (1987). *Remote Sensing Principles and Interpretation*, London: Freeman.

SIMPSON, G., MORRIS, R. and WEBB, K. (1994). *Study on the Application of Artificial Neural Networks for Crop Yield Prediction and Software Substitution: Final Report*, unpublished document EOS-94/097(13000)-REP-001, Earth Observation Sciences Ltd, Farnham, UK.

STRAHLER, A. H. (1980). The use of prior probabilities in maximum likelihood classification of remotely-sensed data, *Remote Sensing of the Environment*, **10**, 135–163.

STUMPF, A. S. (1993). From pixels to polygons: the rule-based aggregation of satellite image classification data using ecological principles, in: *Proc. 7th Annual Symposium on Geographic Information Systems in Forestry, Environment and Natural Resources Management*, Vol. 2, pp. 939–945, Vancouver: Polaris Conferences.

CHAPTER TWENTY

How Many Lakes, Islands and Rivers are there in Finland?

A Case Study of Fuzziness in the Extent and Identity of Geographic Objects

TAPANI SARJAKOSKI

Finnish Geodetic Institute, Department of Cartography and Geoinformatics, Masala, Finland

20.1 Introduction

Promoted by our desire to describe the geographic reality with increased accuracy in geographic information systems (GIS), understanding and modelling the fuzziness of geographic objects has become a topic of considerable interest in geoinformatics. Mathematical methods such as probability theory and fuzzy sets have been successfully applied for handling fuzziness in a number of special applications. But still agreement on a universally applicable theory of the fuzziness of geographic objects evades us.

Here, we approach the problem in the form of a case study about the fuzziness of Finland's lakes, islands and rivers. The Finnish landscape is suitable for such a study in many respects. The Ice Age, which ended about 10 000 years ago, left the landscape heavily fragmented, as is apparent from the map in Figure 20.1. The land is interspersed with many lakes; the lakes are 'filled' with islands, and the coastline against the sea is buffered by an extensive archipelago of many islands. In addition, the land is still in continuous dynamic movement due to land uplift or isostatic rebound.

20.2 Central concepts of objects and of fuzziness in relation to them

Before treating the problem of fuzziness in the existence and extent of geographic objects, we should look at the conventional meanings of some 'object-related' concepts more closely. The following definitions are from *Webster's Electronic Dictionary of Synonyms* and *Webster's Third New International Dictionary*:

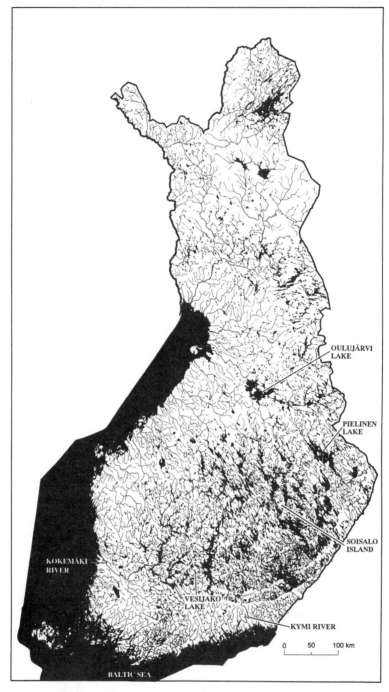

Figure 20.1   Water areas in Finland. (© National Land Survey of Finland, permission no. 297/MAA/94).

Object: (1) Whatever is apprehended as having actual, distinct and demonstrable existence.
                   (2) That which can be known as having existence in space or time.

Entity: (1) One that has real and independent existence.
                   (2) That which can be known as having existence in space or time.

Individual: (1) Whatever is apprehended as having actual, distinct and demonstrable existence.
                   (2) One that has real and independent existence.
                   (3) That which can be known as having existence in space or time.

Existence: The state or fact of having independent reality.

Extent: (1) General or approximate size or amount.
                   (2) The amount of measurable space or area occupied by or comprising a thing.

Identity: The quality of being the same in all the constitutes the objective reality of separate things.

Independent: (1) Not subject to the rule or control of another.
                   (2) Maintaining or able to maintain oneself without outside aid.

Region: A distinguishable extent of surface and especially of the Earth's surface.

Boundary: An enclosing line or margin.

Border: A line or relatively narrow space that marks the outermost bound of something.

Class: (1) A number of individuals thought of as a group because of a common quality or qualities.
                   (2) A unit of subunit of a larger whole made up of members sharing one or more characteristics.

Generalize: (1) To make general
                   (2) To derive or induce a general conception, principle or inference.
                   (3) To make indefinite (as by blurring particular features).
                   (4) To modify or eliminate (nonessential details on a map) for improving legibility or for emphasizing some particular feature (as the location of mountains or the essential character of a coastline).
                   (5) To portray or emphasize in painting general rather than particular features and characteristics.

General: (1) Involving or belonging to the whole of a body, group, class, or type.
                   (2) Involving or belonging to every member of a class, kind, or group.
                   (3) Applicable or pertinent to the majority of individuals involved.
                   (4) Concerned or dealing with universal rather than particular aspects.

Fuzzy: Scarcely or imperfectly perceptible.

The use of these terms and concepts in geography and geoinformatics (often with the prefix geo or geographic) has accorded rather well with the above definitions. In ordinary English, though, the words *object* and *entity* seem to be synonymous.

Existence is a necessary property of an object. An object also has an *identity*. This is the key issue when distinguishing between *field-* and *object-oriented* approaches in describing geographic reality (Frank and Goodchild, 1990; Couclelis, 1992). There is no identity related to fields as such, except the one that renders a field distinguishable from other fields. Objects can be, and also very often are, associated with *regions* in a field; thus fields are indirectly connected with the concept of identity.

Many of the above concepts are closely interwoven. Take the triple existence–extent–identity, for example; it is hard, or even impossible, to speak of one without another. According to our contemporary understanding of the universe, geographic objects can only exist confined in four-dimensional space–time.

How can fuzziness be related to geographic objects? When there appears to be fuzziness, it appears in many of the properties of an object, not in one alone. Some of the extreme cases are pointed out below.

Fuzziness only in 3D space: Fuzziness may be limited to the spatial extent of a geographic object, i.e. the boundary of the object is ill-defined, even though there is no doubt about its identity and temporal extent.

Fuzziness only in time: The temporal extent of a geographic object may be ill-defined, even though the spatial, extent is well-defined, i.e. the spatial boundary is sharp.

Fuzziness in identity: This is closely related to fuzziness in spatial and temporal extent. The issue here is whether or not an object has 'enough identity' to be an individual. With dynamic objects (objects that change in time), the concept is linked to the problem of the extent to which an object can change and still be the same object. With static objects we often face the situation where we do not know whether a small object should be considered part of a larger one instead of being independent (e.g. the Adriatic Sea and the Aegean Sea parts of the Mediterranean Sea), or whether or not two equally large objects should be considered a whole, without being a conceptual aggregate. This problem is obviously related to that of generalization. It is probably true that if a geographic object is truly an object then people will sooner or later name it. Conversely, if something has a name then it definitely is an object – at the chosen level of generalization.

Fuzziness in class definition: Class definitions are sometimes fuzzy and overlap, making it impossible to assign an object to only one class.

Fuzziness in class membership: Even when there are well-defined (axiom- or definition-based) classes, the characteristics of an object may still be such that we cannot be quite sure to which class it belongs.

20.3 How many lakes, islands and rivers are there in Finland?

In the mid-1980s Raatikainen and Kuusisto (1987) carried out a survey to establish the numbers and surface areas of lakes, islands and rivers in Finland. The final figures were: 187 888 lakes, 179 584 islands and 647 rivers. This result immediately raised questions about the existence and extent of geographic objects: how can it be possible to count lakes, islands and rivers so precisely? Or, to put it another way, what is the definition of a lake, an island and a river, and how were the lakes, islands and rivers measured to reach the above numbers? A more detailed look at the survey is enlightening about the notion of fuzzy geographic objects.

20.3.1 How many lakes?

According to the survey, there are 187 888 lakes in Finland, but what are these lakes? The lakes were counted manually from the Finnish 1 : 20 000 scale topographic base maps. In the survey, lakes were defined as waters with a minimum area of 0.05 ha, so that they showed up on the maps with a diameter of 1.3 mm if fully round. Smaller lakes and ponds are also marked on these maps but they were omitted from the count.

The survey had to confront difficult problems of interpretation. Groups of ponds within a bog were interpreted as a single lake; some of them had once been continuous water areas but had filled with vegetation. Artificial reservoirs for hydropower stations were also counted, but the basins of water treatment plants were not.

Lakes in chains of lakes separated by rapid straits were usually considered to be separate, especially if the lakes had different names. Bays separated from a main basin by shallow straits could also have been considered independent lakes, but they seldom were.

Lakes in river basins also presented a problem. They are the wider parts of rivers and often lack individual names. In some places they are very deep basins where the water current slows down and the main stream passes only a narrow section of the water. Such expanses of water can be interpreted as a lake if a major portion of the sediments in the water sink to the bottom of the basin. Neither the depth of the basin nor the accumulation of sediments is shown on the maps used in the survey. In the survey a basin was interpreted as a lake if the current clearly slows down within it.

The change in the number of lakes over time is also problematic. The number is declining naturally, even though uplift is creating some new lakes out of sea bays. The number of lakes has also been reduced by artificial drainage which has produced a large number of new, smaller lakes. In recent decades, many artificial lakes or reservoirs have also been built.

20.3.2 How many islands?

According to the survey, there are 179 584 islands in Finland. The dictionary defines an island as a land area surrounded by water. An island can be natural or a man-made artifact. There is no universally recognized lower limit for the size or area of an island. In the survey an island was defined as in the Finnish 1 : 20 000 scale topographic base maps: it must be larger than about 0.01 ha and is presented on the map with a solid line. According to the survey, identifying Finland's islands was rather easy, but even so, the number of islands is not stable. Uplift and human construction work create new islands. Some islands disappear when uplift narrows straits and causes separate islands to merge. Also, man-made earth fills and bridges erase or at least change the status of islands. Such islands were included in the count because their ecological status is still much as it always was.

Interpretation of some of the large islands was not so straightforward. Soisalo, for example, is commonly considered the largest island in Finland. It fulfils the criterion of an island in that it is a land area surrounded by water. The waters of Lake Kallavesi on the north side of Soisalo run through Lake Unnukka (west) plus Lakes Suvasvesi and Kermajärvi (east). In some places the chains of lakes are so narrow they resemble rivers. On this basis, Soisalo can be interpreted as an area of land between two river branches; such areas are not called islands in the survey.

If no limit is set on the width of the water that has to surround an island, then the area between the basins of the Kymi and Kokemäki rivers would form the largest year-round island in Finland, because the waters of these rivers connect Lakes Lummene and Vesijako in central Finland. During spring floods, there is also a connection between the river basins of Lakes Oulujärvi and Pielinen, forming an island out of the entire southern half of Finland.

20.3.3 How many rivers?

There are 647 rivers in Finland. According to the dictionary definition, 'river' is a general name for a water basin with a current, but which is larger than a ditch or creek. The Finnish water legislation uses a more precise definition: 'A water basin of running water is called a river if it can be travelled by rowboat except during the low-water season, excluding rapids and stones'. A water basin with a mean discharge of 2 m^3/s is, however, always called a river. In the legislation, a creek is smaller than a river.

The names given by people are not always consistent with the legislation; sometimes rivers are called creeks or ditches, and vice versa. Therefore the name as such cannot be used as a basis for determining whether a watercourse is a river or not. The numerous chains of lakes also make the definition of rivers difficult. How long must a stream between two lakes be, to be defined as a river? According to the water legislation, there is not lower limit. In the survey a river was included in the count if its runoff area is over 100 km^2 and it has a continuous stream of 10 km or more with no interruptions by lakes.

20.3.4 Spatial extent of Finland

Finland, as understood in the survey and also otherwise today, is limited by its political or state boundaries. But the spatial extent of Finland has changed many times in history; the current boundaries were defined after the Second World War. Some of Finland's boundaries are natural, like the one with Sweden, which follows the Tornio River; some are largely artificial, like the one with Russia.

20.4 How fuzziness appears in the survey

Factors that make the counting of geographic objects such an ill-defined task are analyzed here more in detail.

20.4.1 Categorical coverage and classification

The problems of counting all of Finland's lakes, islands and rivers can be seen as the formation of *categorical coverage* for the whole country according to a class hierarchy presented in Figure 20.2. To guarantee that the resulting *regions* fulfil the

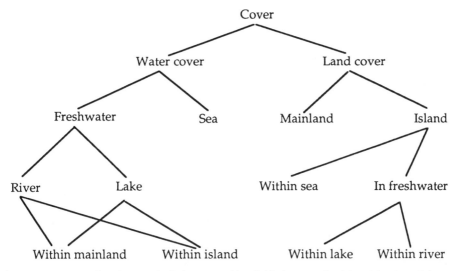

Figure 20.2 Hierarchy of categorical classes used implicitly for counting lakes, islands and rivers.

criteria of the *geographic objects* to be counted, the task is constrained by many conditions. In other words, we have to create a categorical coverage map having land and sea, as well as lakes, rivers and islands with specified properties as *individual regions*. The task is also closely related to the problem of map generalization. *Selection* (Shea and McMaster, 1989) has to be used to omit all the lakes, rivers and islands below a certain *size criterion*. A decision about amalgamation also has to be made, for instance, when considering the independence of bays as lakes, straits in lakes as rivers, and expansions in rivers as lakes.

20.4.2 Difficulties in interpretation

Many of the decisions in the count by Raatikainen and Kuusisto were very heuristic in nature. The salient points set out in Sections 20.3 are elaborated in Tables 20.1–20.3. The classification rules look rather heterogeneous and informal, but this may be justified in view of the overall goal of the survey. It is also possible that the authors did not explicitly express all the knowledge and expertise they actually applied in doing the count, as frequently happens when experts describe their own work. This makes it difficult to formalize such knowledge for computerized expert systems (see e.g. Hayes-Roth *et al.*, 1983). In a recent article, Raatikainen and Kuusisto (1988) define a lake more precisely:

0. In simple terms, a depression in the surface of the earth filled with water can be called a lake. The following clarifications are, however, necessary.

1. Water can fill a depression partly or fully.

2. The water level must be nearly the same all over the lake, except when temporary wind or ice conditions have caused a deviation.

3. The proportion of incoming current is so small that major proportion of the suspended particles sink to the bottom of the basin.

4. The area is more than a specified minimum area.

5. The water level does not continuously follow the variations in sea level.

Table 20.1   Interpretation difficulties with lakes

Quotation	Explanation
The lakes were counted manually from the Finnish 1 : 20 000 scale topographic base maps.	The use of maps as the source of information affects the results of the count very much. How are the lakes defined for the mapping process, in a cartographic sense? When were the maps compiled? What is the overall quality of the maps?
In the survey, lakes were defined as waters with a minimum area of 0.05 ha.	A definition that is based on the relevant disciplines or branches of science (geography, hydrology).
Groups of ponds within bogs were interpreted as a single lake; some of them had once been continuous water areas but had filled with vegetation.	An aggregate operator is here applied, at least on the conceptual level. It is unclear how the area is calculated in this case.
Lakes in chains of lakes separated by rapid straits were usually considered to be separate, especially if the lakes had different names	Here commonsense knowledge (the existence of names) has been applied to direct the identification process.
Bays separated from the main basin by shallow straits could also have been regarded as independent lakes, but they seldom were.	A contradictory explanation that leaves much space for subjective decisions.
In the survey a basin was interpreted as a lake if the current clearly slows down within it	Information on dynamics (current) is used in the interpretation. The linguistic statement *clearly* is inherently fuzzy.
The number of lakes is declining naturally, even though uplift is creating some new lakes out of sea bays.	Another example of dynamics: the number and extent of geographic objects is changing. Thus a count is valid for only a limited time period.

The authors point out that different areas of application would require slightly different definitions. They also conclude that, although the counts were given in the precision of single lakes, the accuracy of the count is no better than a few thousand for lakes of 1 ha or less, and no better than a few hundred for those of 1–100 ha. This is due to (1) the difficulty of the definition, (2) the visual/manual procedure used to determine of the size of a lake, and (3) the variable conditions of aerial photography used to produce of the 1 : 20 000 scale topographic base maps.

20.4.3   Effect of land uplift

Land uplift in Finland is a legacy of the Ice Age. Its effect is most marked on the south and west coasts where the land is rising at a rate of about 1 cm per year in

Table 20.2 Interpretation difficulties with islands

Quotation	Explanation
The dictionary defines an island as an area of land surrounded by water. There is no universally recognized lower limit for the size or area of an island	A dictionary gives a sound definition of an island in the sense of *topology*. It can be recalled, as a curiosity, that Greenland, with its total area of almost 2.2 million km^2, is considered to be the world's largest island. Islands larger than that are called continents.
In the survey an island was defined as in the Finnish 1 : 20 000 scale topographic base maps: it must be larger than about 0.01 ha and is presented on the map with a solid line.	The existing *cartographic* definition of an island was used in this study. Is it truly applicable for this purpose?
According to the survey, identifying Finland's islands was rather easy.	This is a *subjective* statement by the authors of the survey.
Even so, the number of islands is not stable. Uplift and human earth construction work create new islands. Some islands disappear when uplift narrows straits and causes separate islands to merge.	The number of islands, as the number of lakes, is applicable for only a specified time.
If no limit is set on the width of the water that has to surround an island, then the area between the basins of the Kymi and Kokemäki rivers would form the largest year-round island in Finland, because the waters of these rivers connect Lakes Lummene and Vesijako in central Finland. During spring floods, there is also a connection between the river basins of Lakes Oulujärvi and Pielinen, forming an island out of the entire southern half of Finland.	Although the topological definition of an island is clear, it is not sufficient because the width of the surrounding water has to be sufficient for *realistic* interpretation. The *annual variation* must also be considered.

relation to sea level. This is creating new islands and also new lakes. As these changes are very slow, we frequently encounter the problem of fuzziness in the extent and identity of the islands. Figure 20.3 shows how our conceptual model changes from a field view to an object view. To start with, we have a rather ill-defined 'shallow part of the sea'; in the end, a well-defined island.

20.4.4 Fractal nature of the number of lakes and islands

Several studies have demonstrated the fractal nature of geographic phenomena (Mandelbrot, 1975; Lam and Cola, 1993). Of relevance here is probably the often

Table 20.3   Interpretation difficulties with rivers

Quotation	Explanation
According to the dictionary definition river is a general name for a *water basin with a current*, but which is larger than a ditch or a creek.	The dictionary definition uses *water current* but says nothing about the *form* of the water basin.
Finnish water legislation uses a more precise definition: 'A water basin of running water is called a river if it can be travelled by rowboat except during the low-water season, excluding rapids and stones'. However, a water basin with a mean discharge of 2 m³/s is always called a river. In the legislation, a creek is smaller than a river.	As usual, the legal definition is rather precise. Interestingly, it uses a very pragmatic qualifier: *if it can be travelled by rowboat.* From the legal point of view it is essential to know whether or not the water body is part of a water basin. Ditches are no longer treated in a similar way by the water legislation.
The names given by people are not always consistent with the legislation; sometimes rivers are called creeks or ditches, and vice versa.	This indicates that common names are not a reliable source of information in the classification of geographic objects.
The numerous chains of lakes also make the definition of rivers difficult. How long must a stream between two lakes be, to be defined as a river? According to the water legislation, there is no lower limit.	Where to find the answer?
In the survey a river was included in the count if its runoff area is over 100 km² and it has a continuous stream of 10 km or more with no interruptions by lakes.	This size criterion for a river is complex, including the use of the run off area and the length of the river.

used statement that natural shorelines are fractals. This means, broadly speaking, that the irregularity of the shoreline repeats itself on a wide range of scales. This irregularity can be expressed with a single numeric value called the fractal dimension (D). According to the fractal theory and the Korcak–Zipf empirical law for islands (Burrough 1993), the number of islands can be expressed with the formula

$$Nr(A > a) = Fa^{-D/2},$$

where $Nr(A > a)$ is the number of islands larger than a, and F is a constant. The fractal dimension D can be estimated from sample data by plotting $\log Nr(A > a)$ against $\log a$, the slope being $-D/2$.

   To study the fractal behavior of the sizes of lakes and islands in Finland, a limited analysis was done with this method. The results are given in Figures 20.4 and 20.5. Data on islands were available only on the 100 largest islands in lakes and the sea. With the exception of the largest lakes and the very small lakes (which are affected by cartographic generalization) the sample data fit a linear log–log plot over

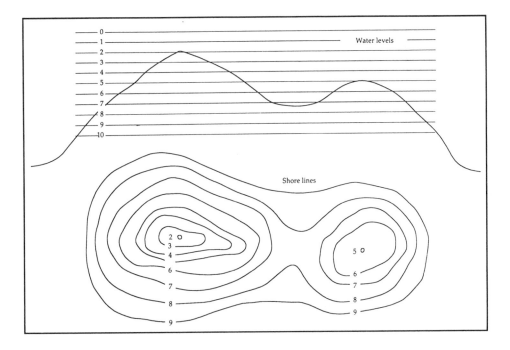

Water at level	Geographic interpretation
0 or above	Shallow part of the sea
1	Shallow part of the sea with a rock under the surface
2	Shallow part of the sea with a rock on the surface
3 and 4	A small island and a rock under the surface next to it
5	A small island and a rock on the surface next to it
6	Two small islands
7	Two small islands that are nearly connected
8	One or two islands
9 or below	One island

Figure 20.3 Example of how an island gradually emerges from the sea as the water level falls.

some four orders of magnitude, which strongly suggests a fractal distribution. The estimated fractal dimension D for the lakes is 1.7, which is close to the value obtained by Korcak (1938) (this coefficient was 0.8 and according to Mandlebrot (1977) $D = 2 \times$ Korcak's coefficient). However Lovejoy and Schertzer (1991, cited in Lavallée et al., 1993, p.169) have pointed out that this method of estimating D yields different results for different threshold values for the area. On the basis of this evidence it is intriguing that the distribution of lake sizes is fractal-like, but there is insufficient evidence to claim that they are ideal fractals.

Because of their fractal behaviour, lakes and islands are uncountable, just as a fractal shoreline has an infinite length. Strictly speaking, then, it is not relevant to ask how many lakes or islands there are in Finland. The question becomes meaningful only when their minimum size is specified. Although common sense tells us this is obvious enough, it is convincing when derived theoretically.

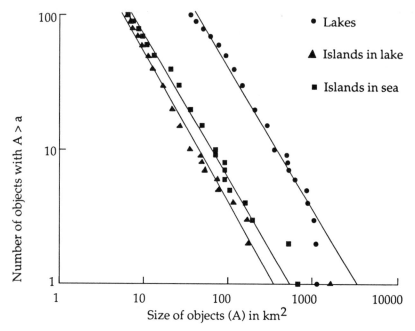

Figure 20.4 Number of lakes and islands as a function of minimum area, represented in a log–log plot for the 100 largest ones in each group.

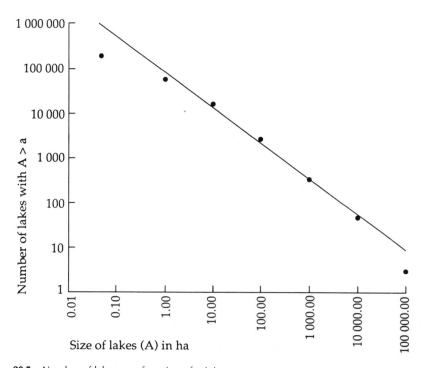

Figure 20.5 Number of lakes as a function of minimum area.

The concept of uncountability must be kept separate from the other problems of the interpretation. These all remain, even when the minimum size of the object is stated exactly. Equally, geographic objects that do not meet the size criteria may be well-defined geographic objects in their own right.

20.5 Counting geographic objects with computerized methods

In the survey, the lakes, rivers and islands were counted manually from 1 : 20 000 scale topographic base maps. Technically, it would be possible to count them by automatic, computerized methods if the necessary information was stored in digital form in a GIS database. The real difficulty is how to model geographic reality to facilitate the task. An extreme alternative would be to incorporate all the relevant basic *physical* information, that is, terrain elevation, water levels, speed of water current, etc., in the model and database. By defining lakes, rivers and islands in terms of these numeric quantities, we would then be able to count them automatically. This would lead to an *objective* count, in the sense that the interpretation of instances would not depend on a human agent.

A more advanced method would utilize the commonsense understanding of place names (e.g. 'Lakes in chains of lakes separated by rapid straits are usually considered separate entities, especially if these lakes have different names'). In that case it would be necessary to add this information to the data model and database. The rules about using place names would then be based on commonsense or heuristics.

Finally, the most 'human-based' approach would be to use a GIS database only for storing the regions of objects already identified. In this approach, automation would be used only for counting regions greater than a specified minimum size. This is nearly the only operational method available with the tools in the current GIS packages.

20.6 Discussion

In principle and by definition, a geographic object will always have an identity and therefore a crisp existence. This implies that a geographic object can *always* have an individual name. This case study demonstrates that, in reality, it is often very difficult to state whether or not something can be modelled as an individual object. The objects on this conceptual borderline can be said to be *fuzzy in existence*, as was done here. In other words, it is not clear whether the object exists at all: Fuzzy existence is always strongly related to fuzziness in extent, whether spatial or temporal. The task of counting the number of geographic objects deals only with their existence; it is in principle unnecessary to know their extent. Nevertheless, because the objects are located in the same geographic field, their spatial extent must be determined in a consistent way.

The use of GIS tools in spatial analytical tasks always presumes a formal definition of the task. As proposed above, Finland's lakes could also be counted by such methods. This would require a rather complex knowledge-based system to reach the same level understanding of geographic reality and the same level of skill in interpretation that a geographer has. The effort involved in building such a system would be of great scientific interest, because it would entail formalizing the interpretation of knowledge. With such a system, the process would be fully repeatable, and the effect of the changes in the knowledge base and input data could be easily studied. It would certainly provide insight into the problems related to the fuzziness of geographic objects and of the basic observables involved in the process.

Acknowledgement

I wish to thank Professor Peter Burrough for his suggestions related to fractal analysis.

REFERENCES

BURROUGH, P. A. (1993). Fractals and geostatistical methods in landscape studies, in: Lam, N. S.-N. and Cola, L. D. (Eds), *Fractals in Geography*, pp. 87–121, Englewood Cliffs, NJ: Prentice Hall.

COUCLELIS, H. (1992). Beyond the raster–vector debate in GIS, in: Frank, A. U., Campari, I. and Formentini, U. (Eds), *Theories and Methods of Spatio-Temporal Reasoning in Geographic Space*, Lecture Notes in Computer Science 639, pp. 65–77, Berlin: Springer.

FRANK, A. U. and GOODCHILD, M. F. (1990). *Two Perspectives on Geographic Data Modelling*, Technical paper 90–11, National Center for Geographic Information and Analysis (NCGIA).

HAYES-ROTH, F., WATERMAN, D. A. and LENAT, D. B. (Eds). (1983). *Building Expert Systems*, Reading, MA: Addison-Wesley.

KORCAK, J. (1938). Deux types fundamentaux de distribution statistique, *Bulletin de l'Institute International de Statistique*, III, 295–299.

LAM, N. S.-N. and COLA, L. D. (Eds). (1993). *Fractals in Geography*, Englewood Cliffs, NJ: Prentice Hall.

LAVALLÉE, D., LOVEJOY, S., SCHERTZER, D. and LADOG, P. (1993). Nonlinear variability of landscape topography: Multifractal analysis and simulation, in: Lam, N. S.-N. and Cola, L. D. (Eds). *Fractals in Geography*, pp. 158–192, Englewood Cliffs, NJ: Prentice Hall.

LOVEJOY, S. and SCHERTZER, D. (1991). Multifractal analysis techniques and the rain and cloud fields from 10^3–10^6 m, in: Schertzer, D. and Lovejoy, S. (Eds), *Nonlinear Variability in Geophysics: Scaling and Fractals*, Dordrecht: Kluwer.

MANDELBROT, B. B. (1975). Stochastic models for the Earth's relief, the shape and the fractal dimension of the coastlines, and the number–area rule for islands, *Proc. National Academy of Sciences.* **72** (10), 3825–3828.

MANDELBROT, B. B. (1977). *Fractals, Form, Chance and Dimension*, San Francisco: Freeman.

RAATIKAINEN, M. and KUUSISTO, E. (1987). *Suomen Kuvalehti*, **4B**, Helsinki (in Finnish).

RAATIKAINEN, M. and KUUSISTO, E. (1988). The number and surface area of the lakes in Finland, *Terra* **102** (2), 97–100 (in Finnish).

SHEA, K. S. and MCMASTER, R. B. (1989). Cartographic generalization in a digital environment: When and how to generalize, *Auto-Carto 9, Proc. 9th Int. Symposium on Computer-Assisted Cartography*, Baltimore, MD, pp. 56–67.

CHAPTER TWENTY ONE

Geoscience Modelling: From Continuous Fields to Entities

MARINOS KAVOURAS

Faculty of Rural and Surveying Engineering, National Technical University of Athens, Athens, Greece

21.1 Introduction

Geoscience applications involve objects of great variety and complexity. Some of these have accurate spatial descriptions with sharp boundaries, and they are 'well-behaved'. Burrough and Frank (1995) also call them *exact entities*. Depending on the characteristics of an exact entity and the indented use, various geometric modelling techniques can be used to describe it at various degrees of detail. Any uncertainty in the boundary definition of 'well-behaved' objects can be due to lack, loss, poor quality, inconsistency of definitional data, or limitations of the selected representation.

In contrast with the exact and 'well-behaved' entities (usually man-made features), other natural objects tend to be highly irregular, fragmented, fuzzier, with variable distributions of attribute values, and uneven sample data – and are thus very difficult to describe with distinct boundaries. Realizations that exhibit a statistical rather than a deterministic character, similar to that of the soil data in GIS applications (Burrough, 1991), are often called *continuous fields* (Burrough and Frank, 1995). In these cases, entities can be defined by the contents and various classification realizations.

The topic of undetermined boundaries relates to data accuracy and quality issues which are active GIS research topics (Goodchild and Gopal, 1989; Chrisman, 1991; Buttenfield, 1993). Geoscience and also environmental modelling have not experienced the same degree of awareness and development as 2D GIS, for they address smaller interest groups with non-standard needs. They also present many difficult problems with regard to data generation, manipulation and visualization. Three special meetings – a NATO Advanced Research Workshop at Santa Barbara, USA, in 1989 (Turner, 1992), a symposium at Freiburg im Breisgau, Germany, in 1990 (Pflug and Harbaugh, 1992), and a European Science Foundation meeting at Il Ciocco, Italy, in 1992, have been held to address and promote research in the field of 3D GIS and geoscientific applications. In this range of activities, applications include oil exploration (Youngmann, 1989; Sims, 1992; Lasseter, 1992), mining (Kavouras, 1987, 1992; Kavouras and Masry, 1987; Bak and Mill, 1989; Houlding, 1992), geological modelling (Denver and Phillips, 1990; Kelk, 1992; Hurni, 1993), sedimentology (Lee and Harbaugh, 1992), hydrogeology (Turner, 1989), environmental monitoring (Smith and Paradis, 1989), and meteorology (Slingerland and Keen, 1992). For the new reader, a number of review articles and textbooks containing a good selection of topics, are presently available (Raper, 1989; Raper and Kelk, 1991; Bonham-Carter, 1991; Turner, 1992; Pflug and Harbaugh, 1992).

In the context of geoscientific data modelling, this chapter focuses on the problem of entity definition from continuous fields (Sections 21.2 and 21.3). The approach introduced here is applicable in several domains, however, our primary application interest has been in geology and mining. Coping with the boundary uncertainty of exact entities is a complex problem, yet much easier to handle, and it is only briefly introduced here (Section 21.4). The concluding Section 21.5 suggests several issues for further research.

21.2 Geoscience data modelling

Some natural entities (geo-objects) are part of a continuum, being characterized by properties which vary continuously in space. The most common strategy for representing a continuous field is a 3D (or *n*-D) domain-partitioning to regular or non-regular elements with variable attributes. Entities can be spatially defined and subsequently modelled (i.e represented or discretized) using a classification of this continuous field. Such entities are also known as '*definition limited*' (Raper and Kelk, 1991), as opposed to '*sampling limited*', a term that is used to describe exact geo-objects. The classification criteria are based on selected thematic values, and therefore control the spatial properties of the resulting entities. Different classification criteria result in different entity representations.

Before presenting the necessary modelling strategy (see Section 21.3), it is important to introduce some background concepts about the geoscientific modelling requirements. The common problem of *ore body definition* provides a good basis for discussion.

An ore body is conceptually defined as a mineral deposit that can be exploited at a profit, under certain market conditions. Thus, as economic conditions and technological capabilities change, the ore body fluctuates (shrinks or expands) in a dynamic manner. The total extent of a mineral deposit and its overall grade are determined during exploration. Some deposits exhibit low variability about their overall grade. Most deposits, however, have lower- and higher-grade zones. When a mine comes near production, it is of great importance to determine these variations with sufficient confidence.

There are several techniques for estimating mineral inventories from sample data. Traditionally, geologists work with drill-holes of cross sections of the deposit and outline the ore-waste contact and zones of certain richness (ore grade). Block models are very suitable structures for both geological interpretation and mine design. The introduction of the theory of *geostatistics* by G. Matheron in the early 1960s, made it possible to formulate a rigorous block estimation as compared to the empirical techniques. The objective is not only to estimate the most likely values of the unknown parameters, but also to assess how accurate these estimates are (their variability). The theory of geostatistics is concerned with variables which are distributed in space, known as *regionalized variables*. The distribution of grades of a particular mineralization is an example of a regionalized variable (Journel and Huijbregts, 1978).

Kriging is a geostatistical procedure, in which the grade (or other) value at a given point (or block), is expressed as a linear combination of the N known (sample) values in its neighbourhood. The procedure requires knowledge of the covariances between the known points and the point (to be estimated). The linear combination uses N weighting coefficients which depend on the structural characteristics of the mineralization and the geometry of the sample data points and the point (or block) to be estimated. The kriging procedure first computes the weighting coefficients and subsequently uses them to estimate the grade value at the point (or block) of interest. The estimation variance is computed as well. The procedure is repeated for all points.

The estimation of kriging is based on some assumptions of homogeneity and isotropy (local estimation). The presence of heterogeneous zones (significant trends), in the deposit, requires more complex treatment (global estimation) (Journel and Huijbregts, 1978). Techniques such as *universal kriging* (Huijbregts and Matheron, 1970), *KTkriging* (Deutsch and Journel, 1992) and other nonlinear estimates (Journel and Huijbregts, 1978) can be used in these cases.

21.3   Entity definition from continuous fields

In the introduction it was stated that in the modelling process, entity definition can be based on classification of continuous fields. Here it is argued that this classification should not only be based on thematic attributes but also on some predefined confidence criteria. Consideration of adjacency, attributes and available accuracy measures results in better entity definitions. This approach is explained in the rest of this section. Details of the aggregation strategy for heterogeneous objects can also be found in Kavouras (1987).

An estimation uses a set of irregularly spaced sample points to estimate the 'best' values (e.g. metal grades) for the block mesh in the continuous field, in the form of a solution vector g and the associated variances s_i^2. The estimation may also provide covariances s_{ij} between any two of the N blocks, in the form of a covariance matrix S of the thematic values, as follows:

$$g^T = (g_1, g_2, \ldots, g_N),$$

$$S = \begin{bmatrix} s_1^2 & s_{12} & s_{13} & \cdots & s_{1N} \\ & s_2^2 & s_{23} & \cdots & s_{2N} \\ & & s_3^2 & \cdots & s_{3N} \\ & \text{symmetric} & & & \vdots \\ & & & & s_N^2 \end{bmatrix}$$

A simple representation scheme can then be specified in the following four steps:

1. By defining a single generic primitive, a cuboid (known as a block, voxel, etc.), containing properties of interest.

2. By associating each block with a tuple $vx(k_i, g_i, s_i)$, where k_i is a locational key describing the spatial properties of the block, g_i is the block attribute, and s_i is the standard deviation of the estimated attribute g_i.

3. By defining the structuring rules needed to build structured representations and support the classification/aggregation scheme.

4. By defining some semantics to ensure that the representation is semantically correct (has desirable properties).

An entity can be described as 'the set of blocks i whose attribute g_i is larger than or equal to a specified value, e.g. current cut-off grade g_t'. In order to attach some confidence to such classification, the covariance matrix information is to be used as well. If, for each block i, a normal probability distribution function is postulated, the blocks satisfy the inequality $g_i - k*s_i \geq g_t$, where k is the coefficient based on a given confidence level e.g. 95%.

Based on this inequality, blocks can be classified as being: (a) clearly inside the entity (*Rich*, high grade), (b) outside the entity (*Poor*, low grade below cut-off), or (c) above cut-off grade, but in which one cannot place high confidence (*Uncertain*). Figures 21.1 and 21.2 exemplify this classification using standardized and actual values (grades). Aggregation of *Uncertain* blocks generally results in larger, more structured representations with higher reliability. The structuring rules can follow a

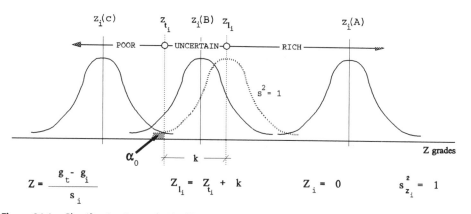

Figure 21.1   Classification intervals R(rich), P(poor), U(uncertain), for the standardized grade Z_i of block i.

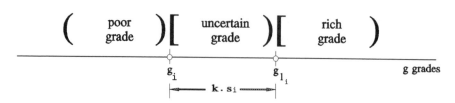

Figure 21.2   Classification intervals R(rich), P(poor), U(uncertain), for the actual grades g.

27.5	31.7	36.9	38.0	37.9	37.9	40.1	40.5
1.9	1.4	3.0	2.8	4.4	4.5	4.1	3.0
29.0	31.7	35.9	36.5	35.2	37.3	41.4	39.8
4.0	3.7	3.8	2.8	3.6	3.3	1.7	1.9
30.5	31.7	33.8	33.6	32.1	35.2	40.0	38.3
2.9	2.9	4.1	2.7	2.2	1.7	1.9	2.7
32.4	32.0	34.1	37.9	34.9	36.8	37.6	33.5
3.0	3.6	3.5	2.4	2.2	1.9	1.1	1.1
31.7	31.3	33.6	37.1	32.7	33.0	31.6	29.4
1.6	2.8	1.3	1.4	2.0	2.5	1.0	1.5
30.3	31.1	31.0	31.8	31.2	34.4	34.5	30.8
1.1	2.7	2.3	2.7	2.6	1.4	1.2	2.5
33.3	33.0	34.9	33.6	29.0	29.1	35.2	32.5
2.2	3.4	3.4	2.3	1.7	1.6	2.5	11.5
33.4	33.7	39.4	38.8	33.8	30.2	32.9	32.2
2.0	2.6	1.6	1.5	1.6	3.1	4.0	3.5

Figure 21.3   A 64-block simulated deposit with estimated grades and standard deviations (from Clark, 1979, p. 112).

simple hierarchical aggregation scheme, which builds larger blocks from smaller ones. When $M = 2^m$ blocks are aggregated ($m > 0$), the resulting block will have a value s_a and a variance s_a^2, which, considering equal weights, are computed from

$$g_a = (1/M) \sum_1^M g_i, \quad \text{and} \quad S_a = a^T S_i\, a,$$

where $a^T = (1/M)(1, 1, 1, \ldots, 1)$ is the design matrix, and S_i is the part of the covariance matrix S which refers to the M blocks. The matrix S_a contains only one element, the variance of the block after aggregation, s_a^2, which is a scalar of the form

$$s_a^2 = (1/M^2)[s_1^2 + s_2^2 + \cdots + s_M^2 + (s_{12} + \cdots + s_{1M} + s_{23} + \cdots + s_{1M}$$
$$+ s_{34} + \cdots + s_{1M} + \cdots s_{(M-2)(M-1)} + s_{(M-2)(M)} + s_{(M-1)(M)})].$$

The aggregation can be loosely viewed as a form of generalization. When blocks of similar attributes (grades) are to be grouped, some restrictions must also be imposed so that there is no significant loss of local detail due to overgeneralization. Thus, certain *cut-off sizes* are selected, in order to specify at what level aggregation should start and where it should end, even if blocks may still be classified as *Uncertain*.

A detailed description of the aggregation procedure can be found in Kavouras (1987). Here, and without loss of generality, the procedure is best illustrated by a two-dimensional example. The sequential nature of the procedure makes extension to the 3D case straightforward. The input data of 64 equal-size blocks with their estimated grades and associated standard deviations (Figure 21.3) are from Clark

(1979). The additional parameters that have to be specified for the aggregation procedure are

(a) the cut-off grade $g_t = 33$ (user specified),
(b) the largest sizes allowed by the aggregation, to prevent overgeneralization (cut-off sizes), and
(c) the level of statistical confidence (e.g. 68%).

Blocks are classified as *Poor* if their grade g_i is less then $g_t = 33$, as *Rich* if $33 + s_i \le g_i$, and as *Uncertain* if $33 \le g_i < 33 + s_i$. The first classification is graphically illustrated in Figure 21.4. Figure 21.5 shows the final result of the aggregation, if a quadtree structuring representation is employed.

The proposed aggregation technique exploits the spatial coherence of nature, also known as spatial homogeneity (Journel and Huijbregts, 1978), which normally characterizes regionalized variables. It is usually accepted (*ibid.*), that in a limited zone of a homogeneous mineralization the correlation that exists between two data values

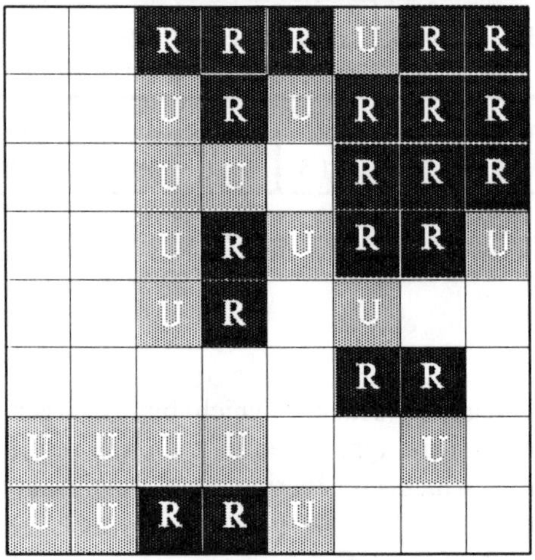

Figure 21.4   Classification of the blocks shown in Figure 21.3.

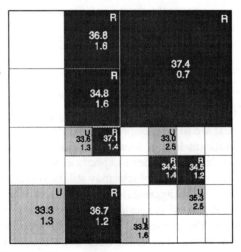

Figure 21.5   Result of the aggregation procedure.

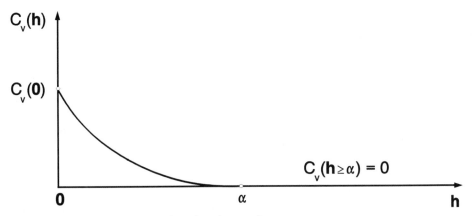

Figure 21.6 Covariance function based on distance h.

$z(x_1)$ and $z(x_1 + h)$ does not depend on their particular positions within the zone, but rather on the distance h which separates them. In order to formulate this spatial structure and make inferences, the range of homogeneity α has to be established. Outside this range, h is considered to be large and $z(x_i)$ and $z(x_1 + h)$ are considered as uncorrelated. A covariance function $C_v(h)$ (Figure 21.6) or a semi-variogram $\gamma(h)$ is often used to model this spatial structure (Journel and Huijbregts, 1978; Clark, 1979).

This *natural correlation* between points, which shows in the attribute values (grades), should not be confused with the covariances shown in the above covariance matrix S. The latter express the *model correlation*, i.e. how the estimation of block i relates to the estimation of block j. If, for example, the block size is large and the distribution of sample data points is dense and homogeneous, the number of data points lying inside each block may be sufficient for the estimation of its metal grade, without having to consider sample points outside the block but still within range α. Such an estimation would produce only variances in the covariance matrix S of the estimated block grades. Of course, the absence of model correlation should not deceive us into claiming that metal grades of adjacent blocks are not naturally correlated, and vice versa.

21.4 Uncertainty and entity-based approaches

In geoscience applications, exact entities, i.e. those that have known and accurate spatial descriptions, can be well-surveyed excavations and sample point locations. Even certain mineralizations, which are quite homogeneous in quality and have sharp, distinct boundaries (e.g. salt domes or perched aquifers), can be accurately located and represented. In modelling (discretizing) this reality, entities are described from selective observations, and are therefore termed *sampling-limited* (Raper and Kelk, 1991). In both cases, any uncertainty in the boundary definition of the entities can be due to

- the lack or loss of definitional data,
- inaccurate surveys, data of poor quality,
- inconsistency of data coming from multiple sources,

- data processing errors,
- uncontrollable generalization operations, and
- limitations of the selected representation scheme.

Currently, there are no systems which estimate, model, propagate and represent uncertainty and quality of geoscientific entities in a systematic and integrated manner, although a significant amount of research has been conducted already (Goodchild and Gopal, 1989; Buttenfield, 1993). What has been recognized for current 2D GIS systems (Burrough and Frank, 1995), is even more valid for geoscientific information systems.

Modelling uncertainty requires some external knowledge and some assumptions about the spatial structure. Such an uncertainty model can build confidence regions (bands or smooth distributions) around the entities, the shapes and sizes of which depend on the accuracy of definitional data and some statistical assumptions about the entities involved. The issue of the visualization of the data quality has been addressed by McGranaghan (1993) and Fisher (1993). Obviously, building such confidence regions increases the dimensionality and therefore the embedded space complexity of the spatial entities. Point uncertainty can be thus expressed by ellipsoids. Space curves are expressed as sweep representations (volumes swept by 2D or 3D shapes as they are moved along a trajectory in space). Solids are expressed as hypersolids.

Drill-holes offer a good example for understanding the uncertainty issue of 'well-behaved' geoscientific objects. Drill-holes can be geometrically represented as straight lines. In real life however, drill-holes can be quite curved, in which case it would be more appropriate to represent them by space curves. Sometimes, due to instrumentation malfunctioning, subsurface drill-hole directional data are lost. This means that the actual location of the drill-hole underground is unknown. Of course, one can assume that the drilling was done vertically and so use the attribute information, or ignore these drill-hole data completely. There are cases, however, when underground activities should not interfere with the drill-holes. In this case, it is important at least to estimate the region in which the drill-hole may lie. In order to tackle this problem, the locational uncertainty of the vector representation of the drill-hole entity is represented by a solid model (Figure 21.7) built with information about the surrounding rock properties, the directional deviation of neighbouring

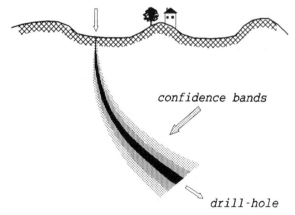

Figure 21.7 Confidence band model of a drill-hole.

drill-holes, and some confidence levels set by the user. Furthermore, this solid can be spatially enumerated with a block model in order to enable individual properties and uncertainty measures to be attached to each point in space.

In addition to the *uncertainty problem in entity representations*, one further issue, which has been almost completely neglected, is that of the *uncertainty of topological relationships*. In general, topological relationships are critical to understand, describe and maintain. Not all relationships, however, carry the same reliability or significance.

Uncertainties in boundary definition can be also due to limitations of the selected representation. Hierarchical data structures, for example, have been proved to be very efficient in modelling irregular geo-solids (Kavouras, 1987, 1992; Bak and Mill, 1989), but pose certain limitations (e.g. they are not invariant under certain transformations) which create topological inconsistencies. Research is under way to develop more robust topological models (Egenhofer *et al.*, 1990; Egenhofer and Herring, 1991; Hadzilakos and Tryfona, 1992; Bundy *et al.*, 1993).

21.6 Suggested topics for further research

In this chapter, an attempt has been made to solve the problem of geoscientific modelling of spatial entities from continuous fields, and a classification/aggregation strategy has been proposed. Exact entities are also complex, yet are much easier to handle. Many issues relating to quality and uncertainty in geoscientific modelling are open for research, some of which are:

- Formalization of uncertainty and fuzziness of spatial entities and relations.
- Formal models of transition zones.
- Estimation of uncertainty and definition of order of topological relationships.
- Quality measures and error propagation during generic operations.
- Visualization of quality and uncertainty.
- Definition of a set of generic dynamic modelling functions.

REFERENCES

BAK, P. R. G. and MILL, A. J. B. (1989). Three-dimensional representation in a geoscientific resource management system of the minerals industry, in: Raper, J. F. (Ed.). *Three-Dimensional Applications in Geographical Information Systems*, pp. 155–182, London: Taylor & Francis.

BONHAM-CARTER, G. (1991). Integration of geoscientific data using GIS, in: Maguire, D. J., Goodchild, M. F. and Rhind, D. W. (Eds), *Geographical Information Systems: Principles and Applications*, Vol. 2, pp. 171–184, London: Longman.

BUNDY, G. L., JONES, C. B. and FURSE, E. (1993). *Holistic Generalization of Large Scale Cartographic Data*, paper presented at the ESF GISDATA Specialist Meeting on Generalization, Compiègne, France.

BURROUGH, P. A. (1991). Soil information systems, in: Maguire, D. J., Goodchild, M. F. and Rhind, D. W. (Eds), *Geographical Information Systems: Principles and Applications*, Vol. 2, pp. 153–169, London: Longman.

BURROUGH, P. A. and FRANK, A. U. (1995). Concepts and paradigms in spatial information: Are current geographical information systems truly generic? *Int. J. Geographical Information Systems*, **9**(2), 101–116.

BUTTENFIELD, P. G. (Ed.) (1993). Mapping data quality: A collection of essays, *Cartographica*, **2**(3), 1–46.

CHRISMAN, N. R. (1991). The error component in spatial data, in: Maguire, D. J., Goodchild, M. F. and Rhind, D. W. (Eds), *Geographical Information Systems: Principles and Applications*, Vol. 1. pp. 165–174, London: Longman.

CLARK, I. (1979). *Practical Geostatistics*, London: Applied Science Publishers.

DENVER, L. E. and PHILLIPS, D. C. (1990). Stratigraphic geocellular modelling, *Geobyte*, **5**(1), 45–47.

DEUTSCH, C. V. and JOURNEL, A. G. (1992). *GSLIB: Geostatistical Software Library and User's Guide*, New York: Oxford University Press.

EGENHOFER, M., FRANK, A. U. and JACKSON, J. (1990). A topological data model for spatial databases, in: Buchmann, A., Gunther, O., Smith T. and Wang, Y. (Eds). *Symposium on the Design and Implementation of Large Spatial Databases*, Zurich, Switzerland pp. 271–286, *Lecture Notes in Computer Science 409*, New York: Springer.

EGENHOFER, M. and HERRING, J. (1991). A mathematical framework for the definition of topological relationships, in: Brazzel, K. and Kishimoto, H. (Eds), *Pro. 4th Int. Symposium on Spatial Data Handling*, Zurich, Switzerland, pp. 803–813.

FISHER, P. F. (1993). Visualizing uncertainty in soil maps by animation, in: Buttenfield, P. B. (Ed.), Mapping data quality: A collection of essays, *Cartographica*, **2**(3), pp. 20–27.

GOODCHILD, M. F. and GOPAL, S. (Eds) (1989). *Accuracy of Spatial Databases*, p. 290, London: Taylor & Francis.

HADZILAKOS, T. and TRYFONA, N. (1992). A model for expressing topological integrity constraints in geographic databases, in: Frank, A. U., Campari, I. and Formentini, U. (Eds). *Theories and Methods of Spatio-Temporal Reasoning in Geographic Space*, pp. 252–268, *Lecture Notes in Computer Science 639*, Berlin: Springer.

HOULDING, S. (1992). The application of new 3-D computer modelling techniques to mining, in: Turner, A. K. (Ed.), *Three-dimensional Modelling with Geoscientific Information Systems*, pp. 303–325, Dordrecht: Kluwer.

HUIJBREGTS, C. and MATHERON, G. (1970). Universal kriging (an optimal method for estimating and contouring in trend surface analysis), *Proc. 9th Int. Symposium on Techniques for Decision-Making in the Mineral Industry*, pp. 159–169, Montreal: Canadian Institute of Mining and Metallurgy.

HURNI, L. (1993). Digital topographic and geological 3D modelling for improved spatial perception, in: Mesenburg, P. (Ed.), *ICA Proc. 16th Int. Cartographic Conference*, Cologne, Germany, Vol. 1, pp. 46–60.

JOURNEL, A. G. and HUIJBREGTS, C. J. (1978). *Mining Geostatistics*, London: Academic Press.

KAVOURAS, M. (1987). *A Spatial Information System for the Geosciences*, unpublished PhD Thesis, Dept. of Surveying Engineering, University of New Brunswick, Fredericton, NB, Canada, p. 229.

KAVOURAS, M. (1992). A spatial information system with advanced modelling capabilities, in: Turner, A. K. (Ed.), *Three-dimensional Modelling with Geoscientific Information Systems*, pp. 59–67, Dordrecht: Kluwer.

KAVOURAS, M and MASRY, S. E. (1987). A spatial information system for the geosciences: design considerations, *Proc. Auto-Carto 8*, Falls Church, VA, ACSM/ASPRS, pp. 336–345.

KELK, B. (1992). 3-D modelling with geographic information systems: The problem, in: Turner, A. K. (Ed.). *Three-Dimensional Modelling with Geoscientific Information Systems*, pp. 29–37, Dordrecht: Kluwer.

LASSETER, T. (1992). An interactive 3-D modelling system for integrated interpretation in hydrocarbon reservoir exploration and production, in: Pflug, R. and Harbaugh, J. W. (Eds), *Computer Graphics in Geology*, pp. 189–198, *Lecture Notes in Earth Sciences 41*, Berlin: Springer.

LEE, Y.-H. and HARBAUGH, J. W. (1992). Standford's SEDSIM project: Dynamic three-dimensional simulation of geologic processes that affect clastic sediments, in: Pflug, R. and Harbaugh, J. W. (Eds), *Computer Graphics in Geology*, pp. 113–127, *Lecture Notes in Earth Sciences 41*, Berlin: Springer.

MCGRANAGHAN, M. (1993). A cartographic view of spatial data quality, in: Buttenfield, P. B. (Ed.), Mapping data quality: A collection of essays, *Cartographica*, 2(3), 8–19.

PFLUG, R. and HARBAUGH, J. W. (Eds). (1992). *Computer Graphics in Geology*, p. 298, *Lecture Notes in Earth Sciences 41*, Berlin: Springer.

RAPER, J. F. (Ed.) (1989). *Three Dimensional Applications in Geographical Information Systems*, London: Taylor & Francis.

RAPER, J. F. and KELK, B. (1991). Three-dimensional GIS, in: Maguire, D. J., Goodchild, M. F. and Rhind, D. W. (Eds), *Geographical Information Systems: Principles and Applications*, Vol. 1, pp. 299–317, London: Longman.

SIMS, D. L. (1992). Applications of 3-D geoscientific modeling for hydrocarbon exploration, in: Turner, A. K. (Ed.), *Three-dimensional Modelling with Geoscientific Information Systems*, pp. 285–289, Dordrecht: Kluwer.

SLINGERLAND, R and KEEN, T. R. (1992). A numerical study of storm driven circulation and 'event bed' genesis, in: Pflug, R. and Harbaugh, J. W., (Eds), *Computer Graphics in Geology* pp. 97–99, *Lecture Notes in Earth Sciences 41*, Berlin: Springer.

SMITH, D. R. and PARADIS, A. R. (1989). Three-dimensional GIS for the earth sciences, in: Raper, J. F. (Ed.), *Three Dimensional Applications in Geographical Information Systems*, pp. 149–154, London: Taylor & Francis.

TURNER, A. K. (1989). The role of 3-D GIS in subsurface characterization for hydrological applications, in: Raper, J. F. (Ed.), *Three Dimensional Applications in Geographical Information Systems*, pp. 115–128, London: Taylor & Francis.

TURNER, A. K. (Ed.) (1992). *Three-dimensional Modelling with Geoscientific Information Systems*. Dordrecht: Kluwer.

YOUNGMANN, C. (1989). Spatial data structures for modelling subsurface features, in: Raper, J. F. (Ed.), *Three-Dimensional Applications in Geographical Information Systems*, pp. 129–136, London: Taylor & Francis.

CHAPTER TWENTY TWO

Geographic Objects, and How to Avoid Them

JOCHEN ALBRECHT

ISPA, University of Osnabrück, Vechta, Germany

22.1 Introduction

People in the West tend to perceive their environment as an agglomeration of objects, and use science to try to systematize and understand the relations among these objects. Eastern philosophies, with their religious belief in reincarnation, provide a completely different perception of the world that has some important implications for cognitive aspects of space as well (Zimmer, 1961). Change (or process as we would call it) plays a far more dominant role than in our object-centred world. The Chinese *I Ch'ing* (Wilhelm, 1923) or the role of 'Chi' in Feng Shui (Walters, 1991) provide good examples of this different perspective. The heated debate between proponents of raster versus vector systems has sometimes included this philosophical aspect of Eastern versus Western thought. In reality, however, even the raster proponents tend to think in terms of objects (or at least fields that, in this respect, could be regarded as a sort of fuzzy objects), rather than in processes occurring in one location. There is no technical constraint forcing them to do so (fields can be represented in the vector structure and objects can be represented in the raster structure), but rather a tradition of thought.

In this chapter it is argued that a consequent use of the possibilities that are inherent to the field model (especially when implemented as irregular points) allows us to find answers to some of the most difficult questions that have been raised in spatio-temporal tasks involving uncertainty. In this chapter the project 'Long-Term Monitoring of the Environmental Situation in the Wadden Sea' is used to illustrate an enormously complex task (Section 22.2), to provide a seemingly simple solution (Section 22.3), to discuss this solution in the light of all the different requirements described (Section 22.4), and finally to reveal the limitations of an imperfect model of an imperfect world.

22.2 The Wadden Sea project

The Wadden Sea is the English name for the tidal mudflats that stretch for 500 km along the North Sea shoreline from the Netherlands to Denmark. It covers an area of about 500 km in length and up to 35 km wide. Due to tidal forces and very flat slopes, it falls dry every 12 hours and displays a constantly changing landscape. The Wadden Sea is one of the most productive bio-regions on Earth. It is a nursery for a substantial proportion of the marine life in the North Sea, and is an important area for tourism. Additional conflicts of interest arise from its location along one of the major sea routes between ports such as London, Rotterdam and Hamburg.

All monitoring, regardless of its purpose (e.g. shipping traffic, oil spill tracking, coastal defence or nature conservation), requires an accurate base map of the topography. However, this is almost impossible to achieve because a 1 cm change in elevation can mean a difference of several square kilometres in the area that is inundated or dry. Severe floods caused by (winter) storms may alter the location of rills that are important for coastal vessels, and may divide islands or give rise to others. All of these fluctuations have a physical basis. On the other hand, the different perceptions of people analyzing this region result in fluctuations in meaning that are expressed by overlapping object definitions. A shoal, for example, means completely different things to a ship's captain, to a geological engineer, and to a biologist studying marine life. A number of geographic information systems (GIS) are used to store spatial data and perform various analyses. The following list gives a short survery of some of the problems faced by the designers of a GIS for the Wadden Sea (Liebig, 1994):

- the objects of investigation all have different life spans, ranging from hours to centuries;
- the dynamics of an object is especially obvious from its shape, suggesting that a fuzzy boundary definition would help to deal with this situation;
- data are gathered from a multitude of sources, including global positioning systems (GPS), satellite and aerial imagery, terrestrial survey, nautical charts, exploring cruises and stationary measuring instruments;
- for the exploring scientist it is difficult to determine whether the object itself has changed or just the estimation of the surveyor, or whether the change is due to a new measuring procedure;
- database updates are made at irregular intervals, resulting in major inconsistencies, and updates to complete coverages are much too expensive;
- common versioning solutions (as used in commercial database management systems) are not appropriate for such a wide range of time scales;
- object definitions depend upon agreement among the many divergent user interests and are therefore difficult to achieve; and
- the systems scientists engaged in research of the Wadden Sea ecosystem are most interested in identifying and analyzing processes, which can occur over a wide range of time frames, and display a variety of dynamics.

22.3   A conceptual solution

The examples listed above give an idea of the complexity of the geography of this region. There are innumerable views on how best to describe geographic information. Simple graphic and image models are clearly inadequate, since conventional maps and pictures are themselves only models of geographic reality. In fact, it is necessary to characterize geographic information by simultaneously capturing the basic spatio-temporal, thematic, and topological aspects of geographic entities and phenomena, for both descriptions and behaviour. These terms need some clarification, as they are used with mixed connotations (Kemp, 1993).

- *Geographic models* are the conceptual models used by environmental modellers as they develop an understanding of the phenomenon being studied and extract its salient features from the background of infinite complexity in nature. Models at this level cannot be completely specified.

- *Spatial data models* are formally defined sets of entities and relationships that are used to discretize the complexity of geographic reality (Goodchild, 1992). The entities in these models can be measured and the models completely specified.

- *Data structures* describe details of specific implementations of spatial data models. A practical example of this differentiation can be found in the SAIF documentation (BCSRMB, 1993).

In SAIF, a 'real-world' geographic entity is described as a geographic object composed of three components: a geometric object, a metadata object,[1] and a set of relationship objects. Geometric objects are composed of multiple geometric primitives which characterize points, areas, images, surfaces, etc. Metadata objects are defined by various descriptors typical of map descriptions, i.e. scale, resolution, data, projection, etc. Finally, relationship objects describe spatial associations with other objects, such as inclusion, proximity, overlap, etc.

All of the difficulties listed in Section 22.2 are in some way related to the notion of 'objects'. Evidently the solution cannot lie in re-education of the users or in forcing them to adhere to one particular object definition. What is needed, therefore, is a database that accommodates the requirements of all users. The Wadden Sea monitoring project is expected to last a couple of decades, so that new, as yet unperceived concepts will need to be integrated in the future. This can more easily be achieved by a new approach suggested here, which completely discards the notion of 'objects' at the time when data are entered into the database. This approach is elaborated in the following.

From the user's perspective, such a database is a set of building blocks, attributes and associations that can be used to describe a particular collection of information in a consistent, methodical manner. All scientists concerned define the relational (in the database sense of the word) view of the attribute values that characterize *their* objects and processes (that is, geographic data structures, spatial types, methods and behaviours). This allows each discipline to keep its own conceptual models. Analyses are performed either directly on the stored attribute values, or, if they require an object model, on a particular geo-object, and the result is re-exported to the database. If scientists from different disciplines want to exchange their results, they would do it preferably via these attributes, or they would discretize the world by using an object catalogue that fits only their conceptual models.

Such a database can be easily implemented with any of the current relational database management systems (DBMS). Each object is defined in terms of its geometric, topological, associative and thematic components, attributes or records (von Bernem, 1990; Farke, 1990). Such a definition is simply a new view of a vast set of attribute tuples. New concepts require only additional attributes (see note 2) to be sampled and entered, or new views to be defined. Finally, objects need to be defined only for visualization and other purposes of communication, both among the participating scientists and the public. No new concepts are required concerning the data model used within the GIS. All spatial analyses are performed on the point data using established interpolation techniques, map algebra and cellular automata (for process simulation). The only measure necessary to communicate[3] about objects is a scheme that guarantees their object identity, giving them a form of persistence.

Technically, the database stores tuples with a threefold key consisting of the project name (domain), the station name (location) and the attribute to be measured. The project name is a pointer to the conceptual (domain) model used within the project. Almost all data (with the exception of remotely sensed images) are gathered at particular stations. Time stamps and metadata information contribute to the overall use. Each geographic feature is distinguished from similar features by some observable or well-defined characteristic defined by the domain specialist, such as name, purpose, owner, physical property, or magnitude. There may be various relationships among these features, either spatial and/or temporal, but these relationships do not necessarily form the basis for the definition of an object; for example, an oil spill may have been caused by a storm (an association or relation) but is does not have to be part of the object definition. Using this simple data model, it becomes easy to include metadata at all levels of object definition, and even to use it to create new views.

Metadata can be used (1) to identify a particular geographic object, through the browsing and searching functions; (2) to evaluate the suitability of an object for the query at hand; and (3) to interpret the data correctly. Conventionally, issues relating to the coordinate system and projection are defined as metadata. This freedom of definition allows objects and processes to be defined in terms of fields. They thus become descriptions of space that may not necessarily correspond to distinct features, but to the distribution of some geographic phenomenon over a given portion of space (Kemp, 1993). In terms of behaviour, a field maps a position in space to a value (as described in the discussion of spatial function below). A Landsat satellite image, for example, can be regarded as a field of the reflectance at various wavelengths from a section of the Earth's surface.

Data describing land use across a broad area may also be regarded as a field (Gardels, 1994). A spatial function is a function defined on positions in space. For example, just as a graph of a quadratic equation provides a value of y for any arbitrary value of x, so a spatial function provides a value for any arbitrary location in space, where the location may be a single point or a larger region. This function can be used to model either continuous phenomena like barometric pressure or altitude, or discrete phenomena like land use or soil type. It may be very simple, such as an assignment of image values to grid positions. Alternatively, the function may be sophisticated, involving for example the interpolation of values at arbitrary locations. In a mathematical context, a field is equivalent to the graph of a function. User-defined fields describe the distribution of a phenomenon over an arbitrary region of space. The geometry of a field represents a continuous region in space. It may, for example, correspond to a set of isolines, a set of velocity vectors at irregularly spaced points, a triangulated irregular network, a regular grid, or a digital elevation model. Because of the nature of a field, estimated values for a phenomenon at arbitrary locations may be determined through interpolation.

22.4   Applications

This new approach addresses many of the difficulties faced by GIS managers as listed in Section 22.2. Object-centred systems require sophisticated techniques for database updates in order to avoid inconsistencies. The approach presented here makes it possible to enter different data independently at different times because what is entered is not the new status of an object (which might affect the status of an associated object) but only the value of an attribute. Since the object definition occurs only later (for communication purposes) this avoids inconsistencies among the objects. Without objects there can be no boundaries.

One of the major problems in current GIS analysis is that errors are propagated if the boundaries are not determined correctly. In the concept of fields, it does not make any sense to discretize them into computable regions. The approach outlined in Section 22.3 does not call for any boundaries; they may be introduced for communication purposes, in which case they become part of the metadata. Usually GIS cause troubles when they are supposed to process heterogeneous sets of data, each with its own confidence rating, coordinate system or classification scheme.

The proposed methodology makes use of the well-established statistical tradition to refer to the source data as much as possible. This does not necessarily imply that knowledge gathered by the analyses of other scientists can not be used. Any computational result can be stored and subsequently used. It is again up to the researchers to choose what kind of aggregated data is used in their analysis. This relates to the problem of changing classification schemes and the fluctuating judgements of field surveyors or image interpreters. One of the most aggravating experiences of the scientists involved in the Wadden Sea project is that they are unable to tell whether an object or just its interpretation has changed.

Questions of spatial topology such as 'adjacent', 'contains', 'surrounds', 'boundedBy', 'on the left side of', and so on, are left to the analyzing model (which could be a commercial GIS). The same holds true for metric associations, including relationships such as 'spatial neighbourhood', 'distance from', distance and angles, and relative linear distance. In the actual database no distinction needs to be made between the different conceptual levels of geographic and spatial data models, since all these definitions are up to the user.

One of the major drawbacks to the use of GIS in a multidisciplinary environment is that most domains have their own (often incompatible) spatial concepts. Burrough and Frank (1995) discuss the pros and cons of attempts to define a universal spatial model that accommodates the compromised needs of everyone. The approach outlined in Section 22.3 circumvents this problem by allowing scientists to work within their own conceptual worlds. Objects need to be defined only (a) for presentations, and (b) for cooperative research. In either case these definitions may be *ad hoc* and need not be universally valid. Those who still wish to use universal concepts may use formal specifications such as ATKIS (AdV, 1989) of the SDTS (USNIST, 1992) profiles (TVP, raster, etc.); however, they then would lose all the advantages described above.

22.5   Conclusions

Since this all is based on traditional GIS techniques, what is new about the presented concept? There are two main answers. First, the GIS is not regarded as an end in itself but is used as a basis for model building. Each cell represents a model that is applied not to an object but to the value of a given attribute. The GIS thus becomes process-based rather than object-based. Second, the discretized world is reassembled into the familiar concept of object at an unusually late stage. Even in its parental domain – image processing – users prefer to work with objects, demanding complicated techniques to classify picture elements for subsequent use as objects. By moving this instance of object definition to the very end of the investigation process, many problems can be avoided (Gahegan, 1994).

One of the remaining problems is well-known from statistics. There is no such thing as an objective measuring of attributes allowing scientists from different domains to share unexamined data. Another disadvantage is the downside of the freedom gained in the procedure described above. All the structural information that is usually inherent in a GIS, rests now with the analytic functions that each scientist has to apply. Each domain needs to formalize conceptual models that include the notion of space, if not geography.[4] While geologists, ecological modellers or meteorologists have no difficulties with this, other disciplines including geography itself have not yet come up with models that could pass the rigid test of practice. The methodology outlined above pertains not only to the Wadden Sea monitoring project. Any region studied by a multidisciplinary team would be a potential candidate for this economic approach allowing the sharing of data.

Notes

[1] It is the information that can be derived from the metadata that allows one to determine the level of confidence in the analysis, map or whatever is produced from the data. The analysis, map, etc., is derivative as well. The GIS analyst would use the metadata to describe mapping functions derived from multiple layers, each of which in turn contains data errors plus their interactions and analytical errors in specifying the functions. In essence, the position along the data stream defines the type of metadata needed. The database developer needs metadata that describe the mechanics of the data, while the end user needs metadata that describe the type of information that can be derived from the data. The term 'new approach' is used here because it has not yet been named; the reader is invited to make suggestions.
[2] Note that attributes are measurements or images or anything else that scientists define to represent their object of investigation, the geographic feature or 'geoobject'.
[3] The stress is on 'to communicate' as this is the only reason for their existence; analyses are to be performed on the lower level of the attributes.
[4] For the field of geography this problem was addressed at a workshop on 'Formalizing Geographic Models for their Subsequent Use in GIS' at the AGIT conference in Salzburg, Austria, July 1994, and will be pursued further by a specialist group that formed there.

REFERENCES

ADV (1989). *Amtlich Topographisch-Kartographisches Informationssystem* (ATKIS), Bonn: Arbeitsgemeinschaft der Vermessungsverwaltungen der Länder der Bundesrepublik Deutschland.

VON BERNEM *et al*. (1990). *Das Wattenmeerinformationssystem WATiS*, Texte Umweltbundesamt 7/90: Ökosystemforschung Wattenmeer, Berlin.

BCSRMB (1993). *Spatial Archive and Interchange Format: Formal Definition*, Release 3.0, Victoria, BC: British Columbia Surveys and Resource Mapping Branch.

BURROUGH, P. A. and FRANK, A. U. (1995). Concepts and paradigms in spatial information: Are current geographical information systems truly generic? *Int. J. Geographical Information Systems*, **9**(2), 101–116.

FARKE, H. (1990). *Ökosystemforschung im Nationalpark Niedersächsisches Wattenmeer*, Texte Umweltbundesamt 7/90: Ökosystemforschung Wattenmeer, Berlin.

GAHEGAN, M. (1996). A model to support the integration of multiple spatial representations from image data into GIS. *Int. J. Geographical Information Systems*, (in press).

GARDELS, K. (1994). Virtual geodata model: The structural view, in: Buehler, K. (Ed.), *The Open Geodata Interoperability Specification*, preliminary report for OGIS meeting at Sun Microsystems in Mountain View, CA, Champaign, IL: USACERL.

GOODCHILD, M. (1992). Geographical data modeling, *Computers and Geosciences*, **18**(4), 401–408.

KEMP, K. (1993). *Environmental Modeling with GIS: A Strategy for Dealing with Spatial Continuity*, NCGIA Technical Report 93-3, Santa Barbara, CA: National Center for Geographic Information and Analysis.

LIEBIG, B. W. (1994). Protecting the environment: GIS and the Wadden Sea, *GIS Europe*, **3**(2), 34–36.

USNIST (1992). *Spatial Data Transfer Standard (SDTS)*, Federal Information Processing Standards Publication 173, Gaithersburg, MD: US National Institute of Standards and Technology.

WALTERS, D. (1991). *The Feng Shui Handbook*, pp. 35–36, London.

WILHELM, H. (1923). *I Ch'ing: The Book of Changes*, (English translation, C. F. Baynes, 1977), Princeton, NJ: Princeton University Press.

ZIMMER, H. (1961). *Philosophie und Religion Indiens*, Zürich.

PART SEVEN

Postscript

Postscript

Practical Consequences of Distinguishing Crisp Geographic Objects

P. A. BURROUGH and H. COUCLELIS

As this book was going to press, a couple of newspaper reports appeared in The Netherlands which illustrate clearly the confusion and absurdities that can occur when people insist on having exact, legally defined geographical objects. The e-mail discussion that ensued prompted Helen Couclelis to explore Wittgenstein's *Philosophical Investigations* (1953); a couple of his observations make pertinent conclusions for a book such as this.

Two articles which appeared in Dutch quality newspapers illustrate the practical problems that can be caused by legally defined or exactly located boundaries, and demonstrate how much ingenuity is needed to obtain a working solution. In both cases the practical solutions involve one or other way of redefining or 'fuzzing' the boundary.

The first story appeared in the Dutch paper *De Volkskrant* on 29 April 1995 under the title 'Dieven zonder grenzen' ('Thieves without boundaries'). It concerns the differences in approach to law and policing along the Dutch–German border in the town of Kerkrade in the southern part of the Dutch province of Limburg. South Limburg is almost cut off from the rest of the Netherlands and is only connected to the rest by a narrow strip of land some 8 km wide along the valley of the River Maas. To the east lies the border with Germany, and to the west Belgium.

In 1963 the police corps in Kerkrade initiated informal coffee meetings with the neighbouring German and Belgian corps in order to deal with trans-border problems. The border between Germany and the Netherlands runs down the length of the High Street, and in the 1960s was marked by a high chainlink fence on concrete poles. By the late 1970s, this fence had been replaced by a symbolic border of knee-high concrete blocks. Although the physical border had been reduced in importance in line with policies of increasing freedom of movement, the legal basis for dealing with border incidents had not. The German police were allowed only to handle crimes that occurred on their side of the border, and the Dutch police were similarly restricted.

On 1 November 1978, the police and the legal authorities were confronted with an incident that rendered them effectively impotent. Four Dutch customs officials were walking along the friendly concrete blocks in the Kerkrade High Street when they saw a man and a woman nonchalantly stepping over the border. When the customs officials asked the pair to come to their office, they both took out automatic pistols and shot two customs officials dead and seriously wounded a third. The attack took place on the boundary, and the woman shot from Germany and the man from Dutch territory. The Dutch police were legally unable to do anything about the shots from Germany – they were legally unable even to touch the gun cartridges that had fallen on the German side of the road. As if this were not enough, the legal status in the Netherlands of reports and charges made in Germany was totally unclear. It was not known if the traces of the incident had equivalent legal status in both countries, and it was not known which information about the case could and could not be exchanged between them. In short, the presence of the boundary caused maximum confusion.

Since 1978, much has happened to remove the uncertainty due to the presence of the border. In 1994 the concrete poles were removed, so that the border is no longer physically marked; effectively it has become 'fuzzied'. Dutch police can cross the road to apprehend criminals on the German side, after which the criminals are dealt with by the German police. Under certain conditions, Dutch police cars can follow criminals throughout Germany, but German police are only allowed to follow criminals up to 10 km from the border in the Netherlands (how do they know when they have reached the limit?); the difference is due to sensitivities that still remain after the 1940–45 war. With electronic communications and with harmonized laws on driving under the influence of alcohol, for example, many anomalies can be resolved. Much of the improvements in communication and cooperation are due to the recent Schengen agreement which covers the legal issues of cooperation and information exchange.

The problems of different laws and different methods of policing on the two sides of the border have not yet been completely removed, however, and incidents that happen on or near the border will clearly continue to cause legal and practical confusion. What is interesting, however, is that the complexities of reality have forced the authorities to realize that conventional ideas of crisp boundaries which had been set up to reduce confusion do not work in practice, and that practical means of dealing with continuous change are essential. Politicians and statesmen in other lands could do well to heed this lesson.

The second story was published a couple of weeks later in the *NRC Handelsblad* on Thursday, 11 May 1995. It concerned the designation of lands that can be considered 'wild', which is to say, terrain on which wild animals such as deer and wild pig can be considered as ranging freely rather than being farmed. The designation is important because the practices that are allowed for managing animal numbers are quite different for 'wild' areas than for farms.

Before 22 January 1993 the managers of the Hoge Veluwe, the main Dutch National Park, were able to cull the numbers of deer and wild pig by simply shooting them. Culling is essential because otherwise the animals breed so successfully that they can double their numbers in a year. On 22 January 1993 the Dutch parliament was presented with a bill on Hunting and Management of Wild Animals, in which it was stated that surplus animals should not be shot but should be caught by lasso on the ground or from helicopters and then be removed for a decent slaughter in an abattoir. To ensure that this law applied to the Hoge Veluwe, the parliamentarians defined the terrain on which shooting and hunting was allowed to be larger than 5000 ha if fenced, and larger than 15 000 ha if unfenced.

Now in its totality the Hoge Veluwe is larger than 5000 ha and it is completely fenced from outside. Inside there is a fenced area of some 3000 ha within which deer and wild pig roam freely. Under the new law the animals in the internal area fall under the law that protects the health and well-being of farm animals, which means that to be culled they must first be rounded up and then transported to an abattoir, or that they must be killed under the supervision of a veterinarian from the National Meat and Livestock Inspection Service. In short, a huge amount of red tape for the managers of the Hoge Veluwe.

The solution for the managers is simple. Remove the internal fences in the Hoge Veluwe and then the area available to the deer and wild pigs is larger than 5000 ha and the problem has been solved! The law would not be broken but the geographic object would be modified to meet the needs of the managers. This solution is not of course acceptable to everyone, and the foresters and ecologists are now up in arms protesting that the animals, in particular the larger deer, will overgraze wooded areas that have until now been protected: this kind of management means the end of the woodland areas and the whole park will in effect become a cattle ranch. The matter has not yet been settled, but it seems likely that the arguments over the designation and delineation of 'wild' areas in the Hoge Veluwe park will continue for some time to come.

Both of these real world examples, so frustrating for the persons concerned, demonstrate human problems with drawing boundaries, or of the legal boundary being inappropriate for the problem in hand. The universality of situations similar to these prompted Wittgenstein (1953) to make the following observations (the numbers refer to paragraphs in his text).

71.   One might say that the concept 'game' is a concept with blurred edges. – 'But is a blurred concept a concept at all?' – Is an indistinct photograph a picture of a person at all? Is it even always an advantage to replace an indistinct picture by a sharp one? Isn't the indistinct one often exactly what we need?

76.   If someone were to draw a sharp boundary I could not acknowledge it as the one that I too always wanted to draw, or had drawn in my mind. For I did not want to draw one at all . . . .

499.   . . . But when one draws a boundary it may be for various kinds of reason. If I surround an area with a fence or a line or otherwise, the purpose may be to prevent someone from getting in or out, but it may also be part of a game and the players be supposed, say, to jump over the boundary; or it may shew where the property of one man ends and that of another begins; and so on. So if I draw a boundary line that is not yet to say what I am drawing it for.

Clearly, geographical objects with indeterminate boundaries not only cause difficult and significant problems for geographers and computer scientists, but also for police, politicians and philosophers. Why have they been ignored for so long?

REFERENCES

WITTGENSTEIN, L. (1953). *Philosophical Investigations*, 3rd edn, New York: MacMillan.

Index

remotely-sensed data 289
soil 277, 278
see also land cover
viewsheds 50–1, 88–9, 92
vlines 146, 150
volumes 4, 197, 273
vpoints 146, 150
vregions 146, 150

Wadden Sea 273, 325–30
wetlands 78

zones 146–7, 227
confidence 273, 320
confusion 3, 26
continuous fields 229, 231–5
see also transition zones